# Effect of Dietary Supplementation on the Growth and Immunity of Fish and Shellfish

# Effect of Dietary Supplementation on the Growth and Immunity of Fish and Shellfish

Editors

**Qiyou Xu**
**Jianhua Ming**
**Fei Song**
**Changle Qi**
**Chuanpeng Zhou**

MDPI

Basel • Beijing • Wuhan • Barcelona • Belgrade • Novi Sad • Cluj • Manchester

*Editors*
Qiyou Xu
Huzhou University
Huzhou
China

Jianhua Ming
Huzhou University
Huzhou
China

Fei Song
South China Normal University
Guangzhou
China

Changle Qi
Huzhou University
Huzhou
China

Chuanpeng Zhou
Chinese Academy of Fishery Sciences
Guangzhou
China

*Editorial Office*
MDPI
St. Alban-Anlage 66
4052 Basel, Switzerland

This is a reprint of articles from the Special Issue published online in the open access journal *Fishes* (ISSN 2410-3888) (available at: https://www.mdpi.com/journal/fishes/special_issues/US6ZN77CB6).

For citation purposes, cite each article independently as indicated on the article page online and as indicated below:

| Lastname, A.A.; Lastname, B.B. Article Title. *Journal Name* **Year**, *Volume Number*, Page Range. |
|---|

**ISBN 978-3-7258-1273-8 (Hbk)**
**ISBN 978-3-7258-1274-5 (PDF)**
**doi.org/10.3390/books978-3-7258-1274-5**

Cover image courtesy of Qiyou Xu

# Contents

# About the Editors

**Qiyou Xu**

Qiyou Xu received his Ph.D. in animal nutrition and feed science from Northeast Agricultural University in 2004. He worked as a Research Assistant and Associate Professor at Animal Nutrition Institute of Northeast Agriculture University (1993–2005), Associate Professor and then Professor at Heilongjiang River Fisheries Research Institute Chinese Academy of Fishery Sciences (2005–2018). In 2018, he moved to Huzhou University as a Professor. His research topics mainly include aquatic animal nutrition, feed, and aquaculture.

**Jianhua Ming**

Jianhua Ming received his Ph.D. in aquatic animal nutrition and feed science from Nanjing Agricultural University in 2011. He worked as Aquatic Engineer in aquatic feed enterprises for 10 years (1996–2006). In 2011, he moved to Huzhou University as Associate Professor and then Professor. His research topics mainly include aquatic animal nutrition and feed science; aquatic animal physiology and molecular immunology; aquatic animal health aquaculture; and metabolic regulation.

**Fei Song**

Fei Song received her Ph.D. in nutrition and feeds for aquatic animals from Ocean University of China in 2018. She worked as Associate Professor at department of life Science at South China Normal University (2018-Now). Her research topic is the nutritional physiology of fish.

**Changle Qi**

Changle Qi received his Ph.D. degree (Zoology) from East China Normal University in 2020. He works in Huzhou University as an associate professor (2020-Current). He is currently working on invertebrate nutrition. His research interests cover gonadal development, molting, and oxidative stress.

**Chuanpeng Zhou**

Chuanpeng Zhou received his Ph.D. in aquaculture at Nanjing Agricultural University in 2012, followed by a postdoctoral fellowship at South China Sea Fisheries Research Institute, Chinese Academy of Fishery Sciences & South China Sea Institute of Oceanology, Chinese Academy of Sciences. Dr. Zhou currently serves as Associate Research at South China Sea Fisheries Research Institute, Chinese Academy of Fishery Sciences, where he has been a scientific researcher since 2013. His research interests include fish nutrition feed technology; carbohydrate metabolism and utilization in fish; and nutrition and immunology in fish.

# Preface

With the development of the aquatic feed industry, new feedstuff and feed additives have been developed. Studies will be needed to ascertain their digestibility along with the composite nutritional value of the formulation to the particular targeted species. This reprint focuses on the application of new feedstuffs and additives in aquaculture species' growth and immunity, especially recently developed feedstuff and additives.

As requirements for aquafeeds increase, shortages are anticipated in various ingredients, especially widely used proteinaceous resources such as fish meal. Technological developments are being explored to enhance the usage of plant and animal feedstuffs in aquafeeds. Examples include chicken meal, enzymatic soybean meal, fermented cottonseed meal, stickwater hydrolysate, and meal. Feed additives can promote growth, feed utilization, and the health of aquatic animals. Several different types of feed additives, such as tryptophan, Myo-Inositol, and Chinese herbal medicine, including *Astragalus membranaceus*, *Coptis chinensis*, *Ginkgo biloba* Leaves, and *Lentinus edodes*, are reported in this reprint. This applied information on a wide range of specific ingredients for use in commercial aquaculture.

We deeply thank all of the authors and reviewers who have participated in this Special Issue. This work was supported by Zhejiang Province R&D Plan (2023C02024) and the Zhejiang Provincial Natural Science Foundation of China under Grant No. LTGN23C190003.

This reprint should be useful to students, researchers, fish nutritionists and feed formulators. We hope this reprint will fulfill its intended purpose of serving as an important and valuable reference resource.

**Qiyou Xu, Jianhua Ming, Fei Song, Changle Qi, and Chuanpeng Zhou**
*Editors*

*fishes*

MDPI

*Editorial*

# Effect of Dietary Supplementation on the Growth and Immunity of Fish and Shellfish

**Changle Qi [1], Qiyou Xu [1,*], Jianhua Ming [1], Fei Song [2] and Chuanpeng Zhou [3]**

1 College of Life Science, Huzhou University, Huzhou 313000, China; qichangle1989@163.com (C.Q.); mingjianhua686@163.com (J.M.)
2 Department of Biology, South China Normal University, Guangzhou 510631, China; sophioe@163.com
3 South China Sea Fisheries Research Institute, Chinese Academy of Fishery Sciences, Guangzhou 510300, China; chpzhou@163.com
* Correspondence: 02655@zjhu.edu.cn

**Citation:** Qi, C.; Xu, Q.; Ming, J.; Song, F.; Zhou, C. Effect of Dietary Supplementation on the Growth and Immunity of Fish and Shellfish. *Fishes* **2024**, *9*, 176. https://doi.org/10.3390/fishes9050176

Received: 7 May 2024
Accepted: 11 May 2024
Published: 13 May 2024



With the rapid development of aquaculture, the shortage of high-quality dietary ingredients has become a critical problem limiting the sustainable development of aquaculture. Therefore, the innovation of new feed materials, especially protein and lipid sources, is of great significance for the healthy development of the aquaculture feed industry. Over the past few decades, a variety of new, high-quality dietary ingredients have been developed successfully with the advent of new technologies. Unfortunately, the database of precise nutritional requirements based on new feed ingredients is not yet perfect, although a preliminary database of nutritional requirements for aquatic animals (based on traditional ingredients) has been established. In addition, feed additives can promote animal growth and improve animal health, which provides important support for the development of functional feed. Therefore, this Special Issue focuses on the application of feed ingredients and additives in aquaculture, especially those more recently developed.

An accurate and reliable method for evaluating the growth performance of aquatic animals is the basic premise of scientific research on the effects of feed ingredients and additives. Traditionally, the condition factor is an empirical indicator and is positively correlated with body weight [1]. It was widely used as a morphological parameter to assess the growth or nutritional status of aquatic animals. However, the growth stages of fish and different feed formulations can lead to deviations from the ideal growth status, which may cause misjudgments [2]. In this Special Issue, the weight–length relationship and condition factor of Gibel carp (*Carassius auratus* CAS V) at different growth stages and feed formulations are investigated. The results indicate that the evaluation of fish growth requires considering diverse indicators such as weight, length, body depth, body width, and carcass ratio, as well as the condition factor, to avoid misjudging the actual growth situation [3]. This study provides a more accurate method for the evaluation of the effects of feed ingredients and additives on the growth of aquatic animals.

Protein is an important nutrient, and fish meal is an ideal high-quality protein source for aquatic animals [4]. Unfortunately, the production of fish meal is limited, and the supply exceeds the demand, leading to the increasing price of fish meal [5]. Therefore, replacing fish meal with plant protein of stable yield and low price is an important direction for aquatic animal nutrition [6,7]. However, plant protein sources cannot completely replace fish meal due to the anti-nutritional factors [8]. Some studies have found that new technological methods such as fermentation, hydrolysis, and enzymatic hydrolysis can eliminate anti-nutritional factors and improve the utilization of feed ingredients [8–10]. In this Special Issue, a study reported that the use of 25% dietary fermented cottonseed meal (FCSM) to partially replace fish meal improved nutrient absorption and reduced intestinal inflammation. However, a high proportion of FCSM negatively affected the intestinal microflora and nutrient absorption [11]. Stickwater hydrolysate (SWH) is obtained by adding hydrolytic enzymes, which can improve the growth performance and feed utilization of fish [12]. In this Special Issue, it is reported that substituting fish meal with SWH or stickwater hydrolysate meal (SWM) does not adversely affect growth or meat quality. Moreover, replacing 15% fish meal with SWH and 5% fish meal with SWM can enhance the immunity of yellow catfish [13]. Similarly, another study in this Special Issue also reported that enzymatic soybean meal can completely replace fish meal in the juvenile Gibel carp's diets without causing adverse impacts on growth performance, antioxidant capacity, or the structure of liver and intestinal tissues [14]. Both of these studies found that partial replacement could even improve the antioxidant capacity and immunity of fish [13,14]. In summary, new technological methods such as fermentation, hydrolysis, and enzymatic hydrolysis can improve the utilization of feed ingredients and enhance the intestinal health of aquatic animals.

The quality of formula feed not only depends on the feed ingredients and their reasonable combination, but is also closely related to feed additives [7]. Feed additives can promote growth, feed utilization, and the health of aquatic animals [4]. Several different types of feed additives, particularly Chinese herbal medicines, are reported in this Special Issue. For example, it has been reported that *Coptis chinensis* supplementation effectively enhances the ability of tilapia to resist *Streptococcus agalactiae* infection by modulating various antioxidant enzymes, immune factors, antimicrobial enzymes, and antimicrobial peptides [15]. Similarly, another two studies in this Special Issue reported that both extracts of *Astragalus membranaceus* (EAms) and extract of *Ginkgo biloba* leaves (EGb) can improve the growth performance and antioxidant capacity of Jian Carp (*Cyprinus carpio* var. Jian) [16,17]. In summary, Chinese herbal medicines can be considered as potential natural antioxidants for aquatic animals.

Nutrients, both as feed ingredients and additives, are closely related to the growth and health of aquatic animals. This Special Issue showcases a collection of original research that highlights the latest discoveries on the applications of new feed ingredients and additives for aquatic animals. It can provide references for the development of high-quality formula feed.

**Author Contributions:** C.Q., writing—original draft and editing. Q.X., supervision. J.M., writing—review and editing. F.S., writing—review and editing. C.Z., writing—review and editing. All authors have read and agreed to the published version of the manuscript.

**Institutional Review Board Statement:** Not applicable.

**Data Availability Statement:** In this editorial, no new data were created. All the result we cited in this manuscript are published articles, which can be found in the database.

**Acknowledgments:** We deeply thank all of the authors and reviewers who have participated in this Special Issue. This work was supported by the Zhejiang Province R&D Plan (2023C02024), the Zhejiang Provincial Natural Science Foundation of China under Grant No. LTGN23C190003, and the Huzhou Natural Science Foundation (2021YZ14).

**Conflicts of Interest:** The authors declare no conflicts of interest.

## References


1. Schulte-Hostedde, A.; Zinner, B.; Millar, J.; Hickling, G. Restitution of mass-size residuals: Validating body condition indices. *Ecology* **2005**, *86*, 155–163. [CrossRef]
2. Cone, R. The need to reconsider the use of condition indices in fishery science. *Trans. Am. Fish. Soc.* **1989**, *118*, 510–514. [CrossRef]
3. Zhang, H.; Xie, S.; Wang, S. Weight–length relationship and condition factor of gibel carp (*Carassius auratus* CAS V) at different growth stages and feed formulations. *Fishes* **2023**, *8*, 439. [CrossRef]
4. Halver, J.; Hardy, R. *Fish Nutrition*; Academic Press: San Diego, CA, USA, 2002.
5. NRC. *Nutrient Requirements of Fish and Shrimp*; National Academies Press: Washington, DC, USA, 2011.
6. Cabral, E.M.; Fernandes, T.J.R.; Campos, S.D.; Castro-Cunha, M.; Oliveira, M.; Cunha, L.M.; Valente, L.M.P. Replacement of fish meal by plant protein sources up to 75% induces good growth performance without affecting flesh quality in ongrowing Senegalese sole. *Aquaculture* **2013**, *380*, 130–138. [CrossRef]
7. Mai, K. *Aquatic Animal Nutrition and Feed Science*; China Agriculture Press: Beijing, China, 2011.

8. Mugwanya, M.; Dawood, M.O.; Kimera, F.; Sewilam, H. Replacement of fish meal with fermented plant proteins in the aquafeed industry: A systematic review and meta-analysis. *Rev. Aquac.* **2023**, *15*, 62–88. [CrossRef]
9. Wang, Z.; Tao, J.; Xie, R.; Zhang, Y.; Zhang, H.; Chen, N.; Li, S. Effects of fish meal replacement with composite mixture of soybean protein hydrolysates and other plant proteins on growth performance, antioxidant capacity and target of rapamycin pathway in Largemouth bass. *N. Am. J. Aquac.* **2023**, *85*, 178–187. [CrossRef]
10. Li, C.; Lin, J. Optimization of soybean meal fermentation process and feeding effect of dietary fish meal partial replacement by fermented soybean meal for large yellow croaker. *Chin. J. Anim. Nutr.* **2022**, *34*, 563–574.
11. Wang, Z.; Liao, S.; Wang, J.; Wang, Y.; Huang, Z.; Yu, W.; Huang, X.; Lin, H.; Luo, M.; Cheng, Z.; et al. Effects of fermented cottonseed meal substitution for fish meal on intestinal enzymatic activity, Iinflammatory and physical-barrier-related gene expression, and intestinal microflora of juvenile golden pompano (*Trachinotus ovatus*). *Fishes* **2023**, *8*, 466. [CrossRef]
12. Shi, Y.; Zhong, L.; Ma, X.; Liu, Y.; Tang, T.; Hu, Y. Effect of replacing fishmeal with stickwater hydrolysate on the growth, serum biochemical indexes, immune indexes, intestinal histology and microbiota of rice field eel (*Monopterus albus*). *Aquac. Rep.* **2019**, *15*, 100223. [CrossRef]
13. Zheng, S.; Shi, Y.; Zhang, J.; Dai, J.; Hu, Y.; Zhong, L. Effects of replacing fish meal with stickwater hydrolysate and meal on the growth, serum biochemical indexes, and muscle quality of Yellow Catfish (*Tachysurus fulvidraco*). *Fishes* **2023**, *8*, 566. [CrossRef]
14. Uyisenga, A.; Liang, H.; Ren, M.; Huang, D.; Xue, C.; Yin, H.; Mi, H. The effects of replacing fish meal with enzymatic soybean meal on the growth performance, whole-body composition, and health of juvenile Gibel carp (*Carassius auratus gibelio*). *Fishes* **2023**, *8*, 423. [CrossRef]
15. Guo, R.; Yu, K.; Huang, K.; Lin, Q.; Liu, T. Immunoprotective effect of coptis chinensis-supplemented diet on *Streptococcus agalactiae* infection in Tilapia. *Fishes* **2023**, *8*, 370. [CrossRef]
16. Xu, J.; Chen, G.; Wu, M.; Yang, Q.; Li, H. The extract of *Astragalus membranaceus* inhibits lipid oxidation in fish feed and enhances growth performance and antioxidant capacity in Jian carp (*Cyprinus carpio* var. Jian). *Fishes* **2023**, *8*, 594. [CrossRef]
17. Chen, G.; Xu, J.; Wu, M.; Li, H.; Yang, Q.; Feng, L. Extract of *Ginkgo biloba* leaves (EGb) decrease lipid oxidation in fish feed and meat and enhance growth and antioxidant capacity in Jian carp (*Cyprinus carpio* var. Jian). *Fishes* **2023**, *8*, 564. [CrossRef]

fishes

Article

# Effects of Partial Substitution of Fish Meal with Soybean Products and Chicken Meal on Growth, Antioxidant Capacity and Intestinal Microbiota of *Penaeus monodon*

**Wanli Yang [1,†], Song Jiang [1,2,3,†], Qibin Yang [1,2,3], Jianhua Huang [1], Jianzhi Shi [2], Yundong Li [2], Yukai Yang [1] and Falin Zhou [2,*]**

1 Shenzhen Base of South China Sea Fisheries Research Institute, Chinese Academy of Fishery Sciences, Shenzhen 518121, China; ywllxt@163.com (W.Y.); tojiangsong@163.com (S.J.); yangqibin1208@163.com (Q.Y.); hjh210440@sina.com (J.H.)

2 Key Laboratory of Aquatic Product Processing, Ministry of Agriculture and Rural Affairs, South China Sea Fisheries Research Institute, Chinese Academy of Fishery Sciences, Guangzhou 510300, China; shijianzhi1989@163.com (J.S.); liyd2019@163.com (Y.L.)

3 Key Laboratory of Efficient Utilization and Processing of Marine Fishery Resources of Hainan Province, Sanya Tropical Fisheries Research Institute, Sanya 572018, China

* Correspondence: zhoufalin@aliyun.com

† These authors contributed equally to this work.

**Abstract:** The aim of this experiment was to investigate the effects of the partial substitution of fish meal with soybean products and chicken meal on the growth performance, antioxidant capacity and intestinal microbiota of *Penaeus monodon*. A total of 450 healthy, consistent shrimp were randomly divided into five groups, with three replicates per group and 30 shrimp per replicate. The proportion of fish meal substituted with soybean products and chicken meal in the five feed groups was 0% (FM), 40% (40SC), 60% (60SC), 80% (80SC) and 100% (100SC). The experiment lasted for 8 weeks. The results showed that, compared to the FM group, the 40SC and 60SC groups had a decrease in WG and SR, but there was no significant difference ($p > 0.05$). In contrast, compared to the FM group, the FCR in the 100SC group was significantly increased ($p < 0.05$), while there was no significant difference among the FM and 40SC, 60SC and 80SC groups ($p > 0.05$). Compared to the FM group, the ACP in the 80SC and 100SC groups significantly increased ($p < 0.05$), while the 40SC and 60SC groups had no significant difference ($p > 0.05$). The AKP in the 100SC group was significantly higher than that in the FM group ($p < 0.05$), while there were no significant differences among the other four groups ($p > 0.05$). There were no significant differences in T-AOC and T-SOD among all the treatment groups ($p > 0.05$). The next-generation sequencing of the intestinal microbiota showed that Proteobacteria was the most abundant phylum in the five groups, accounting for 37.67%, 66%, 40%, 40% and 43.33%, respectively. Compared to the FM group, the Fusobacteriota in the other four groups decreased significantly ($p < 0.05$). The functional prediction of FAPROTAX indicated that no functional components were observed which are harmful to the body. Considering the effects on growth performance, antioxidant capacity and intestinal microbiota, it is feasible to use soybean products and chicken meal to replace 60% of fish meal in the feed of *P. monodon*.

**Keywords:** *Penaeus monodon*; growth performance; antioxidant capacity; intestinal microbiota

**Key Contribution:** A. An appropriate substitution ratio did not affect the growth performance and antioxidant capacity of *P. monodon*, and the intestinal microbiota of *P. monodon* remained stable and did not cause damage to the organism. B. A high substitution ratio can affect the growth and immunoenzyme activity of *P. monodon*.

**Citation:** Yang, W.; Jiang, S.; Yang, Q.; Huang, J.; Shi, J.; Li, Y.; Yang, Y.; Zhou, F. Effects of Partial Substitution of Fish Meal with Soybean Products and Chicken Meal on Growth, Antioxidant Capacity and Intestinal Microbiota of *Penaeus monodon*. *Fishes* **2024**, *9*, 42. https://doi.org/10.3390/fishes9010042

Academic Editor: Kyeong-Jun Lee

Received: 26 December 2023
Revised: 17 January 2024
Accepted: 19 January 2024
Published: 22 January 2024

## 1. Introduction

*Penaeus monodon*, also known as grass shrimp and black tiger shrimp, is a genus of shrimp belonging to the genus Penaeus and family Penaeus. *P. monodon* is the largest individual in the genus Penaeus, with a body mass of 500 g and an average body length of 300–350 mm from maturity. The average body mass is about 350–400 g [1]. *P. monodon* is a valuable edible shrimp with a large body, a delicious taste and strong vitality. So, *P. monodon* is one of the most important marine shrimp breeding varieties in China.

With the rapid development of the aquaculture industry, the demand for aquatic feed is increasing. In aquaculture, feed costs can be as high as 50% [2]. Fish meal has good palatability and has always been an indispensable source of high-quality protein in aquatic feed [3]. It occupies a very high proportion in feed formulas, usually accounting for 20% to 60% of aquatic feed [4]. Fish meal is an unsustainable protein resource. In China, the source of fish meal is mainly imported, resulting in its high cost. The cost of buying fish meal accounts for half of the total breeding cost, so finding new and efficient materials that can replace fish meal protein has become a research hotspot in the aquatic feed industry. At present, many protein sources have been used to replace fish meal in order to achieve the purpose of reducing the use of fish meal in feed, such as poultry by-product meal [5], meat and bone meal [6], soybean meal [7], blood meal [8], cottonseed meal [9], peanut meal [10], rapeseed meal [11], soybean protein concentrate [12] and so on. Soybean meal has been proven to be one of the plant protein sources with a good effect on replacing fish meal. Some studies have found that using a certain amount of soybean meal as a substitute for part of fish meal had no significant effect on the growth performance of *Litopenaeus vannamei* [13]. According to a report by the FAO, currently, poultry meat is ranked in the top category of animal protein in the world. Chicken meal made from the waste, residual mince and offal generated during meat processing has attracted much attention from the animal protein source production market due to its rich nutritional value, high digestibility, amino acid balance and high biological conversion rate [14]. Studies have shown that chicken meal is a high-quality animal protein source for farmed fish [15]. In the process of breeding, differences in feed nutrient composition and raw materials will affect the antioxidant capacity and intestinal microbiota of aquatic animals. Studies have shown that using chicken meal as a substitute for 40% of fish meal could have negative effects on the liver and intestines of *Micropterus salmoides* [16]. The substitution of fish meal with chicken meal can reduce the antioxidant activity of *Trachinotus ovatus* [17]. And the intestinal microbiota of *Procambarus clarkii* can be changed when the additive amount of soybean meal is higher than 40% [18]. There have been many reports which have tested only one of them, while the effect of combining soybean products and chicken mea is unknown.

This study was conducted to study the effects of partially and completely substituting fish meal with soybean products and chicken meal on the growth performance, digestive enzymes, antioxidant capacity of the hepatopancreas, and the intestinal microbiota of *P. monodon*. The results from the current study will enhance our understanding of the effect of soybean products and chicken meal on *P. monodon* and provide basic data and a reference for the formulation and optimization of feed for *P. monodon*.

## 2. Materials and Methods

### 2.1. Experiment Materials

The experiment was conducted at the Shenzhen Base of the South China Sea Fisheries Research Institute, Chinese Academy of Fishery Sciences. The shrimp used in the experiment were of a new variety selected by the research team. The shrimp were taken out of the breeding pond and temporarily raised in an 8 $m^3$ tank for three days; commercial feed (Guangdong Dongteng Feed Co., Ltd., Guangzhou, China) was fed during this period, and feeding was stopped the day before the experiment began. The body mass of the shrimp was 3.44 $\pm$ 0.03 g.

## 2.2. Experiment Feeds

According to the nutritional requirements of *P. monodon*, five kinds of iso-nitrogen and iso-lipid diets were designed, and the substitution ratios of the fish meal were 0% (FM), 40% (40SC), 60% (60SC), 80% (80SC) and 100% (100SC), respectively. Additional crystal amino acids were added to meet the amino acid requirements of the shrimp. Methionine and cystine were hydrolyzed using oxidative acid hydrolysis, and the rest of the amino acids were hydrolyzed via acid hydrolysis. The chicken meal, soybean meal and other materials were purchased through the company Guangdong Kingkey Smart Agri Technology Co., Ltd., in Guangzhou, China. All the ingredients were ground into a powder, sieved through an 80-hole mesh and thoroughly mixed with oil and water. The 1.5 mm-diameter doughs were extruded using a twin screw extruder (F-26, South China University of Technology, Guangzhou, China), cut into pelletized feeds using a pelletizer (G-500, South China University of Technology, Guangzhou, China), steamed in a 90 °C electric oven for 2 h, dried in an air-conditioned room and then stored at −20 °C in a refrigerator until used. The feed formulations and the contents of various nutrients are shown in Table 1.

**Table 1.** Formulation and nutrient levels of the experimental diets (% dry matter).

| | FM | 40SC | 60SC | 80SC | 100SC |
|---|---|---|---|---|---|
| Fish meal | 25.00 | 15.00 | 10.00 | 5.00 | 0.00 |
| Soybean meal | 23.00 | 23.50 | 23.50 | 24.50 | 24.50 |
| Chicken meal | 0.00 | 4.50 | 6.50 | 8.50 | 11.00 |
| Peanut hull | 14.00 | 14.00 | 14.00 | 14.00 | 14.00 |
| Wheatmeal | 23.20 | 22.30 | 22.19 | 21.36 | 21.11 |
| Beer yeast | 3.00 | 3.00 | 3.00 | 3.00 | 3.00 |
| Shrimp med | 3.00 | 3.00 | 4.00 | 4.00 | 4.00 |
| Soy protein concentrate | 0.00 | 4.50 | 6.00 | 8.00 | 10.00 |
| Soybean lecithin | 1.00 | 1.00 | 1.00 | 1.00 | 1.00 |
| Fish oil | 1.55 | 1.65 | 1.70 | 1.80 | 1.85 |
| Soybean oil | 1.55 | 1.65 | 1.70 | 1.80 | 1.85 |
| Vitamin C polyphosphate | 0.10 | 0.10 | 0.10 | 0.10 | 0.10 |
| Cholesterol | 0.50 | 0.50 | 0.50 | 0.50 | 0.50 |
| Vitamin premix (Prawn) | 1.00 | 1.00 | 1.00 | 1.00 | 1.00 |
| Mineral premix (Prawn) | 1.00 | 1.00 | 1.00 | 1.00 | 1.00 |
| $Ca(H_2PO_4)_2$ | 1.00 | 1.75 | 2.05 | 2.45 | 2.85 |
| Lysine hydrochloride (78%) | 0.00 | 0.21 | 0.31 | 0.41 | 0.53 |
| Methionine (99%) | 0.00 | 0.10 | 0.14 | 0.19 | 0.24 |
| Threonine (98%) | 0.00 | 0.09 | 0.13 | 0.18 | 0.23 |
| Carboxymethyl cellulose | 1.00 | 1.00 | 1.00 | 1.00 | 1.00 |
| Taurine (99%) | 0.10 | 0.15 | 0.18 | 0.21 | 0.24 |
| Total | 100.00 | 100.00 | 100.00 | 100.00 | 100.00 |
| Crude protein | 39.92 | 40.00 | 39.91 | 39.93 | 39.89 |
| Crude lipid | 7.26 | 7.22 | 7.19 | 7.25 | 7.26 |
| Methionine | 0.83 | 0.83 | 0.83 | 0.83 | 0.83 |
| Lysine | 2.58 | 2.58 | 2.58 | 2.58 | 2.58 |
| Threonine | 1.63 | 1.63 | 1.63 | 1.63 | 1.63 |
| Taurine | 0.27 | 0.26 | 0.26 | 0.26 | 0.26 |
| Total phosphorus | 1.31 | 1.31 | 1.29 | 1.29 | 1.30 |

Note: vitamin premix: VA, 18 mg/kg; $VD_3$, 5 mg/kg; VE, 150 mg/kg; VC, 500 mg/kg; $VB_1$, 16 mg/kg; $VB_6$, 20 mg/kg; $VB_{12}$, 6 mg/kg; $VK_3$, 18 mg/kg; riboflavin, 40 mg/kg; inositol, 320 mg/kg; calcium-D-pantothenate, 60 mg/kg; niacinamide, 80 mg/kg; folic acid, 5 mg/kg; biotin, 2 mg/kg; ethoxyquin, 100 mg/kg; b. mineral premix: Na, 30 mg/kg; K, 50 mg/kg; Mg, 100 mg/kg; Cu, 4 mg/kg; Fe, 25 mg/kg; Zn, 35 mg/kg; Mn, 12 mg/kg; I, 1.6 mg/kg; Se, 0.2 mg/kg; Co, 0.8 mg/kg.

## 2.3. Feeding Management

Four hundred and fifty shrimp with a uniform size, normal body color and healthy body mass were randomly selected and divided into breeding barrels (500 L). Each group was set up with 3 replicates and 30 shrimp in each replicate. The shrimp were fed three times daily at 8:00, 15:00 and 22:00. The uneaten pellets and feces were removed by a siphon method, and the exuviae and dead shrimp were removed with a dredge. The feed surplus in the pellet tray was observed 1 h and 2 h after feeding, respectively. The feeding amount was increased or decreased according to the condition of the remaining pellets, and the daily feeding amount was 2% to 5% of the body mass of the shrimp. The water was treated by sand filtering. During the feeding experiment, the water temperature was maintained at 27–32 °C, with a salinity of 28–32 ppt, a pH of 7.5–8.0, an ammonia nitrogen concentration of 0–0.2 mg/L and a dissolved oxygen concentration of 6–7 mg/L. The feeding experiment lasted for 8 weeks.

### 2.4. Sample Collection and Index Measurement

After the test, the shrimp were starved for 24 h, and the surface moisture of the shrimp was dried with a towel. The number of shrimp in each glass fiber bucket was calculated, and the total weight (accurate to 0.01 g, Puchun JE2002) was taken to calculate the survival rate, the weight gain rate and the feed conversion rate of shrimp in each feed group. The calculation formula is as follows:

$$\text{Survival rate } (\%) = 100 \times \frac{\text{initial number}}{\text{final number}}$$

$$\text{Weight gain rate } (\%) = 100 \times \frac{\text{final average mass(g)} - \text{initial average mass(g)}}{\text{initial average mass(g)}}$$

$$\text{FCR} = \frac{\text{feed amount(g)}}{\text{final average mass(g)} - \text{initial average mass(g)}}$$

### 2.5. Composition Analysis of Whole Shrimp

After the experiment, five shrimp were taken from each breeding barrel and stored at −20 °C for a later assessment of their whole body composition. The moisture content was measured after the shrimp were oven-dried at 105 °C to a constant weight. And the ash content was measured by burning them in a muffle furnace for at least 5 h at 550 °C. The crude protein content (N × 6.25) was measured by the Kjeldahl method (Kjeltec™8400; FOSS, Hilleroed, Denmark). The crude lipid content was measured by soxhlet extraction using the soxhlet system HT (Soxtec System HT6, Tacator, Sweden) [19].

### 2.6. Determination of Antioxidant Enzymes in Hepatopancreas of Shrimp

After the experiment, five shrimp were taken from each breeding barrel. Hepatopancreas tissue was taken. After freezing treatment with liquid nitrogen, the tissue was stored at −80 °C for the determination of antioxidant enzymes. Acid phosphatase (ACP), alkaline phosphatase (AKP), total superoxide dismutase (T-SOD) and total antioxidant activity (T-AOC) were tested using a commercial test kit (Nanjing Jiancheng Bioengineering Institute, Nanjing, China) according to the manufacturer's instructions.

### 2.7. Determination of Intestinal Microbiota in Shrimp

After the experiment, three shrimp were taken from each breeding barrel. The intestines were taken, frozen with liquid nitrogen and then stored at −80 °C. By extracting the total genomic DNA of a sample using the universal primers 16SrRNA or ITS (Internal Transcribed Spacer) for PCR amplification and then sequencing the highly variable region and identifying the strain, the microbial diversity in the sample can be analyzed through sequencing. We obtained classification tables for species annotation by comparing the current ASV sequences with those in the green genes (16 S rRNA) database. Based on the results of a recent study, the mi-biome diversity of five groups of experimental animals was analyzed by evaluating α-diversity indices as well as β-diversity metrics (using the principal coordinate analysis method). FAPROTAX [20] is a database that was manually constructed by Louca et al. It is more suitable for functional annotation in the biochemical processes of marine environments and lakes (sulfur, nitrogen, hydrogen and carbon cycles).

### 2.8. Data Statistics and Analysis

The statistical analysis was performed using SPSS 21.0 (SPSS Inc., Michigan Avenue, Chicago, IL, USA) for Windows. The effect was tested by a one-way ANOVA. When there were significant differences ($p < 0.05$), the groups' means were further compared by Duncan's multiple range test. The results were presented as the means ± SD ($n = 3$).

## 3. Results

### 3.1. Growth Performance and Feed Utilization of *P. monodon*

The effects of the substitution ratio on the growth performance of *P. monodon* are shown in Table 2. The nutritional composition of *P. monodon* is shown in Table 3. There was no significant difference in the nutritional composition among all the groups ($p > 0.05$). With an increase in the substitution ratio, the WGs in all the treatment groups decreased. Compared to the FM group, the WGs in the 80SC and 100SC groups significantly decreased ($p < 0.05$), while those of the 40SC and 60SC groups had a tendency to decrease, but there was no significant difference ($p > 0.05$). Similarly, the SRs in the 80SC and 100SC groups significantly decreased compared with the FM group ($p < 0.05$), while there was no significant difference in the survival rate among the FM, 40SC and 60SC groups ($p > 0.05$). Compared to the FM group, the FCR in the 100SC group was significantly increased ($p < 0.05$), while there was no significant difference among the FM and 40SC, 60SC and 80SC groups ($p > 0.05$).

**Table 2.** Effects of different substitution ratios on the growth performance of *P. monodon*.

| Items | Diets | | | | |
|---|---|---|---|---|---|
| | FM | 40SC | 60SC | 80SC | 100SC |
| IW (g) | 3.41 ± 0.01 | 3.44 ± 0.01 | 3.47 ± 0.02 | 3.42 ± 0.03 | 3.46 ± 0.03 |
| FW (g) | 13.70 ± 0.45 $^{c}$ | 13.22 ± 0.85 $^{bc}$ | 13.38 ± 1.23 $^{c}$ | 11.52 ± 0.74 $^{ab}$ | 10.44 ± 0.08 $^{a}$ |
| WG (%) | 302.26 ± 12.02 $^{c}$ | 283.69 ± 25.01 $^{bc}$ | 285.98 ± 35.84 $^{bc}$ | 236.65 ± 24.64 $^{ab}$ | 201.90 ± 1.95 $^{a}$ |
| SR (%) | 82.67 ± 0.02 $^{c}$ | 80.67 ± 0.01 $^{bc}$ | 80 ± 0.02 $^{bc}$ | 76.67 ± 0.01 $^{b}$ | 71.33 ± 0.01 $^{a}$ |
| FCR | 1.33 ± 0.08 $^{a}$ | 1.42 ± 0.15 $^{ab}$ | 1.48 ± 0.29 $^{ab}$ | 1.51 ± 0.09 $^{ab}$ | 1.78 ± 0.14 $^{b}$ |

Note: IW: initial body weight; FW: final body weight; SR: survival rate; WG: weight gain rate; FCR: feed coefficient ratio. Data are expressed as mean ± SD ($n = 3$). Means with different superscripts are significantly different ($p < 0.05$).

**Table 3.** Nutritional composition of *P. monodon*.

| Items | Diets | | | | |
|---|---|---|---|---|---|
| | FM | 40SC | 60SC | 80SC | 100SC |
| Moisture | 74.00 ± 0.31 | 74.42 ± 0.71 | 74.81 ± 0.55 | 74.28 ± 0.71 | 72.32 ± 0.47 |
| Crude protein | 73.42 ± 1.06 | 73.58 ± 0.48 | 74.04 ± 0.16 | 73.29 ± 0.77 | 72.83 ± 0.47 |
| Crude fat | 7.13 ± 0.18 | 6.95 ± 0.20 | 5.97 ± 0.56 | 6.16 ± 0.74 | 6.69 ± 0.09 |
| Crude ash | 16.20 ± 0.28 | 15.91 ± 0.71 | 16.60 ± 0.10 | 16.65 ± 0.56 | 14.50 ± 0.47 |

Note: unit: %.

### 3.2. Hepatopancreas Antioxidant Capacity of *P. monodon*

The activities of ACP, AKP, T-AOC and T-SOD in *P. monodon* are shown in Figure 1. Compared to the FM group, the ACP activity (Figure 1a) in the 80SC and 100SC groups significantly increased ($p < 0.05$), while that of the 40SC and 60SC groups had a tendency to increase, but there was no significant difference ($p > 0.05$). The AKP activity (Figure 1b) in the 100SC group was significantly high ($p < 0.05$), while there were no significant differences among the other four groups ($p > 0.05$). There were no significant differences in the T-AOC (Figure 1c) and T-SOD (Figure 1d) activities among all the treatment groups ($p > 0.05$).

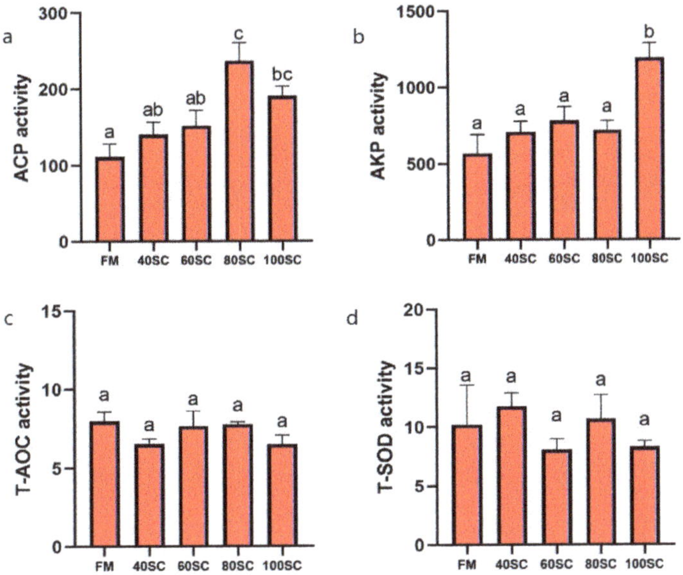


**Figure 1.** Hepatopancreas antioxidant capacity of *P. monodon*. (**a**): ACP activity; (**b**): AKP activity; (**c**): T-AOC activity; (**d**): T-SOD activity. Means with different letters are significantly different ($p < 0.05$).

### *3.3. Composition and Rationality Analysis of Intestinal Microbial ASVs in P. monodon*

As shown in Figure 2a, a total of 354 ASVs were obtained by 16SrDNA high-throughput sequencing. According to the analysis of the Wayne diagram, 42 ASVs were present in the five test groups, no unique ASV was found in the 60SC group and one unique ASV was found in all the other groups. The Shannon–Wiener curve (Figure 2b) [21] considers both species' evenness and richness. When the curve tends to be flat, it indicates that the amount of sequencing data is large enough. A species accumulation curve can show whether a species increased along with an increase in sample size, so it is an effective tool to determine whether a sample size is sufficient. The curve in Figure 2c gradually inclines and then flattens, proving that the species in the samples did not increase with the increase in the sample size or the number of ASVs in the data analysis. The results indicated that the microbial diversity was sufficient to be fully detected, and the sequencing data were reasonable.

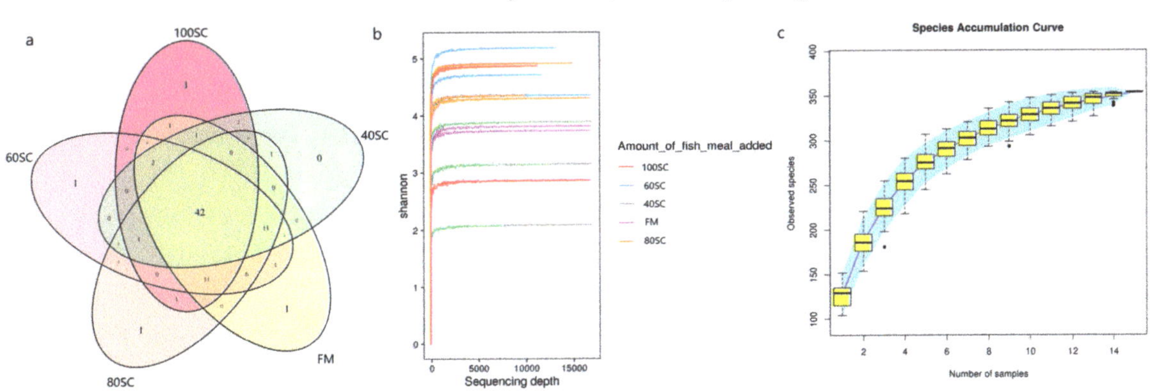


**Figure 2.** Composition and rationality analysis of intestinal microbial ASVs in *P. monodon*. (**a**) Venn diagram; (**b**) Shannon index; (**c**) species accumulation box diagram.

### 3.4. The Richness and Diversity of Intestinal Microbiota in P. monodon

The inter-group difference analysis of the α diversity index was evaluated. A box-type diagram of the inter-group difference analysis is shown in Figure 3a. The Shannon indexes of the 100SC, 80SC and 60SC groups were close to that of the FM group. Based on the Jaccard, bray Curtis, unweighted unifrac and weighted unifrac distances, a PCoA analysis was performed (Figure 3b). The sample distances in the FM and 60SC groups were relatively close. The sample distances of the 100SC and 40SC groups were far away. This proved that the species composition structure in the 60SC group was similar to that of the FM group.

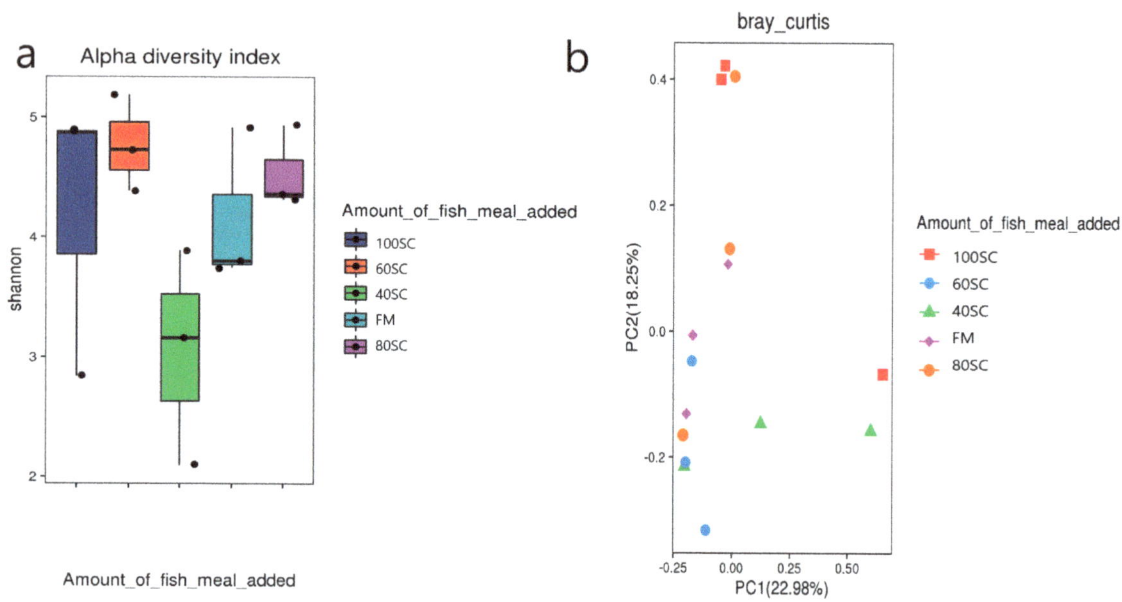


**Figure 3.** The richness and diversity of intestinal microorganisms in *P. monodon*. (**a**) Index group-difference box chart; (**b**) PCoA analysis chart.

### 3.5. Intestinal Microbiota Composition of P. monodon

The analysis of the intestinal contents of *P. monodon* encompassed multiple taxonomic levels, specifically the phylum, order, family and genus. The phylum was chosen as the representative taxonomic level for this study. As shown in Figure 4a,b, at the phylum level, Proteobacteria, Bacteroidota and Actinobacteriota formed the core microbiota. Proteobacteria was the most abundant phyla in the five groups, accounting for 37.67%, 66%, 40%, 40% and 43.33%, respectively (Table 4). Compared to the FM group, the Fusobacteriota in the other four groups decreased significantly ($p < 0.05$). There were no significant differences in the other bacteria among the five groups ($p > 0.05$).

### 3.6. Functional Gene Prediction Analysis of Intestinal Microbiota and Functional Prediction of FAPROTAX

The PCA analysis is presented in Figure 5, which was performed based on the functional abundance table of metabolic pathways predicted by PICRUSt2. The pictures from left to right are the pathway level, protein level and enzyme level. We found that among the three levels, the 100SC and 40SC groups were far apart on the PCA diagram, while on the contrary, the 80SC, FM and 60SC groups were closer together on the PCA diagram, indicating that their functional compositions were similar.

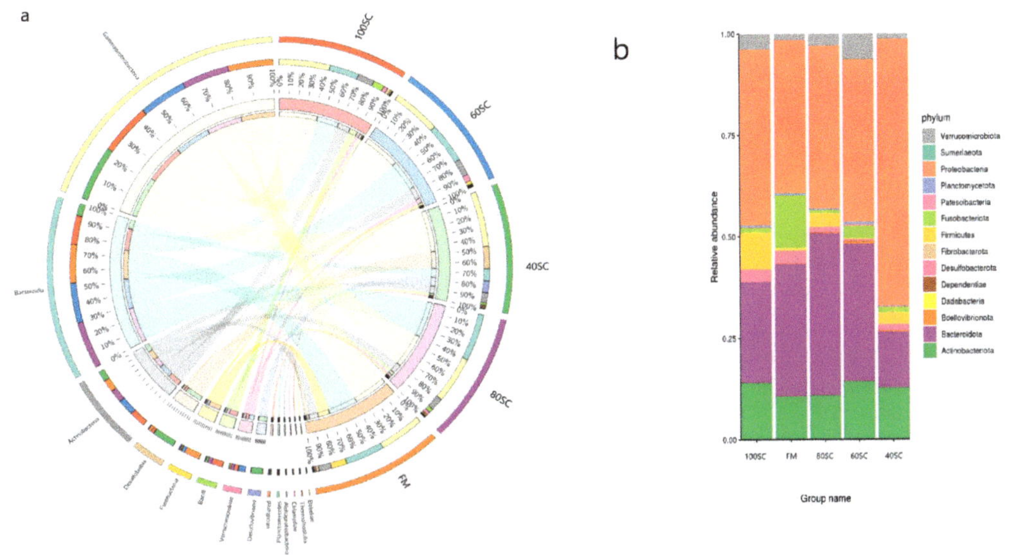


**Figure 4.** Intestinal microbiota composition of *P. monodon*. (**a**) Circos plot; (**b**) histogram of relative abundance of species.

**Table 4.** Distribution of the top 10 microbial phylum levels in the intestinal contents of *P. monodon* in different treatment groups.

| Phylum | Group | | | | |
|---|---|---|---|---|---|
| | **FM** | **40SC** | **60SC** | **80SC** | **100SC** |
| Proteobacteria | 37.67 ± 19.03 | 66.00 ± 9.00 | 40 ± 21.70 | 40 ± 15.10 | 43.33 ± 19.29 |
| Bacteroidota | 33.00 ± 18.36 | 14.00 ± 3.00 | 34.00 ± 7.81 | 40.00 ± 25.06 | 24.67 ± 16.65 |
| Actinobacteriota | 10.67 ± 7.37 | 12.33 ± 11.50 | 14.33 ± 7.23 | 10.67 ± 4.04 | 13.67 ± 10.12 |
| Firmicutes | 0.80 ± 1.04 | 3.02 ± 2.98 | 0.30 ± 0.10 | 3.53 ± 5.60 | 9.06 ± 13.83 |
| Verrucomicrobiota | 1.43 ± 1.40 | 1.10 ± 0.85 | 6.23 ± 6.24 | 2.67 ± 1.52 | 3.50 ± 2.78 |
| Desulfobacterota | 3.08 ± 5.12 | 1.53 ± 1.46 | 0.40 ± 0.53 | 1.17 ± 0.72 | 3.13 ± 2.42 |
| Fusobacteriota | 13.33 ± 10.40 [a] | 1.13 ± 0.81 [b] | 3.03 ± 3.56 [b] | 0.84 ± 10.3 [b] | 10.6 ± 0.90 [b] |
| Planctomycetota | 0.38 ± 0.54 | 0.16 ± 0.14 | 0.63 ± 0.58 | 0.30 ± 0.30 | 0.38 ± 0.53 |
| Patescibacteria | 0.37 ± 0.11 | 0.15 ± 0.05 | 0.37 ± 0.54 | 0.05 ± 0.04 | 0.17 ± 0.21 |
| Dependentiae | 0.01 ± 0.11 | 0.15 ± 0.15 | 0.27 ± 0.30 | 0.08 ± 0.10 | 0.06 ± 0.06 |

Note: data are expressed as mean ± SD ($n$ = 3). Means with different superscripts are significantly different ($p < 0.05$).

FAPROTAX is more complete for analyzing the classification and function of marine microorganisms. The results are shown in Figure 6. In all the groups, the abundances of chemoheterophya and aerobic chemoheterotrophy were high, while the fermentation and nitrate reduction rates were relatively low. Intracellular parasites, aromatic compound degradation and human-associated factors were not found.

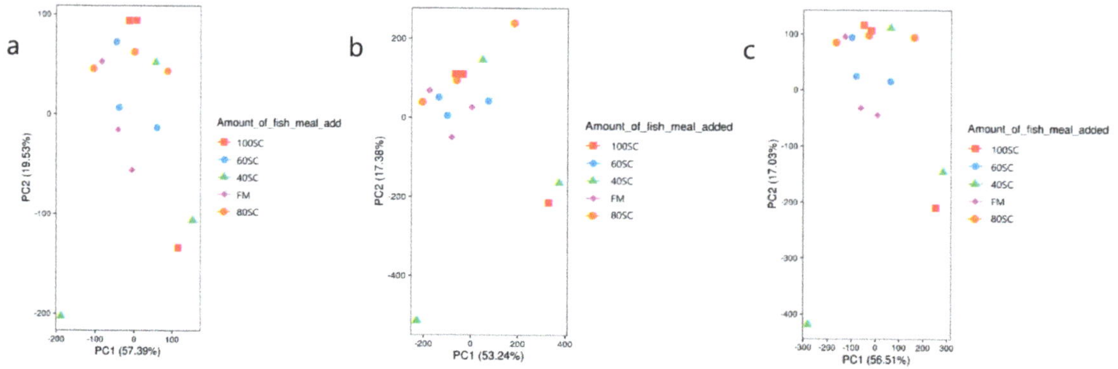


**Figure 5.** PCA dimensionality reduction analysis diagram. (**a**) PCA analysis at pathway level; (**b**) PCA analysis at protein level; (**c**) PCA analysis at enzyme level.

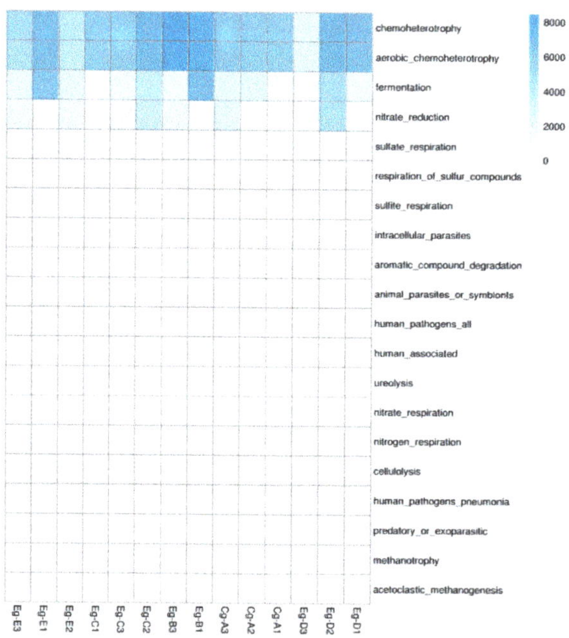


**Figure 6.** FAPROTAX functional abundance heat map.

## 4. Discussion

The growth performance of aquatic animals is generally expressed by the WG, SR, FCR, etc. [22]. Yan et al. [23] found that replacing 50% of fish meal protein with soybean meal did not have a negative impact on the growth of *L. vannamei*, but a higher substitution level would lead to a significant decrease in the WGR and SGR. Zhang et al. also found that the proportional replacement of 20% of fish meal with SPC and peanut bran had no significant effect on the growth performance of *P. monodon* [24]. Studies have shown that a high proportion of plant protein replacing fish meal may inhibit the growth and feed utilization of aquatic animals [25]. In a study by Daniela et al., using soybean meal as a substitute for part of fish meal did not inhibit the growth performance of *Centropomus viridis* [26]. A study by Wu et al. also showed that replacing an appropriate amount of fish meal with chicken meal did not inhibit the growth of *Micropterus salmoides* [27]. In a study by Yu et al., replacing part of fish meal with soy protein concentrate also had no significant effect on the growth performance of *L. vannamei* [28]. Similar conclusions were found in studies by Chen [29] and Wang et al. [30]. In this study, we used chicken meal and soybean products as substitutes for part of fish meal, and with an increase in substitution ratio, the WGs and SRs in all the treatment groups were decreased. The WGs and SRs in the 80SC and 100SC groups significantly decreased compared to those in the FM group, but there was no significant difference, although those in the 40SC and 60SC groups had a slight decrease. The FCR showed a decreasing trend; the FCR in the 100SC group significantly increased ($p < 0.05$), while there was no significant difference between the FM group and the 40SC, 60SC and 80SC groups ($p > 0.05$). These results are similar to those of other studies and prove that it is feasible to partially replace fish meal with soybean products and chicken meal.

Oxidative stress is one of the response mechanisms of animal organisms to environmental stress. Changes in the diet of aquatic animals may affect their antioxidant capacity and immunoenzyme activity. Under environmental stress, many reactive oxygen species will be produced in the body, resulting in organism damage [31]. *P. monodon* has a relatively complete antioxidant system, which can maintain an organism's homeostasis [32]. SOD plays a crucial role in the balance between oxidation and antioxidants in the body and is an important antioxidant oxidase in maintaining a normal metabolism [33]. ACP is an important enzyme in macrophages and plays an important role in the immune response of shrimp [34]. Jiang et al. found that using concentrated dephenolized cottonseed protein (CDCP) as a partial substitute for fish meal did not have a significant effect on the SOD and malondialdehyde (MDA) contents of *P. monodon* [35]. This is consistent with the conclusion of Cui et al. that using soybean meal, fermented soybean meal and cottonseed meal as a substitute for 50% of fish meal did not have a negative impact on the growth performance and antioxidant capacity of juvenile *Eriocheir sinensis* [36]. The study by Wu et al. showed that replacing an appropriate amount of fish meal with chicken meal can improve the antioxidant capacity and immunity of *Micropterus salmoides* [27]. An experiment which used fish soluble pulp as a partial substitute for fish meal to feed *Macrobrachium rosenbergii* did not cause significant differences in the SOD and hepatopancreas ACP contents among all the groups [37]. In our experiment, there was no significant difference between the T-SOD and T-AOC contents among all the groups, indicating that using chicken meal and soybean products as substitutes for part of fish meal would not have a significant effect on the antioxidant capacity of *P. monodon*. The ACP content in the 80SC and 100SC groups was significantly increased compared to that in the FM group ($p < 0.05$), while that in the 40SC and 60SC groups had a slight increase, but there was no significant difference ($p > 0.05$). The AKP content in the 100SC group was significantly high, while there were no significant differences among the other four groups ($p > 0.05$), indicating that a long-term high substitution level would cause the organism to be continuously subjected to stress. The 40SC, 60SC and FM groups did not have significant differences among their ACP and AKP contents, which proved that the substitutions had no effect on the organisms.

Intestinal microbiota refers to the large number of microorganisms present in the intestinal tract of animals, which they rely on to live and help animals carry out a variety of physiological and biochemical functions. Changes in aquatic animal feed could have some impact on the intestinal microbiota [38]. 16SrRNA is located on the small subunit of the ribosome of prokaryotic cells, including ten conserved regions and nine hypervariable regions, among which the conserved regions have little difference among bacteria and the hypervariable regions are specific to the genus or species and vary between them. Therefore, 16SrDNA can be used as a characteristic nucleic acid sequence to reveal biological species and is the most suitable index for bacterial phylogeny and classification identification. Although the OTU clustering method can effectively overcome sequencing errors, it reduces the accuracy of classification, and some sequences below a set threshold cannot be accurately distinguished. The method of de-noising is recommended. OTUs have been given a new name, ASVs (Amplicon Sequence Variants) [39,40]. Intestinal probiotics can promote the body's digestion and absorption of food and competitively inhibit the growth of intestinal harmful microorganisms, thus maintaining the ecological balance in the body and promoting the healthy growth of the body [41]. The functional prediction of FAPROTAX in this paper does not detect the functional components that are harmful to the body. Many studies have shown that Proteobacteria are the dominant bacteria in the gut of shrimp [42,43], and some members of this phyla are involved in the nitrogen cycle and the mineralization of organic compounds [44,45]. In this study, Proteobacteria were the most abundant bacteria in the five groups, accounting for 37.67%, 66%, 40%, 40% and 43.33%, respectively. Compared to the FM group, the abundance of Fusobacteriota in the other four groups decreased significantly. Fusobacteriota are conducive to food digestion and absorption [46]. In the present study, the abundance of Fusobacteriota in the substitution groups decreased significantly, so we speculated that it may be due to the anti-nutritional factors in soybean products, which can hinder the absorption and utilization of proteins. This also explains why the WG in the substitution groups decreased. The results in this study were consistent with the results in a previous study on *L. vannamei* [47].

## 5. Conclusions

Using soybean products and chicken meal as substitutes for 40% and 60% of fish meal did not affect the growth performance and antioxidant capacity of *P. monodon*, and the intestinal microbiota of *P. monodon* remained stable and did not cause damage to the organism. However, a high substitution ratio can affect the growth and immunoenzyme activity of *P. monodon*. Considering the effects on growth performance, antioxidant capacity and intestinal microbiota, it is feasible to use soybean products and chicken meal to replace 60% of fish meal.

**Author Contributions:** S.J. and F.Z. conceived and designed the experiments. W.Y. and J.S. performed the bioinformatics analysis and prepared the manuscript, the table and the figures. Q.Y. and J.H. conducted the experiment. Y.L., S.J. and Y.Y. collected the samples. All authors have read and agreed to the published version of the manuscript.

**Funding:** This study was funded by the National Key Research and Development Plan Project (2022YFD2400104), the Youth Fund of Hainan Natural Science Foundation (321QN351), the earmarked fund for CARS-48, the Central Public-Interest Scientific Institution Basal Research Fund, the South China Sea Fisheries Research Institute, CAFS (2023TD34) and the 2022 Provincial Rural Revitalization Strategy Special Fund Seed Industry Revitalization Project (2022-SPY-00-002, 2022-SPY-00-006).

**Institutional Review Board Statement:** The present study was approved by the Animal Care and Use Committee of the South China Sea Fisheries Research Institute, Chinese Academy of Fishery Sciences (Approval number SCSFRI2021-0731). All procedures were strictly carried out according to the regulations and guidelines approved by the committee.

**Data Availability Statement:** Data are contained within the article.

**Conflicts of Interest:** The authors declare no conflicts of interest.

## References

1. Rahi, L.; Sabbir, W.; Salin, K.R.; Aziz, D.; Hurwood, D.A. Hurwood, Physiological, biochemical and genetic responses of black tiger shrimp (*Penaeus monodon*) to differential exposure to white spot syndrome virus and *Vibrio parahaemolyticus*. *Aquaculture* **2022**, *546*, 737337. [CrossRef]
2. El-Sayed, A.F.M. Long-term evaluation of cotton seed meal as a protein source for *Nile tilapia, Oreochromis niloticus* (Linn.). *Aquaculture* **1990**, *84*, 315–320. [CrossRef]
3. Suárez, J.; Gaxiola, G.; Mendoza, R.; Cadavid, S.; Garcia, G.; Alanis, G.; Suárez, A.; Faillace, J.; Cuzon, G. Substitution of fish meal with plant protein sources and energy budget for white shrimp *Litopenaeus vannamei* (Boone, 1931). *Aquaculture* **2009**, *289*, 118–123. [CrossRef]
4. Watanabe, T. Strategies for further development of aquatic feeds. *Fish. Sci.* **2010**, *68*, 242–252. [CrossRef]
5. Dawood, M.A.O.; Magouz, F.I.; Mansour, M.; Saleh, A.A.; El Asely, A.M.; Fadl, S.E.; Ahmed, H.A.; Al-Ghanim, K.A.; Mahboob, S.; Al-Misned, F. Evaluation of yeast fermented poultry by-product meal in Nile tilapia (*Oreochromis niloticus*) feed: Effects on growth performance, digestive enzymes activity, innate immunity, and antioxidant capacity. *Front. Vet. Sci.* **2020**, *6*, 516. [CrossRef] [PubMed]
6. Moutinho, S.; Peres, H.; Serra, C.; Martínez-Llorens, S.; Tomás-Vidal, A.; Jover-Cerdá, M.; Oliva-Teles, A. Meat and bone meal as partial replacement of fish meal in diets for gilthead sea bream (*Sparus aurata*) juveniles: Diets digestibility, digestive function, and microbiota modulation. *Aquaculture* **2017**, *479*, 721–731. [CrossRef]

7. Nannan, S.; Zhicheng, S.; Cancan, X.; Hua, M.; Xiangning, C.; Hanliang, C.; Jianhe, X.; Yongkai, S.; Chaoqing, W.; Lu, Z. Comparative evaluation of soybean meal vs. extruded soybean meal as a replacer for fishmeal in diets of olive flounder (*Paralichthys olivaceus*): Effects on growth performance and muscle quality. *Aquaculture* **2024**, *578*, 740136.
8. Dominy, W.G.; Ako, H. The utilization of blood meal as a protein ingredient in the diet of the marine shrimp *Penaeus vannamei*. *Aquaculture* **1988**, *70*, 289–299. [CrossRef]
9. Lim, C. Substitution of cottonseed meal for marine animal protein in diets for *Penaeus vannamei*. *J. World Aquac. Soc.* **1996**, *27*, 402–409. [CrossRef]
10. Liu, X.H.; dan Ye, J.; Wang, K.; Yang, W.; Zhou, L. Partial replacement of fish meal with peanut meal in practical diets for the Pacific white shrimp, *Litopenaeus vannamei*. *Aquac. Res.* **2012**, *43*, 745–755. [CrossRef]
11. Liu, Y.L.; Zhong, L.; Chen, T.; Shi, Y.; Hu, Y.; Zeng, J.G.; Liu, H.Y.; Xu, S.D. Dietary sanguinarine supplementation on the growth performance, immunity and intestinal health of grass carp (*Ctenopharyngodon idellus*) fed cottonseed and rapeseed meal diets. *Aquaculture* **2020**, *528*, 735521. [CrossRef]
12. Li, E.; Chen, L.; Cai, Y. Effects of Soy Protein Concentrate as a Dietary Protein Source on Digestive Enzyme Activities of Eriocheir sinensis. *J. Zhanjiang Ocean. Univ.* **2006**, *26*, 14–21.
13. Amaya, E.A.; Davis, D.A.; Rouse, D.B. Replacement of fish meal in practical diets for the Pacific white shrimp (*Litopenaeus Vannamei*) reared under pond conditions. *Aquaculture* **2007**, *262*, 393–401. [CrossRef]
14. Lü, J.X.; Guo, W.D. Dietary nutritional value of poultry by-product meal. *China Feed* **1995**, *13*, 25–26. (In Chinese)
15. Ma, L.; Mu, Y.; Li, W.; Gen, X.; Jiang, M.; Wen, H.; Wang, G.; Li, Y. Effects of Peru Fish Meal Replacement by Pet Grade Chicken Meal on Growth Performance, Serum Biochemical Indexes, intestinal and Liver Histomorphology of Largemouth Bass (*Micropterus salmoides*). *Chin. J. Anim. Nutr.* **2021**, *33*, 4183–4193.
16. Guo, X.; Dai, J.; Zhang, J.; Hu, Y. Research progress on chicken meal and enzymatic hydrolysates of chicken skeleton in aquatic feeds. *Mod. Anim. Husb. Vet. Med.* **2022**, *2*, 80–83.
17. Yi, X.; Chen, R.; Xu, J. Effects of the Poultry By-Product Meal Replacing Partially Dietary Fishmeal on the Growth, Feed Utilization and Anti-Oxidative Capacity of Golden Pompano *Trachinotus ovatus*. *Period. Ocean. Univ. China* **2019**, *49*, 17–24. [CrossRef]
18. Pan, Y.; Huang, Z.; Lv, F.; Lang, Y.; Zhao, Y.; Li, Y. Effect of enzymatic hydrolysis of soybean meal on growth and intestinal microbiota of *Procambarus clarkii*. *Feed. Res.* **2022**, *45*, 44–48. [CrossRef]
19. Horwitz, W. Official Methods of Analysis of the Association of Official Analytical Chemists. *J. Chem. Educ.* **1961**, *38*, 431.
20. Louca, S.; Parfrey, L.W.; Doebeli, M. Decoupling function and taxonomy in the global ocean microbiome. *Science* **2016**, *353*, 1272–1277. [CrossRef]
21. Wong, S.; Waldrop, T.; Summerfelt, S.; Davidson, J.; Barrows, F.; Kenney, P.B.; Welch, T.; Wiens, G.D.; Snekvik, K.; Rawls, J.F.; et al. Aquacultured rainbow trout (*Oncorhynchus mykiss*) possessa large core intestinal microbiota that is resistant to variation in diet and rearing density. *Appl. Environ. Microbiol.* **2013**, *79*, 4974–4984. [CrossRef] [PubMed]
22. Zhang, X.; Liang, M.; Wei, Y.; Liao, Z.; Zhang, Q.; Xu, H. Effects of Dietary Protein Content and Stocking Density on Growth Performance, Nitrogen Excretion, and Relevant Biochemica Parameters of *Juvenile Takifuqu rubripes*. *Prog. Fish. Sci.* **2021**, *42*, 74–83.
23. Yan, Y. Effects of Several Feed Additives on Growth Performance, Immune Response and Intestinal Health of *Litopenaeus Vannamei* Fed Diets with Fish Meal Partially Replaced by Soybean Meal. Master's Thesis, Jimei University, Xiamen, China, 2019.
24. Zhang, J.; Lin, H.; Huang, Z.; Niu, J.; Zhou, F.; Chen, X.; Wang, Y.; Xia, D. Effects of plant proteins supplemented with amino acids on growth and non-specific immunity of *Penaeus monodon*. *South. Fish. Sci.* **2013**, *9*, 44–50.
25. Sun, H.; Tang, J.W.; Yao, X.H.; Wu, Y.F.; Wang, X.; Liu, Y. Effects of replacement of fish meal with fermented cottonseed meal on growth performance, body composition and haemolymph indexes of Pacific white shrimp, *Litopenaeus vannamei* Boone, 1931. *Aquac. Res.* **2016**, *47*, 2623–2632. [CrossRef]
26. Daniela, A.-H.; Crisantema, H.; Emmanuel, M.-M.; Leonardo, I.-C.; Esmeralda, L.-V.; Nayely, L.-L.; Cristina, C.-S.M. Fish meal replacement by soybean products in aquaculture feeds for white snook, *Centropomus viridis*: Effect on growth, diet digestibility, and digestive capacity. *Aquaculture* **2021**, *530*, 735823.
27. Wu, J.; Liao, R.; Kuang, W.; Sun, H.; Chen, Y.; Tan, B.; Lin, S. Effects of Replacing Fish Meal with Domestic Poultry by-Product Meal on Growth, Liver Health Andintestinal Barrier of *Micropterus salmoides*. *J. Fish.* **2023**, *47*, 47–57.
28. Yu, F.; Zhu, L.; Wu, W.; Zhao, S. Research on brown fish meal replacement with soy protein concentrate in the feed of *penaeus vannamel*. *Food Feed. Ind.* **2020**, *5*, 48–52.
29. Chen, J. The Application of Soy Protein Concentrate, Methionine, and Hydroxyl Methionine Calcium in White Shrimp Feed. Master's Thesis, Shanghai Ocean University, Shanghai, China, 2018.
30. Wang, J. Application of Enzymatic Chicken Mash and Soy Protein Concentrate in the Feed of Hybrid Grouper. Master's Thesis, Guangdong Ocean University, Zhanjiang, China, 2021. [CrossRef]
31. Franco, R.; Sánchez-Olea, R.; Reyes-Reyes, E.M.; Panayiotidis, M.I. Environmental toxicity, oxidative stress and apoptosis: Ménage à trois. *Mutat. Res. Genet. Toxicol. Environ. Mutagen.* **2009**, *674*, 3–22. [CrossRef]
32. Mathew, S.; Kumar, K.A.; Anandan, R.; Viswanathan Nair, P.G.; Devadasan, K. Changes in tissue defence system inwhite spot syndrome virus (WSSV) infected *Penaeus monodon*. *Compart. Biochem. Physiol. Toxicol. Pharmacol.* **2007**, *145*, 315–320. [CrossRef]
33. Wu, Z.; You, F.; Wang, Y.; Wen, A.; Ma, D.; Xu, Y.; Zhang, P. The effects of hypoxia and hyperoxia on nucleus anomaly, SOD, CAT activities and MDA content in juvenile turbot *Scophthalmus maximus*. *J. Shanghai Ocean. Univ.* **2011**, *20*, 808–813.

34. Chen, Q.; Chen, S.; Shi, Y.; Zhu, L.; Yan, S. Characterization of Alkaline Phosphatasefrom *Penaeus Penicillatus*. *J. Xiamen Univ. (Nat. Sci.)* **1996**, *35*, 257–261.
35. Jiang, S.; Chen, Z.; Zhou, F.; Yang, Q.; Huang, J.; Yang, L.; Li, Y.; Jiang, S. Study on partial replacement of fish meal with concentrated dephenolized cottonseed protein in feed of *Penaeus monodon*. *Aquac. Res.* **2021**, *52*, 3871–3881. [CrossRef]
36. Cui, Y.; Zhang, N.; Ma, Q.; Chen, Q.; Shen, Z.; Du, Z.; Chen, L. Effects of four plant proteins on growth performance, amino acid deposition rate and antioxidant enzyme activity of juvenile Chinese Mitten crab. *J. Hydrobiol.* **2017**, *41*, 146–154.
37. Luo, Q.; Zhao, Y.; Li, H.; Gu, X.; Ding, G.; Wang, G.; Xu, Q.; Liu, C. Effects of Replacement of Fish Meal by Stickwater on Growth Performance, Body Composition, immune Indexes and Intestinal Digestive Enzyme Activities of *Macrobrachium rosenbergii*. *Chin. J. Anim. Nutr.* **2021**, *33*, 6366–6375.
38. Wang, Z.; Liao, S.; Wang, J.; Wang, Y.; Huang, Z.; Yu, W.; Huang, X.; Lin, H.; Luo, M.; Cheng, Z.; et al. Effects of Fermented Cottonseed Meal Substitution for Fish Meal on Intestinal Enzymatic Activity, Inflammatory and Physical-Barrier-Related Gene Expression, and Intestinal Microflora of Juvenile Golden Pompano (*Trachinotus ovatus*). *Fishes* **2023**, *8*, 466. [CrossRef]
39. Bolyen, E.; Rideout, J.R.; Dillon, M.R.; Bokulich, N.A.; Abnet, C.C.; Al-Ghalith, G.A.; Alexander, H.; Alm, E.J.; Arumugam, M.; Asnicar, F.; et al. Reproducible, interactive, scalable and extensible microbiome data science using QIIME 2. *Nat. Biotechnol.* **2019**, *37*, 852–857. [CrossRef] [PubMed]
40. Knight, R.; Vrbanac, A.; Taylor, B.C.; Aksenov, A.; Callewaert, C.; Debelius, J.; Gonzalez, A.; Kosciolek, T.; McCall, L.I.; McDonald, D.; et al. Best practices for analyzing microbiomes. *Nat. Rev. Microbiol.* **2018**, *16*, 410–422. [CrossRef]
41. El-Saadony, M.T.; Alagawany, M.; Patra, A.K.; Kar, I.; Tiwari, R.; Dawood, M.A.; Dhama, K.; Abdel-Latif, H.M. The functionality of probiotics in aquaculture: An overview. *Fish Shellfish. Immunol.* **2021**, *117*, 36–52. [CrossRef]
42. Rungrassamee, W.; Klanchui, A.; Maibunkaew, S.; Chaiyapechara, S.; Jiravanichpaisal, P.; Karoonuthaisiri, N. Characterization of intestinal bacteria in wild and domesticated adult black tiger shrimp (*Penaeus monodon*). *PLoS ONE* **2014**, *9*, e91853. [CrossRef]
43. Huang, Z.; Li, X.; Wang, L.; Shao, Z. Changes in the intestinal bacterial community during the growth of white shrimp, *Litopenaeus vannamei*. *Aquac. Res.* **2016**, *47*, 1737–1746. [CrossRef]
44. Kersters, K.; De Vos, P.; Gillis, M.; Swings, J.; Vandamme, P.; Stackebrandt, E.R.K.O. *Introduction to the Proteobacteria*; Dworkin, M., Falkow, S., Rosenberg, E., Schleifer, K.-H., Eds.; The Prokaryotes; Springer: New York, NY, USA, 2016; pp. 3–37.
45. Deng, Y.; Xu, X.; Yin, X.; Lu, H.; Chen, G.; Yu, J.; Ruan, Y. Effect of stock density on the microbial community in biofloc water and Pacific white shrimp (*Litopenaeus vannamei*) gut microbiota. *Appl. Microbiol. Biotechnol.* **2019**, *103*, 4241–4252. [CrossRef]
46. Zhang, Z.M. Spatiotemporal Heterogeneity of Microbiome in Southern Catfish. Ph.D. Thesis, Huazhong Agricultural University, Wuhan, China, 2018. (In Chinese).
47. Wang, P.; Chen, S.; Wei, C.; Yan, Q.; Sun, Y.Z.; Yi, G.; Li, D.; Fu, W. Monascus purpureus M-32 improves growth performance, immune response, intestinal morphology, microbiota and disease resistance in *Litopenaeus vannamei*. *Aquaculture* **2021**, *530*, 735947. [CrossRef]

fishes

Article

# The Effects of Replacing Fish Meal with Enzymatic Soybean Meal on the Growth Performance, Whole-Body Composition, and Health of Juvenile Gibel Carp (*Carassius auratus gibelio*)

**Adolphe Uyisenga [1], Hualiang Liang [1,2], Mingchun Ren [1,2], Dongyu Huang [1,2], Chunyu Xue [1,3], Heng Yin [3] and Haifeng Mi [3,*]**

1 Wuxi Fisheries College, Nanjing Agriculture University, Wuxi 214081, China; uyado05@gmail.com (A.U.); lianghualiang@ffrc.cn (H.L.); renmc@ffrc.cn (M.R.); huangdongyu1995@163.com (D.H.); 2022213004@stu.njau.edu.cn (C.X.)
2 Key Laboratory of Integrated Rice-Fish Farming Ecology, Ministry of Agriculture and Rural Affairs, Freshwater Fisheries Research Center, Chinese Academy of Fishery Sciences, Wuxi 214081, China
3 Tongwei Agricultural Development Co., Ltd., Key Laboratory of Nutrition and Healthy Culture of Aquatic, Livestock and Poultry, Ministry of Agriculture and Rural Affairs, Healthy Aquaculture Key Laboratory of Sichuan Province, Chengdu 610041, China; yinh@tongwei.com
* Correspondence: mihf@tongwei.com

**Abstract:** Fish meal (FM) constitutes the main, expensive component in aquatic diets. However, the supply of FM is no longer sufficient to sustain global aquaculture production. This study had the primary goal of assessing if the replacement of FM with enzymatic soybean meal (ESBM) can affect the performance of growth and immunological response in juvenile Gibel carp. Juvenile fish with an initial weight of 45.02 ± 0.03 g were arbitrarily assigned to 18 fish cages of 1 $m^3$ each, then fed with diets of different levels of ESBM (0% (control group), 4%, 8%, 12%, 16%, and 20%) for 159 days. These diets corresponded, respectively, to the replacement levels of 0% (control group), 20%, 40%, 60%, 80%, and 100% FM by ESBM. For the parameters of growth and whole-body composition, no obvious differences were found between the control group and other replacement levels ($p > 0.05$). Similarly, none of the replacement levels showed significant effects for alanine transaminase (ALT), total cholesterol (TC), alkaline phosphatase (ALP), and glucose (GLU) levels ($p > 0.05$). Malondialdehyde (MDA) levels, as well as the activities of superoxide dismutase (SOD) and total antioxidant capacity (T-AOC) in plasma, were not significantly affected at all replacement levels, according to the findings of this study ($p > 0.05$). The replacement level of 60% significantly increased the activities of catalase (CAT), whereas the replacement levels of 20% and 100% markedly decreased the activities of this enzyme ($p < 0.05$). Hepatic and intestinal tissues in this study did not show obvious alterations at all levels of replacement.

**Keywords:** enzymatic soybean meal; Gibel carp (*Carassius auratus gibelio*); growth performance; replacement levels; health

**Key Contribution:** ESBM can completely replace FM meal in the diet of Gibel carp.

**Citation:** Uyisenga, A.; Liang, H.; Ren, M.; Huang, D.; Xue, C.; Yin, H.; Mi, H. The Effects of Replacing Fish Meal with Enzymatic Soybean Meal on the Growth Performance, Whole-Body Composition, and Health of Juvenile Gibel Carp (*Carassius auratus gibelio*). *Fishes* **2023**, *8*, 423. https://doi.org/10.3390/fishes8080423

Academic Editors: Qiyou Xu, Jianhua Ming, Fei Song, Changle Qi and Chuanpeng Zhou

Received: 27 June 2023
Revised: 4 August 2023
Accepted: 15 August 2023
Published: 18 August 2023



## 1. Introduction

Aquatic animal protein significantly contributes to global food security and the supply of animal protein [1]. With an astounding production of 2.75 million tons in 2020, the Gibel carp (*Carassius auratus gibelio*) is among the major freshwater fish species in the world [2]. This fish species is mostly cultivated in China and has a number of desirable qualities, including good growth rates, excellent taste, suitability for culture systems, and a high level of stress tolerance [3].

Feeding costs can account for up to 70% of an aquaculture enterprise's overall operating costs [4]. Protein is the most expensive resource in aquafeeds. Fish meal is the ideal source of protein in aquafeeds because it has a balanced amino acid profile, which is rich in vitamins, minerals, and other nutrients that are vital for fish [5]. Normally, the commercial diet for Gibel carp contains 15–20% fish meal, which is a lot for an omnivorous fish [6]. On the other hand, as fish meal is an expensive ingredient in the commercial diet and a premium source of protein, its price has been rising quickly alongside the development of intensive aquaculture [7]. As a result, the world's fish-meal yield is not sufficient to meet the feeding industry's demand for production. Finding alternative protein sources is therefore essential because it can reduce the cost of fish feeds. Soybean is one of the best and most affordable options to use as a substitute for fish meal in commercial aquafeeds. It has been proven to have the best amino acid profile of all the protein-rich plant feeds for satisfying the needs of fish in terms of essential amino acids [8–10]. However, studies have shown that antinutritional factors (ANFs) like tannins, phytic acid, and trypsin inhibitors can have a negative impact on fish growth performance, intestinal health, and immune response [11–13]. Due to these ANFs, soybean meal needs to be treated to improve its utilization rate by fish. Different processing methods have been employed to minimize or remove ANF components in feeds made from soybean, although some of these procedures often have some defect such as protein loss, commercial viability, and environmental sustainability [14]. With different advantages associated with enzymatic hydrolysis, this treatment method is the most efficient and extensively used technique to upgrade the value of plant-origin proteins [15]. Enzymatic preprocessing of foods can successfully eliminate the antinutritional components in soybean meal and enhance nutrient digestibility [16], with a significant beneficial impact of preprocessing seen on several fish species' growth performance [17].

ESBM has been applied with excellent results to the farming of other animals like poultry and pork; however, little information is available on enzymatic soybean meal when used to replace fish meal in fish diets. Hence, the assessment of the effects of replacing FM with ESBM on the growth performance and immune response in juvenile Gibel carp was the main aim of this feeding experiment.

## 2. Materials and Methods

### 2.1. Experimental Diets

The commercial diet formula design of Gibel carp was adopted in the experimental feed, and the FM replacement levels were 0% (control group), 20%, 40%, 60%, 80%, and 100% with ESBM, which were, respectively, 0%, 4%, 8%, 12%, 16%, and 20% of the diet (Table 1). The raw feed ingredients were combined with additives, calcium dihydrogen phosphate, and amino acids. After that, the mixture was crushed using a 60-mesh sieve. Next, an appropriate amount of water was added, and the expanded feed was granulated into pellets of 1.0–1.5 mm using an expanding feed machine and lastly sun-dried for use.

### 2.2. Fish and Experimental Rearing Conditions

The fish breeding base of the Freshwater Fisheries Research Centre (FFRC) provided the juvenile Gibel carp. The experiment was conducted in the water-flow concrete fish pond of 2000 square meters and 3 m depth, and the water condition was controlled; dissolved oxygen (DO) was maintained at >5.0 mg/L, while pH was kept at 7.5–8.2 and aerators were used to maintain a good level of DO. For acclimatization to the experimental conditions, the fish were kept under observation for two weeks before the start of the experiment. During that period, the fish were fed with commercial feed (obtained from Wuxi Tongwei Feedstuffs Co., Ltd., Wuxi, China; crude protein 35%, crude lipid 7%) to acclimate to the experimental environment, and the feeding frequency was three times (7:30, 12:30, and 5:30 p.m.) daily to apparent satiation each time. The health, size, and weight of the juvenile Gibel carp were the same, and the average initial weight (IW) was 45.02 $\pm$ 0.03 g. Six groups of fish were fed with diets of different FM replacement levels. The feeding frequency was three times (7:30, 12:30, and 5:30 p.m.) daily to apparent satiation each time, and the amount of feed in each cage was recorded to enable the calculation of growth parameters at the end of the experiment. There were three experimental replicates in each group, with 15 test fish for each, and 18 cages (1 m × 1 m × 1 m) were used as outdoor pond cages. The experiment's duration was 159 days.

**Table 1.** Formulation and composition of experimental fish diet.

| Raw Material (%) | Replacing FM with ESBM Levels | | | | | |
|---|---|---|---|---|---|---|
| | 0 | 20% | 40% | 60% | 80% | 100% |
| Fish meal [1] | 15 | 12 | 9 | 6 | 3 | 0 |
| Chicken meal [1] | 5 | 5 | 5 | 5 | 5 | 5 |
| Soybean meal [1] | 15 | 15 | 15 | 15 | 15 | 15 |
| Enzymatic soybean meal [2] | 0 | 4 | 8 | 12 | 16 | 20 |
| Cottonseed meal [1] | 5.5 | 5.5 | 5.5 | 5.5 | 5.5 | 5.5 |
| Rapeseed meal [1] | 24 | 24 | 24 | 24 | 24 | 24 |
| Wheat meal [1] | 14.33 | 14.33 | 14.33 | 14.33 | 14.33 | 14.33 |
| Rice bran [1] | 15.65 | 13.64 | 11.62 | 9.60 | 7.57 | 5.55 |
| Soybean oil | 0.67 | 1.29 | 1.92 | 2.55 | 3.18 | 3.81 |
| Monocalcium phosphate | 2.00 | 2.33 | 2.66 | 3.00 | 3.33 | 3.66 |
| Vitamin premix for omnivorous fish [3] | 0.2 | 0.2 | 0.2 | 0.2 | 0.2 | 0.2 |
| Trace element premix for omnivorous fish [4] | 2 | 2 | 2 | 2 | 2 | 2 |
| Lysine (98.5%) | 0.30 | 0.33 | 0.36 | 0.39 | 0.43 | 0.46 |
| DL-methionine | 0.10 | 0.13 | 0.16 | 0.18 | 0.21 | 0.24 |
| Vitamin C phosphates | 0.05 | 0.05 | 0.05 | 0.05 | 0.05 | 0.05 |
| Choline chloride | 0.2 | 0.2 | 0.2 | 0.2 | 0.2 | 0.2 |
| Proximate Composition (dry basis) | | | | | | |
| Moisture (%) | 8.56 | 8.51 | 8.47 | 8.51 | 8.55 | 8.47 |
| Crude protein (%) | 35.31 | 35.26 | 35.32 | 35.31 | 35.32 | 35.33 |
| Gross energy (MJ/kg) | 15.31 | 15.29 | 15.28 | 15.26 | 15.22 | 15.21 |
| Crude lipid (%) | 7.83 | 7.79 | 7.81 | 7.83 | 7.84 | 7.83 |
| Crude fiber (%) | 7.62 | 7.35 | 7.08 | 6.81 | 6.53 | 6.26 |
| P (%) | 1.47 | 1.45 | 1.46 | 1.48 | 1.46 | 1.48 |

[1] Fish meal, chicken meal, rapeseed meal, cottonseed meal, soybean meal, wheat meal, and rice bran were received from Wuxi Tongwei Feedstuffs Ltd. Fish-meal composition (%, dry matter): arginine, 3.17; histidine, 1.44; isoleucine, 2.55; leucine, 4.22; lysine, 4.32; methionine, 1.66; phenylalanine, 2.36; threonine, 2.43; valine, 2.93; tryptophan, 0.66; moisture, 8.92; crude protein, 60.30; crude lipid, 7.67; crude ash, 14.30; crude fiber, 1.01; phosphorus, 2.88. [2] Enzymatic soybean meal was provided by Jiangsu Fuhai Biotechnology CO., LTD (Nantong, China). ESBM composition (%, dry matter): arginine, 2.96; histidine, 1.25; isoleucine, 2.36; leucine, 3.36; lysine, 3.00; methionine, 0.68; phenylalanine, 2.30; threonine, 1.82; valine, 2.33; tryptophan, 0.61; moisture, 6.01; crude protein, 49.70; crude lipid, 2.41; crude ash, 6.83; crude fiber, 3.21; phosphorus, 0.31. [3] Vitamin premix and trace element premix for omnivore fish were purchased from Wuxi Hanove animal health product Co., Ltd. in Wuxi, China. Vitamin premix (IU or mg/kg of premix): vitamin A, 800,000 IU; vitamin D3, 150,000–250,000 IU; vitamin E, 4500 IU; vitamin K3, 600 mg; thiamin, 800 mg; riboflavin, 800 mg; calcium pantothenate, 2000 mg; pyridoxine HCl, 2500 mg; cyanocobalamin, 8 mg; biotin, 16 mg; folic acid, 400 mg; niacin, 2800 mg; inositol, 10,000 mg; vitamin C, 10,000 mg. [4] Trace element premix (g/kg of premix): magnesium sulphate, 1.0–1.5; ferrous sulphate, 15–30; zinc sulphate, 8–13.5; cupric sulphate, 0.35–0.8; and manganese sulphate, 2–6; rice chaff and zeolite were used as a carrier.

### 2.3. Collection of Samples

Before sampling, fish were kept under starvation for 24 h to allow the evacuation of the digestive system. Prior to the sample collection, the total number of fish in every cage was assessed and the weight was measured to determine the growth parameters. MS-222 (100 mg/L) was used to anesthetize the collected fish. From every cage, five fish were arbitrarily taken for analysis of whole-body composition. Three fish were randomly selected from each cage for sampling and sedated with 100 mg/L MS-222; then, the blood samples were taken through the caudal vein and centrifuged (3000 rpm, 10 min, at 4 °C) for separating from the plasma. The liver and intestine samples were also taken for tissue structure examination. The collected intestine and hepatic samples were kept at −80 °C for further examination.

### 2.4. Chemical Analysis

Moisture content, crude protein, crude lipid, and ash were determined by taking as a reference the protocols established by the Association of Official Analytical Chemists [18]. The methods and testing equipment of plasma biochemical indices and antioxidant indices are shown in Table 2, which displays the specific procedures, kits, and equipment used in this study.

**Table 2.** Main methods and analyzing equipment.

| Item | Methods and Testing Equipment |
| --- | --- |
| Moisture | Dried to constant weight in an oven at 105 °C. |
| Protein | Determined by Hanon K1100 auto Kjeldahl apparatus (Jinan Hanon Instruments Co., Ltd., Jinan, China). |
| Lipid | Determined by Hanon SOX606 auto fat analyzer (Jinan Hanon Instruments Co., Ltd., Jinan, China). |
| Ash | Determined by burning at 550 °C for 5 h in an XL-2A intelligent muffle furnace (Hangzhou Zhuochi Instruments Co., Ltd., Hangzhou, China). |
| Alanine transaminase (ALT)<br>Total cholesterol (TC)<br>Glucose (GLU)<br>Alkaline phosphatase (ALP)<br>Albumin (ALB)<br>Triglyceride (TG)<br>Aspartic transaminase (AST)<br>Total protein (TP) | All plasma parameters were determined by assay kits (Mindray Bio Medical Co., Ltd., Shenzhen, China) with a Mindray BS-400 automatic biochemical analyzer (Mindray Medical International Ltd., Shenzhen, China). |
| Total antioxidant capacity (T-AOC)<br>Superoxide dismutase (SOD)<br>Catalase (CAT)<br>Glutathione (GSH)<br>Glutathione peroxidase (GPx)<br>Malondialdehyde (MDA) | All hepatic antioxidant parameters and MDA levels were tested according to instructions of assay kits purchased from Jian Cheng Bioengineering Institute (Nanjing, China). |

### 2.5. Hematoxylin and Eosin (HE) Staining

Hematoxylin and eosin (HE) staining was used to assess the hepatic and intestinal histology, referring to a previous study [19]. The tissue was first fixed with 4% formaldehyde, washed with water, dehydrated with ethanol, made transparent with methyl salicylate, and immersed in paraffin, then cut on a microtome at a thickness of 5μm, followed by HE staining. Finally, sample photomicrographs were taken before the tissue was examined under a microscope.

For the intestine, the tissue slices were placed on the device using the panoramic slice scanner, and they were then slowly moved in front of the scanner's lens. The tissue information from the tissue slices was scanned and imaged as they moved, creating a folder that comprised all of the tissue information from the tissue slices. CaseViewer2.4 scanning software was used to select the target area of the tissue for imaging. During imaging, the whole field of view was filled with the tissue as far as possible to ensure that the background light of each photo was consistent. Following imaging, the lengths of the villous epithelium and mucosal layer were measured at five different locations within each segment using Image-Pro Plus 6.0 analysis software, and the matching number of goblet cells were counted. The number of goblet cells per unit length = the number of goblet cells/the length of the villous epithelium.

### 2.6. Statistics Analysis

The tissue's target area for imaging was chosen using CaseViewer2.4 scanning software. The lengths of the villous epithelium and mucosal layer were measured at five different locations within each segment using Image-Pro Plus 6.0 analysis software, and the matching number of goblet cells were counted. In addition, data are mean value ± standard deviation. Data analysis was carried out using the statistical software for social sciences (SPSS) for Windows version 22.0. To determine whether there were any significant differences between the means of the various treatments, one-way analysis of variance (ANOVA) was utilized, and Tukey's multiple range statistics was used to show where the means were different ($p < 0.05$).

## 3. Results

### 3.1. Growth Performance

The results (Table 3) reveal that none of the growth parameters at all replacement levels was significantly affected compared to the control group ($p > 0.05$).

**Table 3.** Growth parameters and physical indices of Gibel carp fed with the experimental diets over 159 days [1].

| Parameters | Replacement Levels | | | | | | |
|---|---|---|---|---|---|---|---|
| | 0% | 20% | 40% | 60% | 80% | 100% | *p* Value |
| [2] IW (g) | 45.00 ± 0.23 | 45.02 ± 0.13 | 44.98 ± 0.12 | 45.05 ± 0.13 | 45.05 ± 0.23 | 45.00 ± 0.15 | 0.99 |
| [3] FW (g) | 229.89 ± 4.62 | 231.44 ± 0.51 | 226.62 ± 3.57 | 231.95 ± 2.85 | 230.45 ± 1.17 | 229.03 ± 2.24 | 0.31 |
| [4] FCR | 1.27 ± 0.03 | 1.21 ± 0.02 | 1.22 ± 0.02 | 1.20 ± 0.08 | 1.26 ± 0.08 | 1.30 ± 0.05 | 0.21 |
| [5] SGR (%/d) | 1.03 ± 0.011 | 1.03 ± 0.003 | 1.02 ± 0.010 | 1.03 ± 0.008 | 1.02 ± 0.006 | 1.02 ± 0.008 | 0.61 |
| [6] WGR (%) | 410.86 ± 9.35 | 414.13 ± 2.54 | 403.80 ± 7.93 | 414.87 ± 6.86 | 411.56 ± 4.82 | 408.97 ± 6.18 | 0.41 |
| [7] VSI (%) | 5.77 ± 1.12 | 6.23 ± 1.04 | 6.89 ± 0.66 | 6.60 ± 0.92 | 6.64 ± 0.34 | 6.68 ± 0.81 | 0.27 |
| [8] SR (%) | 97.78 ± 3.85 | 100.00 | 97.78 ± 3.85 | 95.56 ± 7.70 | 97.78 ± 3.85 | 97.78 ± 3.85 | 0.90 |
| [9] HSI (%) | 4.19 ± 0.96 | 4.53 ± 0.71 | 5.36 ± 1.19 | 4.97 ± 0.91 | 5.28 ± 0.40 | 5.21 ± 0.88 | 0.16 |
| [10] CF (g/cm$^3$) | 2.57 ± 0.31 | 2.62 ± 0.07 | 2.74 ± 0.20 | 2.64 ± 0.14 | 2.50 ± 0.11 | 2.47 ± 0.07 | 0.09 |

[1] Data are mean values ± standard deviation (n = 3). [2] IW: initial weight. [3] FW: final weight. [4] Feed conversion ratio (FCR) = total feed intake in dry matter basis (g) ÷ weight gain (g). [5] Specific growth rate (SGR, %/d) = [ln (mean final weight)-ln (mean initial weight) ÷ days] × 100. [6] Weight gain rate (WGR, %) = [final weight (g)-initial weight (g) × 100] ÷ initial weight (g). [7] Viscerosomatic index (VSI, %) = Viscera weight (g) ÷ body weight (g) × 100, [8] Survival rate (SR, %) = (final number of fish ÷ initial number of fish) × 100. [9] Hepatosomatic index (HSI, %) = Hepatopancreas weight (g) ÷ body weight (g) × 100. [10] Condition factor (CF, g/cm$^3$) = (body weight ÷ standard length$^3$) × 100.

### 3.2. Whole-Body Composition

The results (Table 4) indicate that none of the whole-body composition indices at all replacement levels were affected compared to the control group ($p > 0.05$).

**Table 4.** Whole-body composition of Gibel carp fed with the experimental diets over 159 days [1].

| Parameter (% Wet Matter) | Replacement Levels | | | | | | |
|---|---|---|---|---|---|---|---|
| | 0% | 20% | 40% | 60% | 80% | 100% | *p* Value |
| Moisture | 75.39 ± 0.40 | 76.12 ± 0.84 | 75.78 ± 0.59 | 75.59 ± 0.90 | 76.00 ± 0.83 | 75.72 ± 0.94 | 0.87 |
| Crude protein | 16.83 ± 0.78 | 16.61 ± 0.31 | 16.11 ± 0.59 | 17.10 ± 0.84 | 16.69 ± 0.61 | 16.72 ± 0.61 | 0.83 |
| Crude lipid | 1.84 ± 0.10 | 1.70 ± 0.04 | 1.66 ± 0.12 | 1.64 ± 0.16 | 1.61 ± 0.02 | 1.66 ± 0.10 | 0.16 |
| Crude ash | 4.77 ± 0.12 | 4.69 ± 0.24 | 4.65 ± 0.37 | 4.55 ± 0.13 | 4.32 ± 0.12 | 4.58 ± 0.35 | 0.37 |

[1] All data are mean values ± standard deviation (n = 3 × 3).

### 3.3. Plasma Biochemical Indices

Table 5 shows the results of the plasma biochemical parameters obtained at the end of this study. Plasma alanine transaminase (ALT), total cholesterol (TC), glucose (GLU), alkaline phosphatase (ALP), and albumin (ALB) levels of the fish under study were not significantly affected in all treatments ($p > 0.05$). The level of plasma triglyceride (TG) was reduced proportionally as the level of ESBM increased. Compared with the control group, the replacement level of 100% significantly increased the AST activity ($p < 0.05$). In addition, the replacement level of 80% significantly increased the content of TP compared to the control group ($p < 0.05$).

**Table 5.** Biochemical parameters of blood plasma in Gibel carp fed the experimental diets over 159 days [1].

| Parameter | Replacement Levels | | | | | | |
|---|---|---|---|---|---|---|---|
| | 0% | 20% | 40% | 60% | 80% | 100% | *p* Value |
| ALB (g/L) | 11.13 ± 0.82 | 11.11 ± 0.84 | 11.53 ± 1.46 | 10.75 ± 0.94 | 12.66 ± 2.26 | 11.74 ± 1.64 | 0.14 |
| ALT (U/L) | 4.60 ± 1.53 | 4.43 ± 0.69 | 5.15 ± 0.73 | 4.24 ± 0.79 | 4.69 ± 1.25 | 5.19 ± 1.19 | 0.41 |
| AST (U/L) | 100.28 ± 14.17 $^{a}$ | 98.94 ± 12.41 $^{a}$ | 108.85 ± 24.13 $^{ab}$ | 111.81 ± 10.83 $^{ab}$ | 114.61 ± 12.82 $^{ab}$ | 123.19 ± 11.79 $^{b}$ | 0.02 |
| TC (mmol/L) | 5.73 ± 0.63 | 5.50 ± 0.35 | 5.60 ± 0.50 | 5.40 ± 0.59 | 5.73 ± 0.23 | 5.46 ± 0.48 | 0.65 |
| TG (mmol/L) | 1.91 ± 0.23 $^{c}$ | 1.70 ± 0.11 $^{ab}$ | 1.76 ± 0.15 $^{bc}$ | 1.57 ± 0.13 $^{ab}$ | 1.63 ± 0.20 $^{ab}$ | 1.51 ± 0.18 $^{a}$ | 0.01 |
| GLU (g/L) | 6.11 ± 1.64 | 6.92 ± 1.21 | 5.62 ± 1.21 | 5.85 ± 1.04 | 6.64 ± 1.29 | 6.94 ± 0.78 | 0.15 |
| TP (g/L) | 36.74 ± 1.15 $^{a}$ | 36.85 ± 2.02 $^{a}$ | 37.74 ± 3.57 $^{ab}$ | 36.30 ± 2.35 $^{a}$ | 40.92 ± 4.43 $^{b}$ | 38.79 ± 4.63 $^{ab}$ | 0.04 |
| ALP (U/L) | 12.21 ± 4.71 | 9.58 ± 3.57 | 12.15 ± 4.99 | 10.14 ± 3.09 | 12.13 ± 4.38 | 11.88 ± 4.12 | 0.68 |

[1] All data are mean values ± standard deviation (n = 3 × 3). On the same row, different superscripted values represent significant differences ($p < 0.05$).

### 3.4. Results of Plasma Antioxidants Indices

The findings of plasma antioxidant indices are shown in Table 6. Malondialdehyde (MDA) levels, as well as the activities of superoxide dismutase (SOD) and total antioxidant capacity (T-AOC) in plasma, were not significantly affected at all replacement levels, according to the findings of this study ($p > 0.05$). The highest activity of catalase (CAT) was observed in the group at the 60% replacement level, which was significantly higher than the control group ($p < 0.05$). Compared with the control group, 20% replacement showed a significant increase in the content of glutathione (GSH) ($p < 0.05$). Furthermore, the activities of glutathione peroxidase (GPx) were significantly increased in the groups at the 40% and 60% replacement levels compared to the control group ($p < 0.05$).

**Table 6.** Antioxidant parameters in blood plasma of Gibel carp fed with the experimental diets over 159 days [1].

| Parameters | Replacement Levels | | | | | | |
|---|---|---|---|---|---|---|---|
| | 0% | 20% | 40% | 60% | 80% | 100% | *p* Value |
| T-AOC (U/mL) | 0.38 ± 0.02 | 0.40 ± 0.03 | 0.37 ± 0.05 | 0.38 ± 0.02 | 0.39 ± 0.02 | 0.38 ± 0.03 | 0.49 |
| SOD (U/mL) | 22.22 ± 1.07 | 22.80 ± 0.92 | 22.46 ± 0.96 | 22.71 ± 0.46 | 22.82 ± 0.95 | 22.07 ± 1.37 | 0.56 |
| CAT (U/mL) | 21.79 ± 7.49 $^{abc}$ | 13.79 ± 6.65 $^{a}$ | 30.78 ± 7.47 $^{cd}$ | 34.66 ± 9.76 $^{d}$ | 25.90 ± 11.50 $^{bcd}$ | 17.55 ± 10.05 $^{ab}$ | 0.01 |
| GSH (mg/L) | 30.56 ± 9.93 $^{ab}$ | 51.24 ± 20.15 $^{c}$ | 36.84 ± 23.21 $^{abc}$ | 47.92 ± 21.33 $^{bc}$ | 30.28 ± 8.71 $^{ab}$ | 26.30 ± 9.25 $^{a}$ | 0.04 |
| GPx (μmol/L) | 104.70 ± 42.39 $^{a}$ | 118.16 ± 36.44 $^{a}$ | 178.74 ± 10.42 $^{c}$ | 165.08 ± 41.42 $^{bc}$ | 119.09 ± 44.16 $^{a}$ | 128.95 ± 30.09 $^{ab}$ | 0.01 |
| MDA (nmol/mL) | 14.17 ± 5.50 | 12.52 ± 3.34 | 12.77 ± 4.24 | 13.80 ± 2.47 | 13.16 ± 4.17 | 9.88 ± 2.40 | 0.30 |

[1] All data are mean values ± standard deviation (n = 3 × 3). On the same row, different superscripted values represent significant differences ($p < 0.05$).

### 3.5. Results of Hepatic Tissue Structure

As shown in Figure 1, the hepatocytes at 0%, 20%, 40%, 60%, 80%, and 100% replacement levels were arranged in double rows into plates, and the plates were scattered around the central veins. The hepatocytes were irregular polygons with centered nuclei and vacuolated cytoplasm (black arrows). There was no obvious expansion of hepatic sinusoids and no obvious inflammatory cell infiltration.

### 3.6. Results of Intestine Tissue Structure

Among all replacement levels of FM with ESBM, none showed a significantly changed villous epithelium length, number of goblet cells, or number of goblet cells per unit length. In addition, as displayed in the following Figure 2, the structure of each layer of intestinal tissue at 0%, 20%, 40%, 60%, 80%, and 100% replacement levels was clear. The intestinal villi were abundant and arranged regularly, the mucosal epithelial cells were not exfoliated, and many goblet cells were seen without obvious abnormalities (Table 7). This indicates that the structure of intestinal tissue of Gibel carp is not damaged when FM is completely replaced with ESBM in the diet.

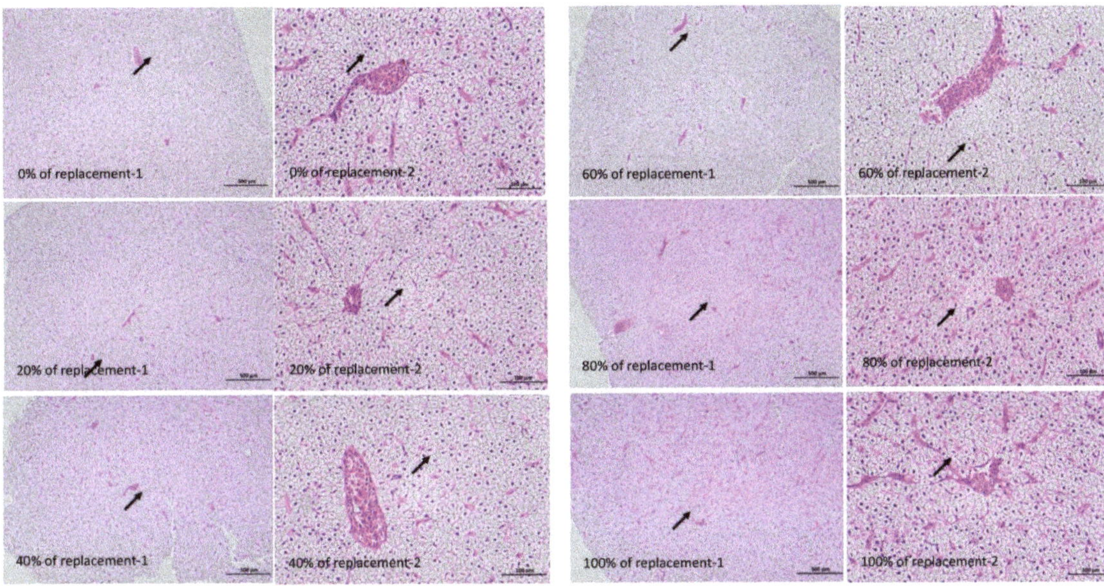


**Figure 1.** Sections of hepatic tissue structure in Gibel carp fed with the experimental diets over 159 days. The hepatocytes were arranged in double rows into plates, and the plates were scattered around the central veins. The hepatocytes were irregular polygons with centered nuclei and vacuolated cytoplasm (black arrow). The number 1 represents 40× and the number 2 represents 200× observation multiples in each image.

**Table 7.** Intestinal tissue structure in Gibel carp fed with the experimental diets over 159 days [1].

| Replacement Levels (%) | Thickness of Mucosal Layer (mm) | Length of the Villous Epithelium (mm) | Number of Goblet Cells | Number of Goblet Cells per Unit Length (Cell/mm) |
|---|---|---|---|---|
| 0 | 0.48 ± 0.14 | 0.54 ± 0.12 | 14.93 ± 7.19 | 27.66 ± 10.52 |
| 20 | 0.33 ± 0.05 | 0.46 ± 0.02 | 12.90 ± 6.65 | 27.81 ± 13.26 |
| 40 | 0.41 ± 0.03 | 0.55 ± 0.03 | 13.00 ± 5.30 | 23.74 ± 8.95 |
| 60 | 0.52 ± 0.22 | 0.69 ± 0.30 | 15.47 ± 5.67 | 22.98 ± 3.33 |
| 80 | 0.40 ± 0.05 | 0.55 ± 0.03 | 11.47 ± 3.65 | 20.70 ± 6.37 |
| 100 | 0.43 ± 0.07 | 0.61 ± 0.13 | 24.13 ± 7.71 | 39.70 ± 9.21 |
| *p* Value | 0.58 | 0.64 | 0.23 | 0.19 |

[1] Data are presented as mean value ± standard deviation (n = 3 × 3).

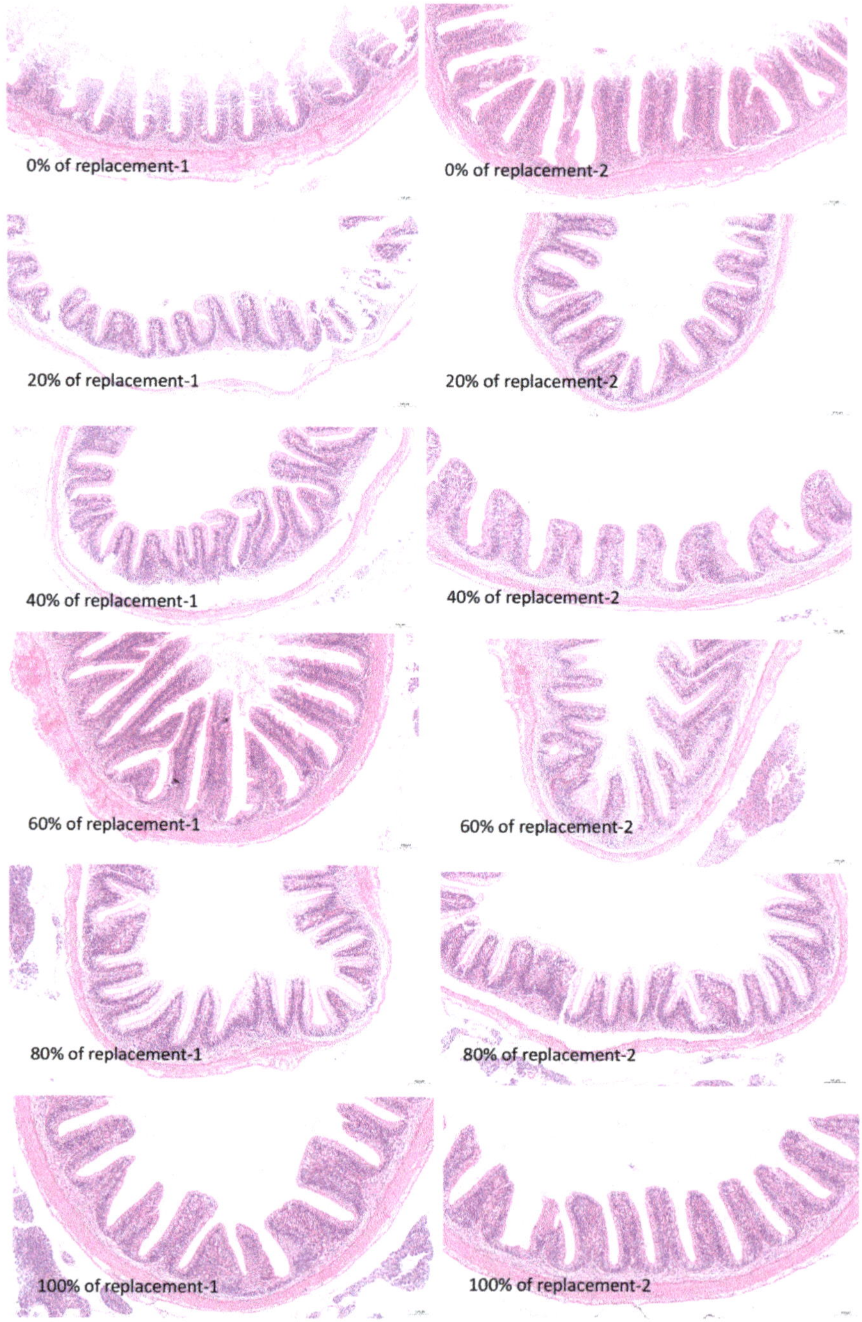


**Figure 2.** Sections of intestinal tissue structure in Gibel carp fed with the experimental diets over 159 days. The 1 and 2 in each group represent different intestines at 10× viewing multiples.

## 4. Discussion

Due to the rising demand for FM and the rapid expansion of the aquafeed industry, the price of FM has climbed significantly during the past years [20]. Cheap plant proteins have typically been employed as FM replacements in aquafeed diets to promote the aquaculture industry [21]. The results of this study show that the growth performance of Gibel carp is not significantly affected, including FW, FCR, SGR, and WGR, when the FM is entirely replaced with ESBM. This indicates that ESBM is a good alternative for FM in the diet of Gibel carp. Liu et al. [22] also provided a similar suggestion after studying the effect of the substitution of FM with ESBM in the feed of juvenile largemouth bass (*Micropterus salmoides*). The capacity of ESBM to replace FM in aquatic diets is correlated to the increase in nutritional status and reduction in antinutritional factors when soybean is treated with enzymes. However, the study performed on rainbow trout (*Oncorhynchus mykiss*) fry suggests that the replacement level of FM with ESBM should be 50% [23]. This dissimilarity may come from different fish species on which the studies were conducted. Furthermore, the SR was not significantly affected at all the replacement levels. Samira et al. [23] found similar results in the study carried out on rainbow trout fry fed with ESBM (HP310) as a replacement for FM in the diet. Fan et al. [24] also gave the same suggestion in their experiment conducted on Pacific white shrimp (*Litopenaeus vannamei*), where FM was replaced with ESBM. The important indicators that represent the body lipid, lean, and growth of fish include the CF, HSI, and VSI [25]. The fish's nutritional status can be determined using the HSI, which is an indirect indicator of hepatic glycogen content. Lower values may indicate stress, and both the HSI and VSI are used to determine whether energy is being diverted away from organ or tissue growth in order to evade stress [26]. In the current study, no significant differences in the HSI and VSI were found at all replacement levels; therefore, total replacement of FM with ESBM did not disturb glycogen and carbohydrate energy reserves. Liu et al. [22] also suggested similar results in juvenile largemouth bass.

Whole-body composition can indicate directly and indirectly an animal's growth [27,28]. The results of this study showed that the moisture, crude lipids, protein, and ash contents of juvenile Gibel carp were not significantly affected by any replacement levels. Similar results for moisture, crude protein, and ash contents after replacing FM with SBM were also found in the hybrid of *Carassius auratus gibelio* ♀ × *Cyprinus carpio* ♂ [29]. Furthermore, Samad et al. [30] suggested that replacement of FM with fermented soybean meal had no significant impact on the crude protein and ash contents of Japanese seabass (*Lateolabrax japonicus*). However, previous studies have stated that soybean can reduce crude lipid content in rainbow trout fry [23]. Additionally, the study of Duan et al. [31] confirmed that the replacement of FM with fermented soybean lessens the level of crude lipid in the fish body of hybrid snakehead (*Channa argus* × *Channa maculate*). These dissimilarities could be caused by differences in fish species, experimental setups, and methods of soybean processing. Plasma biochemical parameters reflect fish metabolism and health status [32]. Information on the state of fish health and protein metabolism is detected using plasma TP and ALB [33]. In addition, fish immunity is also correlated with plasma TP and ALB [34]. The replacement of FM with graded levels of ESBM did not significantly affect the plasma ALB level compared to the control group. However, the 80% replacement level of FM with ESBM was significantly improved compared to the control group. In contrast, it was reported by Ajani et al. [35] that the level of plasma TP is reduced when the FM is completely replaced by soybean with or without methionine fortification in the diet of Nile tilapia (*Oreochromis niloticus*). This difference can be caused by the presence of ANFs in raw soybean and the difference in fish species studied; however, enzymolysis was shown to effectively reduce the levels of these ANFs in this study. Fish plasma GLU is regarded as a stress indicator, and its level reflects the energy produced by a stressor [36]. In this study, the level of plasma glucose was not significantly modified at all replacement levels, which indicates that total replacement of FM with ESMB does not significantly disturb the energy homeostasis of Gibel carp. Ye et al. [37] also reported the same results of plasma glucose in juvenile obscure puffer (*Takifugu obscurus*). The findings of this study also demonstrated that, except at the 40% replacement level, other replacement levels significantly decreased the level of TG in comparison to the control group. On the other hand, TC was not significantly modified by different replacement levels. As the dietary

ESBM level grew, the body's overall crude lipid content decreased, which is correlated to the declining trend of plasma TC and TG. These results agree with the previous report that replacement of fish meal with defatted and fermented soybean meals could lower the level of TC compared to FM in pompano (*Trachinotus blochii*) [38], which is mainly caused by the presence of high concentrations of estrogenic isoflavones [39]. ALT and ALP are enzymes that contribute to the metabolism of protein; the level of these enzymes may indicate fish health status [40]. At all replacement levels, there were no significant modifications in the level of plasma ALT and ALP compared to the control group, which showed that the liver and protein metabolism of Gibel carp were not impaired by the entire replacement (100%) of FM with ESBM. T-AOC is a frequently used analyte to assess the antioxidant defense for combatting against the free radicals caused by a given disease [41]. There was no significant difference in the activity of T-AOC between the control group and all replacement levels. Lin et al. [25] also found similar results in juvenile pompano fish (*Trachinotus ovatus*). The level of MDA in fish plasma can be utilized as an indication of oxidative damage because it is one of the byproducts of cell membrane lipid peroxidation and is regarded as an indirect biomarker of lipid peroxidation [42]. All replacement levels of FM with ESBM in this study had no significant influence on the plasma MDA level compared to the control group. This indicated that the replacement (FM with ESBM) of these ingredients did not cause damage to the health of Gibel carp. Different results have been observed in studies of other fish species like redlip mullet (*Liza haematocheila*) [7], where FM was replaced with soybean, and Japanese seabass [31], where FM was replaced with fermented soybean. These contradictory findings could be the result of different fish species and various soybean-processing techniques. SOD, CAT, and GPx are crucial indicators of the body's capacity to defend against oxidative cell damage [27]. Catalase is a typical enzyme that may be found in almost all living things that are exposed to oxygen, including bacteria, plants, and mammals. It helps hydrogen peroxide break down into water and oxygen. It is a crucial enzyme in preventing reactive oxygen species (ROS) from oxidatively harming the cell.

In this study, no significant differences were observed in SOD activity between the control group and all other replacement levels. This indicated that the full replacement (100%) of FM with ESBM did not significantly disturb the mitochondrial oxidative metabolism of fish cells and did not cause oxidative stress in Gibel carp. GPx eliminates hydrogen peroxide and lipid peroxides generated during metabolism; in addition, it also catalyzes the reduction in glutathione. In this study, the 40 and 60% replacement levels of FM by ESBM significantly improved the activities of GPx; however, lower and higher levels of substitution did not affect the activities of GPx. Huaxing et al. [25] also reported the same results in juvenile pompano fed on soy protein peptide. Furthermore, at the replacement level of 60%, CAT activity was significantly improved compared to the control group. Liu et al. [22] also found the change in CAT activity in juvenile largemouth bass when the FM was replaced by enzyme-treated soybean in diets. As a result, ESBM is capable of producing small peptides with features that increase the body's antioxidant capacity by neutralizing free radicals, preventing lipid peroxidation, and chelating metal ions [22]. A perfect indicator of the fish's good nutrition is the liver and intestine's histological condition [43]. The findings of this feeding experiment showed that none of the replacement levels affected the hepatic and intestinal tissue structure in Gibel carp. These results indicated that ESBM could be used as an FM replacement without causing problems in food digestion, nutrient absorption, and metabolism in Gibel carp. Muhammad et al. [43] and Fotini et al. [44], respectively, stated that fermented soybean meal may contribute to the histological, morphological, and functional alterations in African catfish (*Clarias gariepinus*) liver tissues, and processed soybean resulted in the accumulation of lipids within hepatocytes of gilthead sea bream (*Sparus aurata*). The dissimilarities may be due to the difference in fish species and the low content of antinutritional factors in ESBM [23,28,45].

## 5. Conclusions

In general, ESBM could completely replace FM in ration formulation for juvenile Gibel carp's diets without causing adverse impacts on growth performance, antioxidant capacity, and the structure of liver and intestine tissues. Furthermore, the partial replacement could even improve the antioxidant capacity of juvenile Gibel carp.

**Author Contributions:** Conceptualization, H.M. and M.R.; data curation, A.U. and D.H.; funding acquisition, M.R. and H.L.; investigation, A.U., D.H., H.L., H.Y. and C.X.; methodology, A.U., D.H., H.Y. and H.L.; project administration, H.M. and M.R.; supervision, M.R. and H.L.; writing—original draft, A.U.; writing—review and editing, H.M. and M.R. All authors have read and agreed to the published version of the manuscript.

**Funding:** This research was funded by the National Key R & D Program of China (2022YFD2400904), the earmarked fund for CARS (CARS-46), the National Natural Science Foundation of China (32102806), and Transformation of Technological Innovation Engineering Achievements in the Aquatic Industry (2019ZHCG0133).

**Institutional Review Board Statement:** The study was conducted according to Standardization Administration of China protocols and guidelines (GB/T 35892-2018) and approved by the Institutional Animal Care and Ethics Committee of Nanjing Agricultural University (Permit number: SYXK (Su) 2011–0036).

**Informed Consent Statement:** Not applicable.

**Data Availability Statement:** The authors confirm that the data supporting the findings of this study are available within the manuscript, tables, and figures.

**Acknowledgments:** The authors would like to express sincere thanks to the personnel of Freshwater Fisheries Research Centre, Chinese Academy of Fishery Sciences, for their kind assistance.

**Conflicts of Interest:** Authors Haifeng Mi and Heng Yin are employed by Tongwei Agricultural Development Co., Ltd. The remaining authors declare that the research was conducted in the absence of any commercial or financial relationships that could be construed as potential conflicts of interest. Haifeng Mi played a major role in the design of the study, data analysis, and supervision of the published results. Heng Yin played a role in data analysis and experimental culture.

## References


1. Boyd, C.E.; McNevin, A.A.; Davis, R.P. The contribution of fisheries and aquaculture to the global protein supply. *Food Secur.* **2022**, *14*, 805–827. [CrossRef] [PubMed]
2. Zhang, Y.J.; Deng, P.; Dai, C.J.; Wu, M.K.; Liu, X.Q.; Li, L.J.; Pan, X.Y.; Yuan, J.F. Investigation of putative antimicrobial peptides in *Carassius gibel*, revealing a practical approach to screening antimicrobials. *Fish Shellfish Immunol.* **2022**, *121*, 254–264. [CrossRef]
3. Zhou, Q.L.; Xie, J.; Ge, X.P.; Habte-Tsion, H.M.; Liu, B.; Ren, M.C. Growth performance and immune response of gibel carp, *Carassius auratus gibelio*, fed with graded level of rare earth-chitosan chelate. *Aquacult. Int.* **2016**, *24*, 453–463. [CrossRef]
4. Thompson, K.R.; Muzinic, L.A.; Engler, L.S.; Morton, S.R.; Webster, C.D. Effects of feeding practical diets containing various protein levels on growth, survival, body composition, and processing traits of Australian red claw crayfish (*Cherax quadricarinatus*) and on pond water quality. *Aquac. Res.* **2004**, *35*, 659–668. [CrossRef]
5. Ji, K.; He, J.Y.; Liang, H.L.; Ren, M.C.; Ge, X.P.; Masagounder, K. Response of gibel carp (*Carassius auratus gibelio*) to increasing levels of dietary lysine in zero fish meal diets. *Aquacult. Nutr.* **2021**, *27*, 49–62. [CrossRef]
6. Ren, M.C.; Liang, H.L.; He, J.Y.; Masagounder, K.; Yue, Y.; Yang, H.; Ge, X.P.; Xi, B. Effects of DL-methionine supplementation on the success of fish meal replacement by plant proteins in practical diets for juvenile gibel carp (*Carassius auratus gibelio*). *Aquacult. Nutr.* **2017**, *23*, 934–941. [CrossRef]
7. Liu, T.; Han, T.; Wang, J.T.; Liu, T.; Bian, P.; Wang, Y.B.; Cai, X.J. Effects of replacing fish meal with soybean meal on growth performance, feed utilization and physiological status of juvenile redlip mullet *Liza haematocheila*. *Aquacul. Rep.* **2021**, *20*, 100756. [CrossRef]
8. Dersjant-Li, Y. The use of soy protein in aquafeeds. *Oilseeds Focus* **2021**, *7*, 29–31. Available online: https://hdl.handle.net/10520/ejc-vp_oilseeds_v7_n2_a10 (accessed on 26 June 2023).
9. Porter, M.A.; Jones, A.M. Variability in soy flour composition. *J. Am. Oil Chem. Soc.* **2003**, *80*, 557–562. [CrossRef]
10. Tan, C.; Zhou, H.H.; Wang, X.; Mai, K.S.; He, G. Resveratrol attenuates oxidative stress and inflammatory response in turbot fed with soybean meal based diet. *Fish Shellfish Immunol.* **2019**, *91*, 130–135. [CrossRef]
11. Gatlin, D.M.; Barrows, F.T.; Brown, P.; Dabrowski, K.; Gaylord, T.G.; Hardy, R.W.; Wurtele, E. Expanding the utilization of sustainable plant products in aquafeeds: A review. *Aquac. Res.* **2007**, *38*, 551–579. [CrossRef]

12. Hernández, M.D.; Martínez, F.J.; Jover, M.; Garcia, B.G. Effects of partial replacement of fish meal by soybean meal in sharpsnout seabream (*Diplodus puntazzo*) diet. *Aquaculture* **2007**, *263*, 159–167. [CrossRef]
13. Wang, P.; Zhou, Q.C.; Feng, J.; He, J.J.; Lou, Y.D.; Zhu, J.Q. Effect of dietary fermented soybean meal on growth, intestinal morphology and microbiota in juvenile large yellow croaker, *Larimichthys crocea*. *Aquac. Res.* **2019**, *50*, 748–757. [CrossRef]
14. NRC (National Research Council). *Nutrient Requirements of Fish and Shrimp*; National Academic Press: Cambridge, MA, USA, 2011.
15. Sun, X.D. Enzymatic hydrolysis of soy proteins and the hydrolysates utilisation. *Int. J. Food Sci. Technol.* **2011**, *46*, 2447–2459. [CrossRef]
16. Jacobsen, H.J.; Kousoulaki, K.; Sandberg, A.S.; Carlsson, N.G.; Ahlstrom, O.; Oterhals, A. Enzyme pre-treatment of soybean meal: Effects on non-starch carbohydrates, protein, phytic acid, and saponin biotransformation and digestibility in mink (*Neovison vison*). *Anim. Feed Sci. Technol.* **2018**, *236*, 1–13. [CrossRef]
17. Tibaldi, E.; Hakim, Y.; Uni, Z.; Tulli, F.; Francesco, M.D.; Luzzana, U.; Harpaz, H. Effects of the partial substitution of dietary fish meal by differently processed soybean meals on growth performance, nutrient digestibility and activity of intestinal brush border enzymes in the European sea bass (*Dicentrarchus labrax*). *Aquaculture* **2006**, *261*, 182–193. [CrossRef]
18. Association of Official Analytical Chemists (AOAC). *Official Methods of Analysis of the Association of Official Analytical Chemists*, 15th ed.; Association of Official Analytical Chemists Inc.: Arlington, TX, USA, 2003.
19. Liang, H.L.; Wu, L.H.; Chama, M.K.H.; Ge, X.P.; Ren, M.C.; Chen, X.R.; Pan, L.K.; Xia, D. Tributyrin Plays an Important Role in Regulating the Growth and Health Status of Juvenile Blunt Snout Bream (*Megalobrama amblycephala*), as Evidenced by Pathological Examination. *Front. Immunol.* **2021**, *12*, 652294. [CrossRef]
20. Gui, D.; Liu, W.B.; Shao, X.P.; Xu, W.N. Effects of different dietary levels of cottonseed meal protein hydrolysate on growth, digestibility, body composition and serum biochemical indices in crucian carp (*Carassius auratus gibelio*). *Anim. Feed Sci. Technol.* **2010**, *156*, 112–120. [CrossRef]
21. Wang, Y.; Kong, L.K.; Li, C.; Bureau, D.P. Effect of replacing fish meal with soybean meal on growth, feed utilization and carcass composition of Cuneate drum (*Nibea miichthioides*). *Aquaculture* **2006**, *261*, 1307–1313. [CrossRef]
22. Liu, X.; Chi, S.Y.; Li, S.; Cheng, X.L.; Gao, W.H.; Xu, Q.Q.; Zhang, W.B.; Zhou, X.Q. Substitution of fish meal with enzyme-treated soybean in diets for juvenile largemouth bass (*Micropterus salmoides*). *Aquacult. Nutr.* **2021**, *27*, 1569–1577. [CrossRef]
23. Haghbayan, S.; Mehrgan, M.S. The Effect of Replacing Fish meal in the Diet with Enzyme-Treated Soybean Meal (HP310) on Growth and Body Composition of Rainbow Trout Fry. *Molecules* **2015**, *20*, 21058–21066. [CrossRef] [PubMed]
24. Fan, Y.; Luo, K.; Guo, Y.L.; Gao, W.H.; Xu, Q.Y.; Zhang, W.B.; Mai, K.S. Replacement of fish meal by enzyme-treated soybean on the growth performance, intestine microflora, immune responses and disease resistance of Pacific white shrimp (*Litopenaeus vannamei*). *Aquac. Res.* **2021**, *52*, 4619–4628. [CrossRef]
25. Lin, H.; Tan, B.; Ray, G.W.; Zeng, M.; Li, M.; Chi, S.; Yang, Q. A Challenge to Conventional Fish Meal: Effects of Soy Protein Peptides on Growth, Histomorphology, Lipid Metabolism and Intestinal Health for Juvenile Pompano *Trachinotus ovatus*. *Front. Mar. Sci.* **2022**, *8*, 815323. [CrossRef]
26. Voorhees, J.M.; Barnes, M.; Chipps, S.; Brown, M. Direct substitution of fishmeal with bioprocessed soybean meal in brown trout diets. *J. Fish. Aqua. Dev.* **2018**, *1*, 100043. [CrossRef]
27. Liang, H.; Xu, G.; Xu, P.; Zhu, J.; Li, S.; Ren, M. Dietary Histidine Supplementation Maintained Amino Acid Homeostasis and Reduced Hepatic Lipid Accumulation of Juvenile Largemouth Bass, *Micropterus Salmoides*. *Aquac. Nutr.* **2022**, *2022*, 4034922. [CrossRef]
28. Li, S.L.; Ding, G.T.; Song, F.; Sang, C.Y.; Wang, A.; Chen, N.S. Comparison of dehulled, fermented and enzyme-treated soybean meal in diets for largemouth bass, *Micropterus salmoides*: Effects on growth performance, feed utilization, immune response and intestinal morphology. *Anim. Feed Sci. Technol.* **2020**, *267*, 114548. [CrossRef]
29. Liu, X.R.; Han, B.; Xu, J.; Zhu, J.Y.; Hu, J.T.; Wan, W.L.; Miao, S.Y. Replacement of fish meal with soybean meal affects the growth performance, digestive enzymes, intestinal microbiota and immunity of *Carassius auratus gibelio*♀ × *Cyprinus carpio*♂. *Aquacult. Rep.* **2020**, *18*, 100472. [CrossRef]
30. Rahimnejad, S.; Lu, K.L.; Wang, L.; Song, K.; Mai, K.S.; Davis, D.A.; Zhang, C.X. Replacement of fish meal with *Bacillus pumillus* SE5 and *Pseudozyma aphidis* ZR1 fermented soybean meal in diets for Japanese seabass (*Lateolabrax japonicus*). *Fish Shellfish Immunol.* **2019**, *84*, 987–997. [CrossRef]
31. Duan, Z.P.; Zhang, C.Y.; Huang, L.L.; Lan, Q.; Hu, J.; Li, X.Q.; Leng, X.J. An Evaluation of Replacing Fish Meal with Fermented Soybean Meal in Diet of Hybrid Snakehead (*Channa argus* × *Channa maculata*): Growth, Nutrient Utilization, Serum Biochemical Indices, Intestinal Histology, and Microbial Community. *Aquac. Nutr.* **2022**, *2022*, 2964779. [CrossRef]
32. Roques, S.; Deborde, C.; Richard, N.; Skiba-Cassy, S.; Moing, A.; Fauconneau, B. Metabolomics and fish nutrition: A review in the context of sustainable feed development. *Rev. Aquacult.* **2020**, *12*, 261–282. [CrossRef]
33. Sardar, P.; Abid, M.; Randhawa, H.S.; Prabhakar, S.K. Effect of dietary lysine and methionine supplementation on growth, nutrient utilization, carcass compositions and haemato-biochemical status in Indian Major Carp, Rohu (*Labeo rohita* H.) fed soy protein diet. *Aquac. Nutr.* **2009**, *15*, 339–346. [CrossRef]
34. Zhang, Q.L.; Liang, H.L.; Xu, P.; Xu, G.C.; Zhang, L.; Wang, Y.L.; Ren, M.C.; Chen, R. Effects of Enzymatic Cottonseed Protein Concentrate as a Feed Protein Source on the Growth, Plasma Parameters, Liver Antioxidant Capacity and Immune Status of Largemouth Bass (*Micropterus salmoides*). *Metabolites* **2022**, *12*, 1233. [CrossRef]

35. Ajani, E.K.; Orisasona, O.; Omitoyin, B.O.; Osho, E.F. Total Replacement of Fish meal by Soybean Meal with or Without Methionine Fortification in the Diets of Nile Tilapia, *Oreochromis niloticus*. *J. Fish. Aquat. Sci.* **2016**, *11*, 238–243. [CrossRef]
36. Martin, J.L.K.; Black, M.C. Biomarker assessment of the effects of coal-strip mine contamination on channel catfish. *Ecotoxicol. Environ. Saf.* **1998**, *41*, 307–320. [CrossRef]
37. Ye, H.Q.; Xu, M.L.; Liu, Q.Y.; Sun, Z.Z.; Zou, C.Y.; Chen, L.L.; Su, N.N.; Ye, C.X. Effects of replacing fish meal with soybean meal on growth performance, feed utilization and physiological status of juvenile obscure puffer, *Takifugu obscurus*. *Comp. Biochem. Phys. C* **2019**, *216*, 75–81. [CrossRef] [PubMed]
38. Nguyen, H.P.; Do, T.V.; Tran, H.D.; Nguyen, T.T. Replacement of fish meal with defatted and fermented soybean meals in pompano *Trachinotus blochii* (Lacepède, 1801) diets. *Ann. Anim. Sci.* **2021**, *21*, 575–587. [CrossRef]
39. Setchell, K.D.; Cassidy, A. Dietary isoflavones: Biological effects and relevance to human health; 1999; Symposium on Phytochemicals: Biochemistry and Physiology. *J. Nutr.* **1999**, *129*, 758S–767S. [CrossRef]
40. Moure, A.; Domínguez, H.; Parajo, J.C. Antioxidant properties of ultrafiltration-recovered soy protein fractions from industrial effluents and their hydrolysates. *Process Biochem.* **2006**, *41*, 447–456. [CrossRef]
41. Del, R.D.; Stewart, A.J.; Pellegrini, N. A review of recent studies on malondialdehyde as toxic molecule and biological marker of oxidative stress. *Nutr. Metab. Cardiovasc. Dis.* **2005**, *15*, 316–328. [CrossRef]
42. Kong, F.K.; Liu, T.; Liu, Y.J.; Yu, Z.K.; Zhang, W.; Fan, D.F.; Fan, J.Q.; Kong, L.C.; Li, B.; Chen, S.J.; et al. Dietary effects of enzymolytic soybean meal inclusion on antioxidant capacity, intestinal morphology and caecal microbiota of Rex rabbits. *Ital. J. Anim. Sci.* **2022**, *21*, 1220–1231. [CrossRef]
43. Zakaria, M.K.; Kari, Z.A.; Doan, H.A.; Kabir, M.A.; Harun, H.C.; Sukri, S.A.M.; Khang Wee, K.W.G.W.; Khoo, M.I.; Wei, L.S. Fermented Soybean Meal (FSBM) in African Catfish (*Clarias gariepinus*) Diets: Effects on Growth Performance, Fish Gut Microbiota Analysis, Blood Haematology, and Liver Morphology. *Life* **2022**, *12*, 1851. [CrossRef] [PubMed]
44. Kokou, F.; Sarropoulou, E.; Cotou, E.; Rigos, G.; Henry, M.; Alexis, M.; Kentouri, M. Effects of Fish meal Replacement by a Soybean Protein on Growth, Histology, Selected Immune and Oxidative Status Markers of Gilthead Sea Bream, *Sparus aurata*. *J. World Aquacult. Soc.* **2015**, *46*, 115–128. [CrossRef]
45. Jacobsen, H.J.; Samuelsen, T.A.; Girons, A.; Kousoulaki, K. Different enzyme incorporation strategies in atlantic salmon diet containing soybean meal: Effects on feed quality, fish performance, nutrient digestibility and distal intestinal morphology. *Aquaculture* **2018**, *491*, 302–309. [CrossRef]

fishes

MDPI

*Article*

# Effects of Fermented Cottonseed Meal Substitution for Fish Meal on Intestinal Enzymatic Activity, Inflammatory and Physical-Barrier-Related Gene Expression, and Intestinal Microflora of Juvenile Golden Pompano (*Trachinotus ovatus*)

**Zhanzhan Wang [1,2], Shuling Liao [3], Jun Wang [1,4], Yun Wang [1,4], Zhong Huang [4,5], Wei Yu [4,5], Xiaolin Huang [4,5], Heizhao Lin [4,5], Maoyan Luo [3], Zhenyan Cheng [2,*] and Chuanpeng Zhou [1,4,*]**

1 Key Laboratory of Aquatic Product Processing, Ministry of Agriculture and Rural Affairs, South China Sea Fisheries Research Institute, Chinese Academy of Fishery Sciences, Guangzhou 510300, China; wangyun@scsfri.ac.cn (Y.W.)
2 School of Fisheries, Tianjin Agricultural University, Tianjin 300384, China
3 School of Life Science, Guangzhou University, Guangzhou 510006, China
4 Key Laboratory of Efficient Utilization and Processing of Marine Fishery Resources of Hainan Province, Sanya 572426, China
5 Shenzhen Base of South China Sea Fisheries Research Institute, Chinese Academy of Fishery Sciences, Shenzhen 518121, China
* Correspondence: chengzhenyan2005@126.com (Z.C.); chpzhou@163.com (C.Z.)

**Citation:** Wang, Z.; Liao, S.; Wang, J.; Wang, Y.; Huang, Z.; Yu, W.; Huang, X.; Lin, H.; Luo, M.; Cheng, Z.; et al. Effects of Fermented Cottonseed Meal Substitution for Fish Meal on Intestinal Enzymatic Activity, Inflammatory and Physical-Barrier-Related Gene Expression, and Intestinal Microflora of Juvenile Golden Pompano (*Trachinotus ovatus*). *Fishes* **2023**, *8*, 466. https://doi.org/10.3390/fishes8090466

Academic Editor: Fei Ling

Received: 31 August 2023
Revised: 14 September 2023
Accepted: 15 September 2023
Published: 18 September 2023



**Abstract:** The present study was conducted to investigate the effects of dietary fermented cottonseed meal (FCSM) substitution for fish meal on intestinal enzymatic activity, inflammatory and physical-barrier-related gene expression, and intestinal microflora of juvenile golden pompano. The 375 golden pompanos were divided into 15 groups of 25 fish each, with three replicates for each experimental group. The fish were fed five experimental diets (0 (FM), 12.5% (CSM12.5), 25% (CSM25), 50% (CSM50), and 100% (CSM100) substitution levels) for 8 weeks. The fish were reared and fed the experimental diets under a natural-day light cycle. Compared with the control group, the activities of AMY (amylase) enzymes in the CSM12.5 group and all other groups were elevated ($p < 0.05$). The CSM25 group exhibited a considerable up-regulation of *IL-10* (Interleukin-10) expression relative to the FM group ($p < 0.05$). With an increase in dietary FM substitution with FCSM from 0 to 25%, the relative expressions of *NF-κB* (Nuclear factor kappa-B), *IL-1β* (Interleukin-1 beta), and IL-8 (Interleukin-8) were down-regulated. In this study, the relative expressions of ZO-1 (zonula occluden-1) and *Occludin* were up-regulated, and those of *Claudin-3* and *Claudin-15* significantly up-regulated, when the FCSM substitution ratio was 25%. The results of high-throughput sequencing of the intestinal microflora showed that ACE indices the lowest in the CSM25 group, which was significantly different from those in the CSM100 group ($p < 0.05$). The CSM50 group had the highest Shannon and Simpson indices and the highest community diversity. In addition, replacing a high percentage of fish meal with FCSM can negatively affect the intestinal flora of fish. In this study, the 25% substitution ratio improved nutrient absorption, reduced intestinal inflammation, improved intestinal physical barrier damage, did not affect intestinal microecology, and had no adverse effects on fish. However, substitution of a high proportion of FM with FCSM negatively affects the intestinal microflora and nutrient absorption capacity of fish.

**Keywords:** fermented cottonseed meal; substitution; intestinal enzymatic activity; gene expression; intestinal microflora

**Key Contribution:** A. Appropriate substitution ratio could improve nutrient absorption, reduce intestinal inflammation, and improve intestinal physical barrier damage while not affecting intestinal microecology. B. Replacing a high percentage of fish meal with fermented cottonseed meal can negatively affect the intestinal flora of fish.

## 1. Introduction

Golden pompano (*Trachinotus ovatus*) is a valuable edible fish with thornless flesh, a tender texture, and a delicious taste, with the unique aroma of the trevally family. It is distributed in the Atlantic, Indian, and Pacific Oceans' tropical and subtropical seas. In China, the coasts of Guangdong, Guangxi, Hainan, and Fujian are main growing regions [1]. The demand for protein raw materials (e.g., fishmeal or soybean meal) is rising as aquaculture production increases. Import policies have a significant impact on the aquaculture industry, since fishmeal and soybean meal are import-dependent. To fulfill the rising domestic demand for protein raw materials, guarantee a continuous supply of aquatic products, and maintain global food security, it is crucial to discover and develop novel, effective, sustainable, and environmentally friendly protein sources.

The ability of the intestines to digest and absorb nutrients is crucial for fish, especially stomachless fish, and is considered to be an important indicator of the overall health of the body [2]. There are large amounts of microflora in the intestines of fish which live in symbiosis with the fish and are known as the normal intestinal flora. Fish with normal intestinal flora can have better digestion and receive essential nutrients which strengthen their natural defenses and immunity [3]. Dysbacteriosis has the potential to trigger enteritis, leading to diminished appetite, stunted growth, and potentially fatal outcomes in fish [4]. An effective and properly controlled intestinal barrier is crucial for safeguarding the organism against food antigens and its indigenous intestinal bacteria [5]. As a result, research on the structure and changes of fish intestinal microflora has recently gained attention in the aquaculture industry.

Cottonseed meal (CSM) serves as a valuable source of protein of superior quality. The utilization of this nutrient-dense resource as animal feed, however, encounters obstacles due to the presence of free gossypol (FG) [6]. FG-containing diets can have negative effects on animal growth, digestive health, and reproduction. Fermentation technology allows for efficient separation of cottonseed phenols in cottonseed meals, raising the crude protein level of the raw material and producing small peptides and growth-promoting factors that enrich the nutrition of cottonseed protein [7]. The application of fermented cottonseed protein in mackerel culture was investigated, and it was found that the survival and weight gain rates of mackerel in the group with fermented cottonseed protein were significantly higher than those in the control group [8]. It was discovered that using 23% FCSM to replace 9% soybean meal and 15% cottonseed meal promoted the growth of grass carp, reduced the feed coefficient, and improved nonspecific immunity [9]. Most current research on fermented cottonseed protein is gathered in livestock and poultry, with little focus on aquatic animals. Based on these findings, this study was conducted to investigate the effects of the replacement of fish meal (FM) with FCSM on the intestinal enzymatic activity, inflammatory and physical-barrier-related gene expression, and intestinal microflora of juvenile *T. ovatus*.

## 2. Materials and Methods

### 2.1. Preparation of Experimental Diets

Fish meal, casein, soybean protein concentrate, and soybean meal were added as protein sources, and fish oil and soybean lecithin were added as lipid sources. Five isoproteic and isolipidic experimental diets were formulated with different levels of replacement of FM with FCSM, specifically, 0 (FM), 12.5% (CSM12.5), 25% (CSM25), 50% (CSM50), and 100% (CSM100), respectively; the feed formulations and the contents of various nutrients are shown in Table 1. To eliminate the effect of limiting amino acids, lysine and methionine were added to each group, respectively, so that the amino acid content of each group of feeds was balanced; the amino acid composition of the experimental feeds is shown in Table 2. All the ingredients were ground into powder, sieved through 60 mesh, and thoroughly mixed with oil and water; the 2.5 mm and 3.0 mm diameter long doughs were extruded using a twin screw extruder (F-26, South China University of Technology, Guangzhou, China), cut into pelletized feeds using a pelletizer (G-500, South China University of Technology, Guangzhou, China), and then stored at −20 °C in a refrigerator until used.

**Table 1.** Formulation and nutrient levels of the experimental diets (% dry matter).

| Ingredients | Fermented Cottonseed Meal Substitution Percentage/% | | | | |
|---|---|---|---|---|---|
| | FM | CSM12.5 | CSM25 | CSM50 | CSM100 |
| Fish meal | 40 | 35 | 30 | 20 | 0 |
| Soy protein concentrate | 16 | 16 | 16 | 16 | 16 |
| Soybean meal | 4 | 4 | 4 | 4 | 4 |
| Fermented cottonseed meal | 0 | 5.1 | 10.2 | 20.5 | 40.9 |
| Corn starch | 17.7 | 16.2 | 14.7 | 11.7 | 5.8 |
| Porcine blood cell protein powder | 2 | 2 | 2 | 2 | 2 |
| Beer yeast powder | 2 | 2 | 2 | 2 | 2 |
| Fish oil | 7.3 | 7.7 | 8.1 | 8.9 | 10.5 |
| Vitamin and mineral premix [1] | 1 | 1 | 1 | 1 | 1 |
| $Ca(H_2PO_4)_2$ | 0.5 | 0.5 | 0.5 | 0.5 | 0.5 |
| Choline chloride | 0.5 | 0.5 | 0.5 | 0.5 | 0.5 |
| Lecithin | 1 | 1 | 1 | 1 | 1 |
| Microcrystalline cellulose | 7.5 | 8.4 | 9.1 | 10.7 | 13.8 |
| Betaine | 0.5 | 0.5 | 0.5 | 0.5 | 0.5 |
| Lysine | 0 | 0.1 | 0.3 | 0.5 | 1 |
| Methionine | 0 | 0 | 0.1 | 0.2 | 0.5 |
| Nutrient levels | | | | | |
| Ash | 12.08 | 12.03 | 12.01 | 10.35 | 7.99 |
| Crude protein | 43.89 | 43.65 | 43.31 | 43.62 | 43.49 |
| Crude lipid | 11.30 | 11.54 | 11.71 | 11.85 | 12.02 |
| Crude fiber | 7.29 | 7.83 | 8.46 | 10.04 | 12.93 |
| Nitrogen free extract | 24.71 | 24.95 | 24.52 | 24.13 | 23.57 |
| Free gossypol (mg/kg) | 64.7 | 95.9 | 88.6 | 171.0 | 190.0 |

[1] Vitamin and mineral premix provided by Shenzhen Jingji Zhinong Times Co., Ltd. (mg $kg^{-1}$ diet). The formulation includes the following amounts of vitamins and minerals per kilogram: vitamin A at a minimum of 450,000 IU, vitamin B1 at a minimum of 1000 mg, vitamin B2 at a minimum of 1000 mg, vitamin B6 at a minimum of 1500 mg, vitamin B12 at a minimum of 5 mg, vitamin K3 at a minimum of 800 mg, inositol at a minimum of 12,000 mg, D-Pantothenic acid at a minimum of 3500 mg, nicotinic acid at a minimum of 2000 mg, folic acid at a minimum of 500 mg, D-Biotin at a minimum of 5 mg, vitamin D3 at a range of 300,000 to 400,000 IU, vitamin E at a minimum of 8000 IU, $Na_2SeO_3$ at 20 mg, $CuSO_4 \cdot 5H_2O$ at 24 mg, $FeSO_4 \cdot H_2O$ at 266.65 mg, $ZnSO_4 \cdot H_2O$ at 100 mg, $MnSO_4 \cdot H_2O$ at 120 mg, Ca $(IO_3)_2$ at 50 mg, $CoSO_4 \cdot 7H_2O$ at 10 mg, Mg at 20 g, and zeolite at 4380.55 mg.

### 2.2. Fish and Experimental Conditions

The feeding trial was conducted in a seawater pond at the Shenzhen Base of the South China Sea Fisheries Research Institute of the Chinese Academy of Fishery Sciences (Shenzhen, China). For two weeks, juvenile golden pompanos were acclimated to the experimental system and fed commercial diets (Guangdong Yuequn Biotechnology Co., Ltd., Guangzhou, China). At the outset of the feeding experiment, the fish were fasted for 24 h and then weighed. The fish were randomly assigned into 18 cages, with 25 uniformly sized fish per cage (5.6 ± 0.14 g) for 8 weeks. Experimental fish were fed different experimental diets twice daily, at 6:00 and 18:00, until they appeared to be satiated. During the feeding-trial period, the water temperature was maintained at 28.3–33.3 °C. Dissolved oxygen was higher than 6.0 mg/L. The salinity and ammonia were in the range of 20–25‰ and 0.05–0.1 mg/L, respectively. The photoperiod was the natural-day light cycle throughout the experimental period. The protocols for all fish were approved by the Ethical Committee of the South China Sea Fisheries Research Institute.

**Table 2.** Proximate amino acid profiles of experimental diets (g/100 g).

| Amino Acids | Fermented Cottonseed Meal Substitution Percentage/% | | | | |
|---|---|---|---|---|---|
| | FM | CSM12.5 | CSM25 | CSM50 | CSM100 |
| Aspartic acid | 2.73 | 2.71 | 2.72 | 2.72 | 2.54 |
| Threonine | 1.15 | 1.14 | 1.13 | 1.08 | 0.95 |
| Serine | 1.04 | 1.07 | 1.11 | 1.15 | 1.10 |
| Glutamic acid | 3.92 | 4.07 | 4.26 | 4.47 | 4.30 |
| Glycine | 1.63 | 1.49 | 1.44 | 1.25 | 0.83 |
| Alanine | 1.72 | 1.72 | 1.67 | 1.54 | 1.28 |
| Cystine | 0.24 | 0.25 | 0.26 | 0.26 | 0.26 |
| Valine | 1.42 | 1.41 | 1.38 | 1.34 | 1.26 |
| Methionine | 0.54 | 0.51 | 0.57 | 0.52 | 0.59 |
| Isoleucine | 1.09 | 1.08 | 1.04 | 0.98 | 0.87 |
| Leucine | 2.17 | 2.17 | 2.13 | 2.09 | 1.88 |
| Tyrosine | 0.69 | 0.68 | 0.72 | 0.71 | 0.72 |
| Phenylalanine | 1.13 | 1.19 | 1.23 | 1.30 | 1.37 |
| Lysine | 2.02 | 1.98 | 1.98 | 1.89 | 1.59 |
| Histidine | 0.72 | 0.74 | 0.75 | 0.76 | 0.74 |
| Arginine | 1.69 | 1.77 | 1.90 | 2.19 | 2.60 |
| Proline | 1.44 | 1.41 | 1.35 | 1.30 | 1.18 |
| Total | 25.3 | 25.4 | 25.6 | 25.6 | 24.1 |

### 2.3. *Collection of Samples*

By the end of the feeding trial, all the fish were deprived of diets for 24 h. Fish that were put under anesthesia were treated with diluted MS-222 (Sigma, St. Louis, MO, USA). Three fish per cage were anesthetized and sampled. The digestive contents of four fish from each cage were collected and were then quickly frozen in liquid nitrogen, followed by storage at −80 °C. To reduce the impact of interindividual differences, intestinal contents from each treatment were mixed for the analysis of intestinal microbiota. Mid-intestines from three fish in each cage were frozen in liquid nitrogen and then stored at −80 °C until total RNA was extracted. Three fish in each cage had a section of their gut frozen in liquid nitrogen, which was subsequently kept at −80 °C until the enzyme activity was determined.

### 2.4. *Measurement and Analysis*

#### 2.4.1. Intestinal Enzymes Activities Measurements

Intestinal samples were homogenized in sterilized physiological saline (0.86%, pH = 7.4; 1:9, *w/v*) by a handheld homogenizer. Then the samples were centrifuged for 15 min (2000 r/min, 4 °C) and the supernatant removed for the quantification of chymotrypsin, lipase, and α-amylase using commercial kits (Beijing Huaying Biotechnology Research Institute, Beijing, China).

#### 2.4.2. Free Cotton Phenol Content Measurements

The free cotton phenol content analysis of diets was determined according to the standard methods of the American Oil Chemists Society (AOCS 2009; method Ba 7b-96) [10]. All samples were analyzed with an 8453 ultraviolet-visible spectrophotometer (Agilent Technologies Co., Ltd., Qingdao, China).

#### 2.4.3. Quantitative Real-Time PCR

Total RNA was extracted from intestinal tissues using the Animal Total RNA Isolation Kit (FOREGENE Co., Ltd., Chengdu, China), and the integrity and quality of the RNA were detected using the NanoDropOne Micro Spectrophotometer (Thermo Scientific, Waltham, MA, USA). The cDNA was obtained by reverse transcription using the PrimeScriptTM RT reagent kit with a gDNA Eraser kit (Takara, Kusatsu City, Japan). The reaction conditions: 37 °C for 15 min, 85 °C for 5 s, and 4 °C. The obtained cDNA was stored at −20 °C. The primer sequences of the target gene and the internal reference gene β-Actin for real-time fluorescence quantitative PCR are shown in Table 3. All primer pairs required for real-time fluorescence quantitative PCR were synthesized by Sangon Biotech (Shanghai) Co., Ltd. (Shanghai, China). Real-time fluorescence quantitative PCR was performed on the Quant Studio Dx PCR instrument (ABI, Foster City, CA, USA). The expression of these genes was quantified by the $2^{-\Delta\Delta CT}$ method [11].

**Table 3.** The primers for real-time fluorescence quantification PCR.

| Gene | Sequence | Reference |
|---|---|---|
| β-Actin-qF | TACGAGCTGCCTGACGGACA | [12] |
| β-Actin-qR | GGCTGTGATCTCCTTCTGCA | |
| il-1β-qF | CGGACTCGAACGTGGTCACATTC | [12] |
| il-1β-qR | AATATGGAAGGCAACCGTGCTCAG | |
| IL-8-qF | TGCATCACCACGGTGAAAAA | [12] |
| IL-8-qR | GCATCAGGGTCCAGACAAATC | |
| TNF-α-qF | CGCAATCGTAAAGAGTCCCA | [12] |
| TNF-α-qR | AAGTCACAGTCGGCGAAATG | |
| il-10-qF | CTCCAGACAGAAGACTCCAGCA | [12] |
| il-10-qR | GGAATCCCTCCACAAAACGAC | |
| NF-κB-qF | TGCGACAAAGTCCAGAAAGAT | [12] |
| NF-κB -qR | CTGAGGGTGGTAGGTGAAGGG | |
| ZO-1-qF | TTTGTGGCAGGAGTTCT | [12] |
| ZO-1-qR | TTCTTGTTGGGGATGAT | |
| Occludin-qF | TACGCCTACAAGACCCGCA | [12] |
| Occludin-qR | CACCGCTCTCTCTGATAAA | |
| Claudin-3-qF | CTCCTCTGCTGCTCCTGTCC | [12] |
| Claudin-3-qR | CGTAGTCTTTCCTTTCTAACCCTG | |
| Claudin-15-qF | AAGGTATGAAATAGGAGAAGGGC | [12] |
| Claudin-15-qR | TGGTTTGATAAGGCAGAGGGTA | |

2.4.4. Gut Microflora

The microbial DNA extraction was performed using Hi Pure Soil DNA Kits (Magen, Guangzhou, China). The 16S rDNA V4 region of the ribosomal RNA gene was amplified using the polymerase chain reaction (PCR) technique. The primers used for this amplification were Arch519 (CAGCMGCCGCGGTAA) and Arch915R (GTGCTCCCCCGCCAATTCCT). In a duplicate 50 μL combination, the components included 5 μL of 10 × KOD buffer, 5 μL of 2.5 mM dNTPs, 1.5 μL of each primer (5 μM), 1 μL of KOD Polymerase, and 100 ng of template DNA for PCR reactions. The AxyPrep DNA Gel Extraction Kit, manufactured by Axygen Biosciences in Union City, CA, U.S., was employed to extract amplicons from 2% agarose gels. The ABI Step One Plus Real-Time PCR System, manufactured by Life Technologies in Foster City, USA, was employed for the purpose of quantification. In accordance with established techniques, the purified amplicons were combined in equimolar proportions and subjected to paired-end sequencing (2 × 250) using an Illumina platform. The raw data underwent splicing and filtering processes in order to create a refined dataset. The construction of operational taxonomic units was carried out, and the final feature table and feature sequences were created using the Divisive Amplicon Denoising Algorithm method.

*2.5. Statistical Analysis*

The experimental data were presented as mean ± SEM. Statistical analysis was performed using SPSS 26.0 software ( IBM Corporation, Somers, NY, USA) for Windows. Before performing analysis of variance (ANOVA), the normality and homogeneity of experimental data were tested using the Kolmogorov–Smirnov test and Levene's test, respectively. After passing the test, the experimental data were subjected to a one-way analysis of variance. When there were significant differences, the group means were further compared with Duncan's multiple-range test, and a probability of $p < 0.05$ was considered significant.

## 3. Results

### 3.1. Intestinal Enzymatic Activity of *T. ovatus*

The effects of dietary fish meal substitution by fermented cottonseed meal on the intestinal enzymatic activity of golden pompano are shown in Table 4. The LPS content levels of the FM, CSM12.5, and CSM25 groups were not significantly different ($p > 0.05$), with the highest content being found in the CSM12.5 group. There was a tendency towards an increase in chymotrypsin activity in the intestine of golden pompano at a substitution rate of 25%, which was significantly lower in the CSM12.5 and CSM50 groups ($p < 0.05$). As the substitution level of a fermented cottonseed meal increased, the AMY activity in the substitution group was significantly higher than that in the FM group ($p < 0.05$), and the highest activity of AMY in the intestine of golden pompano was observed when the substitution rate reached 25%.

**Table 4.** Effects of fish meal substitution by fermented cottonseed meal on an index of intestinal digestive enzymes of *T. ovatus*.

| Enzyme | Fermented Cottonseed Meal Substitution Percentage/% | | | | |
|---|---|---|---|---|---|
| | FM | CSM12.5 | CSM25 | CSM50 | CSM100 |
| LPS (U/mg) | $26.26 \pm 0.66^{a}$ | $26.3 \pm 1.43^{a}$ | $26.02 \pm 1.76^{a}$ | $21.53 \pm 0.32^{b}$ | $21.2 \pm 1.25^{b}$ |
| Chymotrypsin (U/mg) | $49.92 \pm 2.17^{ab}$ | $42.23 \pm 0.93^{c}$ | $51.69 \pm 3.99^{a}$ | $43.03 \pm 0.3^{c}$ | $46.24 \pm 1.59^{bc}$ |
| AMY(U/mg) | $236.33 \pm 6.45^{d}$ | $290.11 \pm 3.63^{c}$ | $405.72 \pm 10.24^{a}$ | $304 \pm 6.84^{b}$ | $290.22 \pm 3.98^{c}$ |

LPS: Lipase; AMY: Amylase. Data are expressed as means ± SEM ($n = 3$). Means with different superscripts are significantly different ($p < 0.05$).

### 3.2. Intestinal Immune-Related Gene Expression of *T. ovatus*

The gene expressions of the intestinal NF-κB-related signaling pathway in golden pompano after ingestion of different levels of experimental diets are shown in Figure 1. Compared with FM and CSM25, the expression of *IL-1β* and *TNF-α* of the fish in the group fed with CSM12.5, CSM50, and CSM100 was notably increased ($p < 0.05$). Compared with the FM group, the relative expression of the IL-8 gene in the intestine of golden pompano was not significantly different between the FM, CSM25, and CSM50 groups ($p > 0.05$), with the highest expression being found in the CSM100 group, which was significantly different from the other groups ($p < 0.05$). The expression of *IL-10* of the fish in the CSM12.5 dietary group was significantly upregulated ($p < 0.05$). The relative expression of the *NF-κB* gene was significantly lower in the intestine of golden pompano at different levels in the FCSM substitution group compared to the FM group ($p < 0.05$), with the highest expression found in the CSM100 group.

### 3.3. Intestinal Physical-Barrier-Related Gene Expression of *T. ovatus*

The expression of physical-barrier-related genes in golden pompano after ingestion of different levels of FCSM are shown in Figure 2. The relative expression of the *ZO-1* gene was lowest in the CSM50 group, which was lower than the FM group ($p < 0.05$), while the relative expression was significantly higher in the CSM12.5 and CSM100 groups. The relative expression of the *Occludin* gene was highest in the CSM100 group, and the levels of the CSM12.5 and CSM100 groups were significantly higher than that of the FM group ($p < 0.05$); additionally, there was no significant difference between the CSM25 and FM groups ($p > 0.05$). The relative expression of the *Claudin-3* gene was significantly affected by the substitution of FM by FCSM ($p < 0.05$), with the highest relative expression found in the CSM25 group. When the level of FCSM substitution was 50%, the relative expression of the *Claudin-15* gene in the intestine of golden pompano was significantly ($p < 0.05$) lower than in all other groups. Meanwhile, the CSM12.5 group, CSM25 group, and CSM100 group were significantly higher than the FM group ($p < 0.05$).

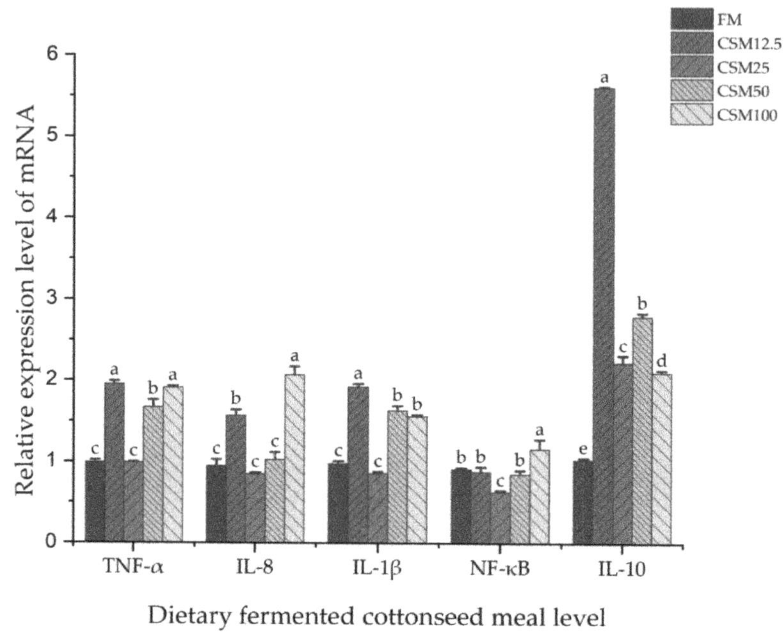


**Figure 1.** Effect of fermented cottonseed meal substitution for fish meal on immune-related gene expression of *T. ovatus*. Data are expressed as means ± SEM ($n$ = 3). Means with different superscripts are significantly different ($p < 0.05$).

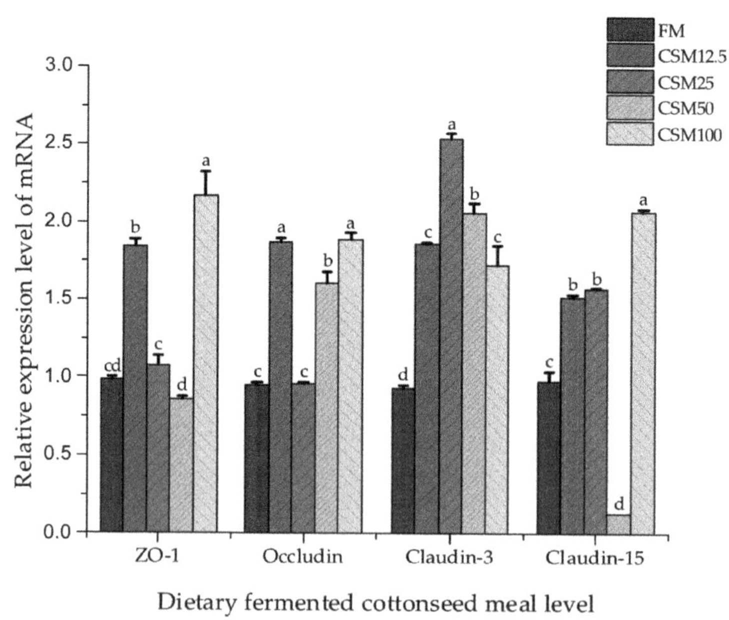


**Figure 2.** Effects of fish meal substitution by a fermented cottonseed meal on the barrier-related gene expression. Data are expressed as means ± SEM ($n$ = 3). Means with different superscripts are significantly different ($p < 0.05$).

### 3.4. Intestinal Microbiota

#### 3.4.1. Analysis of Microbial OTUs and Alpha Diversity of Intestinal Flora

The curve of intestinal microflora dilution (Figure 3) of juvenile golden pompano in different treatment groups tends to be flat, with sequencing coverage ≥ 99.96%. This indicates that the sequencing depth of the intestinal microflora of each group of golden pompano is reasonable and can cover the vast majority of species in the sample. A grand number of 1,549,208 sequencing reads of good quality were acquired.

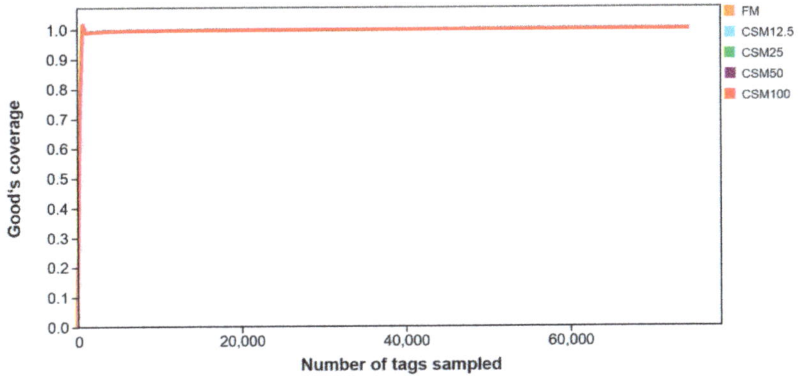


**Figure 3.** Effects of fish meal substitution by a fermented cottonseed meal on the dilution curve of the intestinal flora of *T. ovatus*.

From the Venus plot of OTUs of golden pompano in different treatment groups (Figure 4), the total number of OTUs between groups was 71. The alpha diversity metrics were computed by analyzing the rarefaction curves at the operational taxonomic units (OTUs) level for each experimental meal. Table 5 displays the Chao1, ACE, Shannon, and Simpson indexes. The CSM100 group had the highest ACE indices and the largest community richness, which was not significantly different from the other groups ($p > 0.05$). Compared with the remaining groups, the CSM25 group had the lowest ACE and Chao1 indices and the lowest community richness, while the CSM100 group had the highest ACE and Chao1 indices and the highest community richness. The CSM50 group had the highest Shannon index and Simpson index, and the highest community diversity. The Shannon index and Simpson index in the CSM50 group differed significantly from those of the other groups ($p < 0.05$) (Table 5).

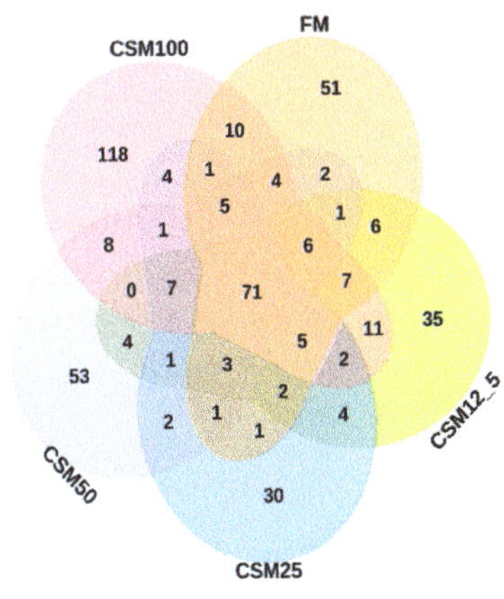


**Figure 4.** Venus map of OTUs of the intestinal flora of *T. ovatus* in different treatment groups.

**Table 5.** Diversity statistics of intestinal samples of juvenile *T. ovatus* in different treatment groups.

| Item | Group | | | | |
|---|---|---|---|---|---|
| | **FM** | **CSM12.5** | **CSM25** | **CSM50** | **CSM100** |
| OTU | 166.7 ± 11.09 | 158.25 ± 8.06 | 139.25 ± 7.85 | 154.75 ± 27.92 | 195.00 ± 59.57 |
| Ace | 189.67 ± 15.13 | 176.73 ± 1.76 | 161.73 ± 14.21 | 165.01 ± 9.89 | 191.77 ± 22.51 |
| Chao1 | 183.38 ± 18.21 | 169.74 ± 3.1 | 157.15 ± 9.73 | 156.37 ± 9.01 | 187.16 ± 21.13 |
| Shannon | 1.86 ± 0.08 [bc] | 1.68 ± 0.11 [c] | 2.34 ± 0.29 [ab] | 2.71 ± 0.26 [a] | 1.88 ± 0.29 [bc] |
| Simpson | 0.61 ± 0.03 [b] | 0.57 ± 0.06 [b] | 0.7 ± 0.1 [ab] | 0.78 ± 0.05 [a] | 0.64 ± 0.06 [ab] |

ACE: The ACE index is utilized as a means to approximate the abundance of operational taxonomic units (OTUs) within a given ecological community. It has been observed that higher ACE values exhibit a positive correlation with the overall diversity and richness of the microbial community. Chao1: The number of OTU in the community was estimated by Chao1 algorithm, and the value of Chao1 was positively correlated with the total number of species. Shannon: The index considers both the abundance and evenness of the community, and there is a positive correlation between its value and the richness and evenness of the community. Simpson: The index was used to estimate microbial diversity in samples, and the values were negatively correlated with community diversity. Data are expressed as means ± SEM ($n = 4$). Means with different superscripts are significantly different ($p < 0.05$). [a,b,c] Means (±SEM) values within the row unlike superscript letters were significantly different ($p < 0.05$).

### 3.4.2. Analysis of Intestinal Flora Composition and Relative Abundance

The analysis of the intestinal contents of juvenile golden pompano encompassed multiple taxonomic levels, specifically, phylum, order, family, and genus. Phylum and genus were chosen as representative taxonomic levels for this study.

As shown in Figure 5, at the phylum level, Proteobacteria and Firmicutes formed the core microflora. Proteobacteria were the most abundant phyla in the control group and CSM25 group, accounting for 49.4% and 52.16%, respectively. Firmicutes was the most abundant phylum in CSM12.5, CSM50, and CSM100, accounting for 61.46%, 51.71%, and 52.83%, respectively. There were no significant differences among the five groups, except for Fusobacteria and Planctomycetes ($p > 0.05$) (Table 6).

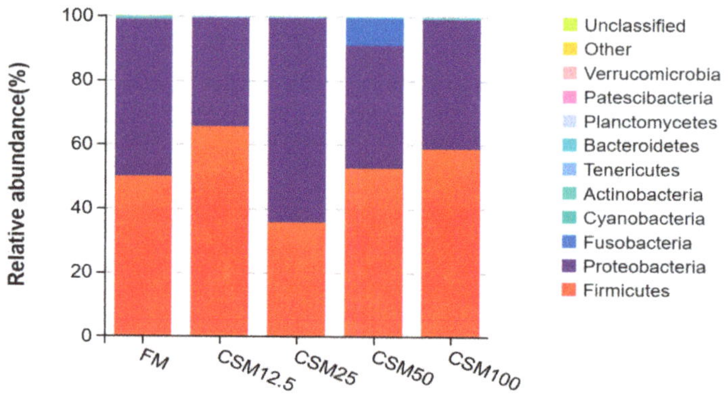


**Figure 5.** Distribution of the top 10 microbial phylum levels in the intestinal contents of *T. ovatus* in different treatment groups.

As shown in Figure 6, at the genus level, Photobacterium and Paraclostridium formed the core microflora. The dominant genera in the intestinal flora of *T. ovatus* were Photobacterium and Paraclostridium, which accounted for 52.5% and 44.86% of the total species, respectively. Photobacterium is the most abundant genus in the FM group and CSM25. There were significant differences among the five groups, except for Lysinibacillus, Bacillus, and Aerococcus ($p < 0.05$) (Table 7). The Fusobacteria levels of CSM100 were significantly lower than those of all other groups ($p < 0.05$). There was no significant difference in Planctomycetes levels for CSM12.5, CSM25 and CSM50 ($p > 0.05$), and the FM group and CSM100 were significantly higher than all other groups ($p < 0.05$).

**Table 6.** Distribution of the top 10 microbial phylum levels in the intestinal contents of *T. ovatus* in different treatment groups.

| Phylum | Group | | | | |
|---|---|---|---|---|---|
| | FM | CSM12.5 | CSM25 | CSM50 | CSM100 |
| *Firmicutes* | 49.36 ± 6.73 | 61.46 ± 30.38 | 47.14 ± 23.64 | 51.71 ± 14.17 | 52.83 ± 13.50 |
| *Proteobacteria* | 49.40 ± 6.44 | 37.84 ± 30.28 | 52.16 ± 23.64 | 41.16 ± 12.34 | 43.82 ± 10.15 |
| *Fusobacteria* | 0.11 ± 0.03 [a] | 0.10 ± 0.01 [a] | 0.10 ± 0.04 [a] | 0.07 ± 0.04 [a] | 0.01 ± 0.00 [ab] |
| *Cyanobacteria* | 0.45 ± 0.13 | 0.19 ± 0.02 | 0.20 ± 0.02 | 0.26 ± 0.12 | 0.40 ± 0.38 |
| *Actinobacteria* | 0.22 ± 0.08 | 0.14 ± 0.02 | 0.16 ± 0.02 | 0.21 ± 0.13 | 0.30 ± 0.29 |
| *Tenericutes* | 0.17 ± 0.09 | 0.14 ± 0.10 | 0.11 ± 0.03 | 0.09 ± 0.07 | 0.29 ± 0.42 |
| *Planctomycetes* | 0.05 ± 0.03 [b] | 0.03 ± 0.01 [ab] | 0.01 ± 0.00 [ab] | 0.02 ± 0.02 [ab] | 0.06 ± 0.02 [a] |
| *Bacteroidetes* | 0.10 ± 0.06 | 0.02 ± 0.00 | 0.04 ± 0.02 | 0.03 ± 0.02 | 0.10 ± 0.13 |
| *Patescibacteria* | 0.02 ± 0.01 | 0.01 ± 0.00 | 0.02 ± 0.01 | 0.02 ± 0.02 | 0.13 ± 0.17 |
| *Verrucomicrobia* | 0.01 ± 0.00 | 0.00 ± 0.00 | 0.00 ± 0.00 | 0.00 ± 0.01 | 0.09 ± 0.14 |

Data are expressed as means ± SEM ($n$ = 4). Means with different superscripts are significantly different ($p < 0.05$).

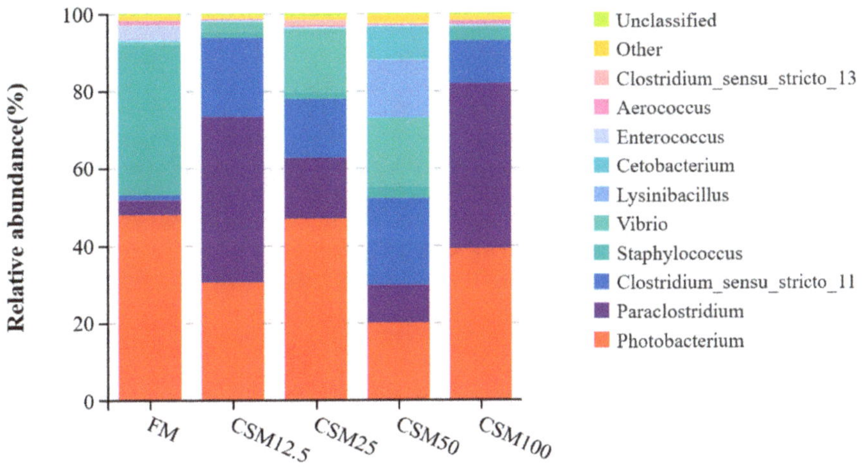


**Figure 6.** Distribution of the top 10 microbial genus levels in the intestinal contents of *T. ovatus* in different treatment groups.

**Table 7.** Distribution of the top 10 microbial genus levels in the intestinal contents of juvenile *T. ovatus* in different treatment groups.

| Genus | Group | | | | |
|---|---|---|---|---|---|
| | FM | CSM12.5 | CSM25 | CSM50 | CSM100 |
| *Photobacterium* | 48.34 ± 6.37 [a] | 41.44 ± 24.84 [ab] | 39.41 ± 18.01 [ab] | 21.91 ± 3.84 [b] | 42.26 ± 9.75 [ab] |
| *Paraclostridium* | 4.07 ± 1.85 [b] | 53.97 ± 1.43 [a] | 17.58 ± 3.88 [ab] | 13.16 ± 8.79 [ab] | 32.97 ± 19.24 [ab] |
| *Clostridium_sensu_stricto_11* | 2.16 ± 1.42 [b] | 29.93 ± 10.68 [ab] | 31.67 ± 20.52 [a] | 19.98 ± 10.36 [ab] | 8.58 ± 9.41 [ab] |
| *Staphylococcus* | 34.36 ± 10.65 [a] | 1.44 ± 0.77 [b] | 1.88 ± 1.48 [b] | 3.87 ± 2.35 [b] | 5.54 ± 5.45 [b] |
| *Vibrio* | 0.71 ± 0.13 [ab] | 4.70 ± 4.47 [ab] | 12.28 ± 8.86 [ab] | 18.62 ± 11.27 [a] | 0.63 ± 0.30 [b] |
| *Lysinibacillus* | 0.23 ± 0.04 | 0.30 ± 0.32 | 0.21 ± 0.03 | 0.15 ± 0.12 | 0.02 ± 0.01 |
| *Enterococcus* | 3.92 ± 1.66 [a] | 0.11 ± 0.05 [b] | 0.13 ± 0.03 [b] | 0.23 ± 0.09 [b] | 0.48 ± 0.29 [b] |
| *Bacillus* | 0.20 ± 0.05 | 0.13 ± 0.07 | 0.19 ± 0.08 | 0.33 ± 0.31 | 0.14 ± 0.23 |
| *Cetobacterium* | 0.11 ± 0.03 [a] | 0.10 ± 0.0 [a] | 0.10 ± 0.04 [a] | 0.07 ± 0.04 [a] | 0.01 ± 0.00 [b] |
| *Aerococcus* | 0.41 ± 0.20 | 0.05 ± 0.03 | 0.02 ± 0.01 | 0.74 ± 0.37 | 1.58 ± 0.79 |

Data are expressed as means ± SEM ($n$ = 4). Means with different superscripts are significantly different ($p < 0.05$).
[a,b] Means (±SEM) values within the row unlike superscript letters were significantly different ($p < 0.05$).

3.4.3. Analysis of β Diversity of Intestinal Flora

Non-measured multidimensional analysis (NMDS) is a commonly used β diversity analysis method. As the basis for community structure research, NMDS analysis is often used to compare the differences between different ecosystems and reflect the heterogeneity of biological species due to the environment. The points in the graph represent the sample, and the distance between the points reflects the degree of sample difference; the closer the distance, the higher the similarity. Points in the same circle represent no significant difference between the samples, and points in the non-intersection circle represent significant differences between the samples. As can be seen from Figure 7, the FM group intersects with the CSM12.5 group, but does not intersect with any other groups. The CSM25 group intersects with the CSM12.5 group and the CSM50 group and does not intersect with any other groups. The CSM50 group intersects with the CSM25 group and does not intersect with any other groups. The CSM100 group intersects with the CSM12.5 group but does not intersect with any other groups. The CSM12.5 group is not intersected with the CSM50 group and is intersected with all other groups. This means that there is no significant difference between the FM group and the CSM12.5 group ($p > 0.05$), while there is a significant difference with all other groups ($p < 0.05$). There is no significant difference between the CSM25 group and the CSM12.5 group or the CSM50 group ($p > 0.05$), and there is a significant difference with all other groups ($p < 0.05$). There is no significant difference between the CSM50 group and the CSM25 group ($p > 0.05$), and the former is significantly different from each of the other groups ($p < 0.05$); there is no significant difference between the CSM100 group and the CSM12.5 group ($p > 0.05$), and the former is significantly different from each of the other groups ($p < 0.05$). There is a significant difference between the CSM12.5 group and the CSM50 group ($p < 0.05$), and the former is not significantly different from any of the other groups ($p > 0.05$). There was an intersection between the CSM25 group and the CSM50 group, indicating that when the amount of FCSM in the feed was 25%, there was no significant difference in the intestinal microbial composition than with the addition of 50% in *T. ovatus* ($p > 0.05$). There was an intersection among the CSM25 group, the CSM12.5 group, and the CSM100 group, indicating that when the amount of FCSM in the feed was 25%, there was no significant difference in the intestinal microbial composition than with the addition of 12.5% and 100% in *T. ovatus* ($p > 0.05$).

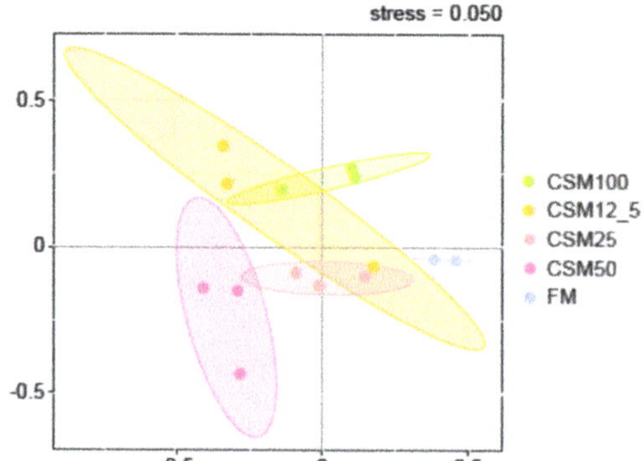


**Figure 7.** NMDS analysis at the OTU level of *T. ovatus* in different treatment groups.

## 4. Discussion

Fish meal (FM) is widely utilized as a protein source in commercial aquafeeds due to its well-balanced composition of essential amino acids, vitamins, and minerals, and its palatability [10]. In recent years, the increasing demand for animal nutrition has led to a search for sustainable protein sources of both plant and animal origin, due to the limited availability of fishmeal [13]. Cottonseed meal is digested relatively well by fish and crustaceans [14]. When markets are favorable, CSM is an economical alternative protein source for use in aquatic animal feeds. Furthermore, it is generally highly palatable to most aquatic animals [14].

The intestinal tracts of animals serve as complex microbial ecosystems that harbor dynamic communities of microorganisms. These microorganisms play various roles, such as absorbing nutrients, improving energy production, and maintaining immune homeostasis [15]. The process of digestion is related to the activity of digestive enzymes, including chymotrypsin and α-amylase. These enzymes play a crucial role in breaking down nutrients, as they facilitate the digestion process [16]. In the present study, the intestinal amylase content levels in the CSM12.5 and CSM25 groups were higher than that in the FM group. The chymotrypsin and AMY activity levels in the intestine were higher than in the FM group when the substitution rate reached 25%. AMY activity reflects the absorption and utilization of nutrients by fish [17]. Protease is an important proteolytic enzyme in fish intestine, and its activity can reflect the ability of fish to decompose feed protein [18]. These findings indicated that 25% substitution of FCSM for FM boosted the activity of digestive enzymes and improved nutrient uptake, which in turn improved growth performance of *T. ovatus*. In this study, when 50–100% of FM was replaced by FCSM, the activity levels of LPS and chymotrypsin were lower than those of the control group. This may be due to the lower digestion and absorption ability of FCSM as compared to fish meal. A similar phenomenon was also observed in a previous study by Sillago sihama Forsskål [19]. Research findings indicate that the American redfish species (*Sciaenops ocellatus*) has demonstrated proficient ability in digesting cottonseed meal [20].

The significant function of pro- and anti-inflammatory cytokines in the maintenance of tissue and immunological homeostasis has been well established [21]. The pro-inflammatory cytokines, which include *TNF-α* and *IL-8*, have a variety of effects that cause inflammation [22]. Interleukin-1β (*IL-1β*) plays a crucial role in both the onset and amplification of the inflammatory response, which ultimately leads to intestinal damage [23]. Inflammatory cytokine *IL-8* is a component of the immune response, and disruption of the balance of these cytokines is key to the pathophysiology of inflammatory bowel disease [24]. In epithelial tissues, *NF-κB* signaling is crucial for preserving immunological homeostasis. Thus, *NF-κB* appears to exhibit two sides in chronic inflammation: on the one hand, *NF-κB* activation increases and continues to induce inflammation and tissue damage, but on the other hand, inhibition of *NF-κB* signaling also disrupts immune homeostasis and triggers inflammation and disease [22]. In the present study, *TNF-α*, *IL-1β*, *IL-8*, and *NF-κB* gene expression levels were all significantly decreased at the level of 25% FCSM replacement. In contrast, excessive substitution levels significantly increased *TNF-α*, *IL-1β*, *IL-8*, and *NF-κB* gene expression levels, indicating that higher substitution levels of FCSM can induce inflammatory responses in the intestine. *IL-10* is an anti-inflammatory factor which plays a role in down-regulating inflammatory response and antagonizing inflammatory mediators [25]. Gene expression of the pro-inflammatory factor *IL-10* was significantly higher than that of the control group at 12.5%, 25% and 50% substitution levels, suggesting that intestinal inflammation improves at appropriate substitution ranges.

The intestinal tract in fish is widely recognized as a significant physical barrier, and maintaining its structural integrity is essential for effectively resisting foreign antigens [26]. The presence of compromised or diminished physical barrier function within the intestinal tract of fish is associated with an elevated susceptibility to pathogen infection and a hindered ability to suppress pathogen growth [12]. *ZO-1* is involved, in its function as a scaffolding protein, by determining the tightness of the epithelial barrier and contributing to the polarity of the epithelial cells [5]. The extracellular loop of *Occludin* goes straight into the TJ complex and interacts with the transmembrane domain to alter the permeability of extracellular selectivity [27]. In the present study, the mRNA transcript levels of *ZO-1* were up-regulated, but the relative expression levels of the *Occludin* gene in the CSM25 and FM groups were not significantly different. *Claudin-3* is a tightening TJ protein that aids in sealing the paracellular intestinal barrier [28]. In this study, the *Claudin-3* gene had the highest relative expression in the CSM25 group, indicating that the intestinal barrier closure was strongest at the 25% substitution level. *Claudin-15* is widely distributed in the tight junctions of the villi and crypt cells of the small and large intestines. In this study, the relative expression of *Claudin-15* in the CSM12.5, CSM25, and CSM100 groups was significantly higher than that in the FM group, suggesting that the structural integrity of intestinal intercellular structures was better in the CSM12.5, CSM25, and CSM100 groups. The results of the above studies suggest that appropriate substitution ratios can ameliorate golden pompano intestinal physical barrier damage induced by FCSM. This is similar to the results of a study on the genes related to the improvement of intestinal tight junction in yellow catfish by fermented rapeseed meal [29]. Therefore, FCSM may alleviate the damage, associated with cottonseed meal, to the intestinal structure of golden pompano by degrading anti-nutritional factors, thereby improving the intestinal health of fish.

The microbial ecosystem of animals consists of microbial communities and their host microenvironment. The intestinal microenvironment consists of intestinal microorganisms and is the most critical microenvironment of the animal body [30]. The intestinal microbiota play a crucial role in various physiological processes, including metabolism, nutrient absorption, growth and development, and immune function. Therefore, it is very important for body health to investigate changes in fish intestinal flora. In the present study, the OTU Wayne diagram found that 71 OTU were common in all groups, indicating that there are still inherent core flora at different levels of FCSM addition in the same culture environment. These results suggested that the tested variations in diet did not exert large, long-term alterations on the intestinal microbiota of golden pompano.

Intestinal bacterial microflora level is closely related to digestive function and overall gastrointestinal health [31]. In terms of community composition, we found that the dominant intestinal bacteria of the golden pompano belonged to three phyla: Proteobacteria, Firmicutes, and Tenercutes [32]; this has also been reported for the turbot [33]. *Proteobacteria* can catabolize feed ingredients [34]. The increased abundance may allow *T. ovatus* to absorb nutrients from the feed. Proteobacteria, a class of bacteria characterized by their Gram-negative cell wall structure, serves a significant function in the breakdown and fermentation of polysaccharides, proteins, and various organic substances. Moreover, they constitute the prevailing microbial population within the intestinal ecosystem of numerous fish species [35]. It is well known that many kinds of Vibrio, Photobacterium, and Mycoplasma are pathogenic bacteria [19,36]. These microorganisms are found within the gastrointestinal tract. The degradation of the aquatic environment leads to an escalation in nutrient deficits and physiological dysfunctions, hence heightening the vulnerability of fish to these infections [37]. Research shows that the balance of the intestinal microflora was altered primarily by probiotic supplementation and, to a lesser extent, by the energy content of the diet [33]. The majority of Fusobacteriales strains identified in fish intestinal samples are attributed to Cetobacterium, a prevalent and extensively distributed species found within the gastrointestinal tracts of fish [38]. In this experiment, the dominant intestinal flora of the golden pompano was composed of *Firmicutes* and *proteobacteria*. However, with the change in the proportion of FCSM instead of fishmeal in feed, the absolute dominant flora between the groups also changed. At the phylum level, the dominant phylum of the control group for the CSM12.5 group, CSM50 group, and CSM100 group was firmicutes, while the dominant phylum of the CSM25 group was proteobacteria. The results of the above studies showed that the substitution of FM by FCSM did not lead to changes in the dominant strains of the intestinal flora of golden pompano.

At the genus level, the dominant genera in the intestinal flora of the *T. ovatus* are *Photobacterium* and *Paraclostridium*. *Photobacterium*, a bacterium classified as Gram-negative, has a combination of anaerobic and aerobic characteristics, displaying both respiratory and fermentative metabolic pathways. The inclusion of Bacillus probiotics in the diet has been found to promote the proliferation of lactic acid bacteria within the gastrointestinal system, leading to a decrease in pH levels. Consequently, this creates an unfavorable environment for the growth and survival of pathogenic bacteria, such as *E. coli* and *Salmonella* spp. [39]. Photobacterium, a Gram-negative bacterium, is partly anaerobic and has respiratory and fermentative metabolism types. Photobacterium can infect the shrimp Litopenaeus vannamei to cause its mortality, and may be a common pathogen of shrimps [19]. The Photobacterium levels in the CSM25 group and CSM50 group were significantly lower than that in the control group, indicating that the abundance of Photobacterium decreased when the amount of FCSM added to the feed was 25–50% [40].

Beta diversity analysis is mainly used to compare differences in the overall structure of microbial communities in each sample. Non-metric multidimensional scaling (NMDS) is a data analysis method that simplifies the research objects of multidimensional space to low-dimensional space for locating, analysis, and classification while retaining the original relationship between objects [41]. In this experiment, we found that the gut microbial composition of the FM group was significantly different from that of the CSM12.5, CSM25, CSM50, and CSM100 groups, and that there were significant differences in the composition of the CSM12.5 and CSM50 groups and the CSM25 and CSM100 groups, indicating that varying levels of fermented cotton meal substitution in the feeds resulted in significant differences in gut microbial composition. Complex microbiota interactions may explain the fact that fermented fishmeal increased the diversity of the gut microbiota of the vanabin carp, a finding which is similar to the results of our study [42].

## 5. Conclusions

In conclusion, by analyzing the intestinal enzyme activity, inflammation, physical barrier-related gene expression, and intestinal microecology of juvenile golden pompano (*T. ovatus*), this investigation demonstrated in its results that FCSM replacement of FM affected the intestinal health status of the fish. Replacing 25% of FM with FCSM had no significant effect on LPS and chymotrypsin in juvenile golden pompano, but AMY activity was significantly increased. However, when 50–100% was replaced, LPS and chymotrypsin were significantly reduced, indicating that the substitution of 25% FM by FCSM did not affect intestinal digestion. It was demonstrated that an appropriate substitution ratio could improve nutrient absorption, reduce intestinal inflammation, and improve intestinal physical barrier damage, while not affecting intestinal microecology. However, substitution of a high proportion of FM with FCSM negatively affects the intestinal microflora and nutrient absorption capacity of fish.

**Author Contributions:** Conceptualization, Z.W. and C.Z.; methodology, Z.W., C.Z. and Z.C.; validation, C.Z. and J.W.; investigation, Z.W., S.L., J.W., Y.W., Z.H., W.Y., X.H., H.L. and M.L.; Formal analysis, Z.W. and C.Z.; data curation, Z.W.; writing—original draft preparation, Z.W.; writing—review and editing, Z.W. and C.Z.; supervision and project administration, C.Z. and Z.C.; funding acquisition, C.Z. All authors have read and agreed to the published version of the manuscript.

**Funding:** This work was supported by the Central Public-Interest Scientific Institution Basal Research Fund, South China Sea Fisheries Research Institute, CAFS (2021XK02); the Central Public-Interest Scientific Institution Basal Research Fund, CAFS (2022XT0404; 2023TD58); the project was subsidized by Special Funds for Science, Technology, Innovation and Industrial Development of Shenzhen Dapeng New District (Grand No. PT202101-26); the National Natural Science Foundation of China (32172984); and the Fund of Southern Marine Science and Engineering Guangdong Laboratory (Zhanjiang) (ZJW-2019-06).

**Institutional Review Board Statement:** The experimental design was approved by the Ethics Committee of the South China Sea Fisheries Institute. Approval Code: 2022040401. Approval Date: 12 May 2022.

**Informed Consent Statement:** Not applicable.

**Data Availability Statement:** The data are available upon request.

**Conflicts of Interest:** The authors declare no conflict of interest.

## References


1. Liu, C.; Chen, C. The biology and cultured technology of Pompano (*Trachinotus ovatus*). *Shandong Fish.* **2009**, *26*, 32–33.
2. Zhou, X. Nutrients in relation to fish gut health. In *Advances in Animal Nutrition Research*; Chinese Society of Animal Husbandry and Veterinary Medicine, Animal Nutrition Branch: Beijing, China, 2012; pp. 253–267.
3. DeMeo, M.T.; Mutlu, E.A.; Keshavarzian, A.; Tobin, M.C. Intestinal Permeation and Gastrointestinal Disease. *J. Clin. Gastroenterol.* **2002**, *34*, 385–396. [CrossRef]
4. Li, C.; Wang, L.; Wang, H. Study on the effects of Aeromonas sobria and probiotics on the diversity of intestinal flora in carp. *Heilongjiang Anim. Sci. Vet. Med.* **2017**, 171–174. [CrossRef]
5. Akbari, P.; Braber, S.; Varasteh, S.; Alizadeh, A.; Garssen, J.; Fink-Gremmels, J. The intestinal barrier as an emerging target in the toxicological assessment of mycotoxins. *Arch. Toxicol.* **2017**, *91*, 1007–1029. [CrossRef] [PubMed]
6. Wang, X. Effect of *Bacillus cereus* Br on bacterial community and gossypol content during fermentation in cottonseed meal. *Afr. J. Microbiol. Res.* **2012**, *6*, 6537–6544. [CrossRef]
7. Wu, X.; Chen, J. The utilization of microbes to break down FG in cottonseed meal. *Chin. J. Agric. Sci.* **1989**, *4*, 82–86.
8. Jin, H. Research on Microbial Fermentation of Detoxified Cotton Meal and Its Utilization on Mackerel Culture. Master's Thesis, Hunan Agricultural University, Changsha, China, 2011.
9. Chen, D.; Leng, X.; Xiao, X. Effect of fermented cotton meal on the growth of grass carp. *Mag. Introd.* **2010**, 40–42. [CrossRef]
10. Sun, H.; Tang, J.; Yao, X.; Wu, Y.; Wang, X.; Liu, Y. Effects of replacement of fish meal with fermented cottonseed meal on growth performance, body composition and haemolymph indexes of Pacific white shrimp, *Litopenaeus vannamei* Boone, 1931. *Aquac. Res.* **2016**, *47*, 2623–2632. [CrossRef]
11. Livak, K.J.; Schmittgen, T.D. Analysis of Relative Gene Expression Data Using Real-Time Quantitative PCR and the $2^{-\Delta\Delta CT}$ Method. *Methods* **2001**, *25*, 402–408. [CrossRef]
12. Zhou, C.; Huang, Z.; Lin, H.; Ma, Z.; Wang, J.; Wang, Y.; Yu, W. Rhizoma curcumae Longae ameliorates high dietary carbohydrate-induced hepatic oxidative stress, inflammation in golden pompano *Trachinotus ovatus*. *Fish Shellfish Immunol.* **2022**, *130*, 31–42. [CrossRef] [PubMed]
13. Henry, M.A.; Gai, F.; Enes, P.; Peréz-Jiménez, A.; Gasco, L. Effect of partial dietary replacement of fishmeal by yellow mealworm (*Tenebrio molitor*) larvae meal on the innate immune response and intestinal antioxidant enzymes of rainbow trout (*Oncorhynchus mykiss*). *Fish Shellfish Immunol.* **2018**, *83*, 308–313. [CrossRef] [PubMed]
14. Li, M.H.; Robinson, E.H. Use of Cottonseed Meal in Aquatic Animal Diets: A Review. *N. Am. J. Aquac.* **2006**, *68*, 14–22. [CrossRef]
15. Wang, X.; Luo, H.; Zheng, Y.; Wang, D.; Wang, Y.; Zhang, W.; Chen, Z.; Chen, X.; Shao, J. Effects of poultry by-product meal replacing fish meal on growth performance, feed utilization, intestinal morphology and microbiota communities in juvenile large yellow croaker (*Larimichthys crocea*). *Aquac. Rep.* **2023**, *30*, 101547. [CrossRef]
16. Yan, L.; Xiao, Q.-Z. Dietary glutamine supplementation improves structure and function of intestine of juvenile Jian carp (*Cyprinus carpio* var. Jian). *Aquaculture* **2006**, *256*, 389–394. [CrossRef]
17. Qin, Q.; Chen, X.; Jiang, G.; Wang, M.; Zhang, S.; Zhong, L.; Liu, W.; Pan, J. Growth, body composition and digestion of juveniles in five yellow catfish families. *Genom. Appl. Biol.* **2018**, *37*, 768–773. [CrossRef]
18. Zhang, W.; Tan, B.; Deng, J.; Yang, Q.; Chi, S.; Pang, A.; Xin, Y.; Liu, Y.; Zhang, H. Soybean protein concentrate causes enteritis in juvenile pearl gentian groupers (*Epinephelus fuscoguttatus* ♀× *Epinephelus lanceolatus* ♂). *Anim. Nutr.* **2023**, *12*, 171–185. [CrossRef] [PubMed]
19. Liu, H.; Dong, X.; Tan, B.; Du, T.; Zhang, S.; Yang, Y.; Chi, S.; Yang, Q.; Liu, H. Effects of fish meal replacement by low-gossypol cottonseed meal on growth performance, digestive enzyme activity, intestine histology and inflammatory gene expression of silver sillago (*Sillago sihama* Forsskál) (1775). *Aquac. Nutr.* **2020**, *26*, 1724–1735. [CrossRef]
20. Gaylord, T.G.; Gatlin, D.M. Determination of digestibility coefficients of various feedstuffs for red drum (*Sciaenops ocellatus*). *Aquaculture* **1996**, *139*, 303–314. [CrossRef]
21. Mantovani, A.; Dinarello, C.A.; Molgora, M.; Garlanda, C. Interleukin-1 and Related Cytokines in the Regulation of Inflammation and Immunity. *Immunity* **2019**, *50*, 778–795. [CrossRef]
22. Zhou, C. Effects of dietary leucine levels on intestinal antioxidant status and immune response for juvenile golden pompano (*Trachinotus ovatus*) involved in Nrf2 and NF-κB signaling pathway. *Fish Shellfish Immunol.* **2020**, *107*, 336–345. [CrossRef]
23. Ahmed, I. Dietary amino acid l-threonine requirement of fingerling Indian catfish, *Heteropneustes fossilis* (Bloch) estimated by growth and biochemical parameters. *Aquac. Int.* **2007**, *15*, 337–350. [CrossRef]
24. Hossain, M.S.; Lee, S.; Small, B.C.; Hardy, R.W. Histidine requirement of rainbow trout (*Oncorhynchus mykiss*) fed a low fishmeal-based diet for maximum growth and protein retention. *Aquac. Res.* **2021**, *52*, 3785–3795. [CrossRef]
25. She, D.; Dong, Y.; Qin, Z.; Li, J.; Huang, D.; Yang, K.; Xu, H.; Chen, Y. Effects of chicken spleen transfer factor on the secretion of anti-inflammatory factors and pro-inflammatory factors in chicken intestine. *Chin. J. Vet. Med.* **2017**, *53*, 3–5.
26. Zhao, L.; Liang, J.; Chen, F.; Tang, X.; Liao, L.; Liu, Q.; Luo, J.; Du, Z.; Li, Z.; Luo, W.; et al. High carbohydrate diet induced endoplasmic reticulum stress and oxidative stress, promoted inflammation and apoptosis, impaired intestinal barrier of juvenile largemouth bass (*Micropterus salmoides*). *Fish Shellfish Immunol.* **2021**, *119*, 308–317. [CrossRef]
27. Wu, X.; Chen, X.; Wang, G.; Niu, X.; Kong, Y.; Li, M. Research progress of tight junction proteins in fish intestinal barrier injury. *Feed Ind.* **2023**, *44*, 56–62. [CrossRef]

28. Patel, R.M.; Myers, L.S.; Kurundkar, A.R.; Maheshwari, A.; Nusrat, A.; Lin, P.W. Probiotic Bacteria Induce Maturation of Intestinal Claudin 3 Expression and Barrier Function. *Am. J. Pathol.* **2012**, *180*, 626–635. [CrossRef]
29. Jia, B.; Zou, F.; Xu, J.; Zhao, T.; Liu, S.; Luo, Z. Effects of fermented rapeseed meal on apparent digestibility, liver and intestinal health of yellow catfish. *Acta Hydrobiol. Sin.* **2023**, *48*, 1–12.
30. Xie, Y.; Zuo, Y.; Li, Q.; Li, Y. Recent advances in the interactions between animal gut microbiota and host intestinal immune system. *Chin. J. Cell Biol.* **2017**, *39*, 1467–1472.
31. Bonos, E.; Giannenas, I.; Sidiropoulou, E.; Stylianaki, I.; Tzora, A.; Skoufos, I.; Barbe, F.; Demey, V.; Christaki, E. Effect of *Bacillus pumilus* supplementation on performance, intestinal morphology, gut microflora and meat quality of broilers fed different energy concentrations. *Anim. Feed Sci. Technol.* **2021**, *274*, 114859. [CrossRef]
32. Xun, P.; Lin, H.; Wang, R.; Huang, Z.; Zhou, C.; Yu, W.; Huang, Q.; Tan, L.; Wang, Y.; Wang, J. Effects of dietary vitamin B1 on growth performance, intestinal digestion and absorption, intestinal microflora and immune response of juvenile golden pompano (*Trachinotus ovatus*). *Aquaculture* **2019**, *506*, 75–83. [CrossRef]
33. Xing, M. Analysis of Microbial Diversity and Function in Intestinal Tract of Cultured *Scophichthys maximus*. Ph.D. Thesis, Graduate School of the Chinese Academy of Sciences (Institute of Oceanology), Beijing, China, 2013.
34. Roeselers, G.; Mittge, E.K.; Stephens, W.Z.; Parichy, D.M.; Cavanaugh, C.M.; Guillemin, K.; Rawls, J.F. Evidence for a core gut microbiota in the zebrafish. *ISME J.* **2011**, *5*, 1595–1608. [CrossRef] [PubMed]
35. Gao, Y.; He, J.; He, Z.; Li, Z.; Zhao, B.; Mu, Y.; Lee, J.-Y.; Chu, Z. Effects of fulvic acid on growth performance and intestinal health of juvenile loach *Paramisgurnus dabryanus* (Sauvage). *Fish Shellfish Immunol.* **2017**, *62*, 47–56. [CrossRef] [PubMed]
36. Niu, J.; Xie, S.-W.; Fang, H.-H.; Xie, J.-J.; Guo, T.-Y.; Zhang, Y.-M.; Liu, Z.-L.; Liao, S.-Y.; He, J.-Y.; Tian, L.-X.; et al. Dietary values of macroalgae Porphyra haitanensis in Litopenaeus vannamei under normal rearing and WSSV challenge conditions: Effect on growth, immune response and intestinal microbiota. *Fish Shellfish Immunol.* **2018**, *81*, 135–149. [CrossRef] [PubMed]
37. Zhao, W.; Xie, J.-J.; Fang, H.-H.; Liu, Y.-J.; Tian, L.-X.; Niu, J. Effects of corn starch level on growth performance, antioxidant capacity, gut morphology and intestinal microflora of juvenile golden pompano, *Trachinotus ovatus*. *Aquaculture* **2020**, *524*, 735197. [CrossRef]
38. Meng, X.; Hu, W.; Wu, S.; Zhu, Z.; Lu, R.; Yang, G.; Qin, C.; Yang, L.; Nie, G. Chinese yam peel enhances the immunity of the common carp (*Cyprinus carpio* L.) by improving the gut defence barrier and modulating the intestinal microflora. *Fish Shellfish Immunol.* **2019**, *95*, 528–537. [CrossRef]
39. Reis, M.P.; Fassani, E.J.; Júnior, A.A.P.G.; Rodrigues, P.B.; Bertechini, A.G.; Barrett, N.; Persia, M.E.; Schmidt, C.J. Effect of Bacillus subtilis (DSM 17299) on performance, digestibility, intestine morphology, and pH in broiler chickens. *J. Appl. Poult. Res.* **2017**, *26*, 573–583. [CrossRef]
40. Li, W.-J.; Zhang, L.; Wu, H.-X.; Li, M.; Wang, T.; Zhang, W.-B.; Du, Z.-Y.; Zhang, M.-L. Intestinal Microbiota Mediates Gossypol-Induced Intestinal Inflammation, Oxidative Stress, and Apoptosis in Fish. *J. Agric. Food Chem.* **2022**, *70*, 6688–6697. [CrossRef]
41. Gu, Y.-G.; Huang, H.-H.; Liu, Y.; Gong, X.-Y.; Liao, X.-L. Non-metric multidimensional scaling and human risks of heavy metal concentrations in wild marine organisms from the Maowei Sea, the Beibu Gulf, South China Sea. *Environ. Toxicol. Pharmacol.* **2018**, *59*, 119–124. [CrossRef]
42. Han, F.; Qian, J.; Qu, Y.; Li, Z.; Chen, H.; Xu, C.; Zhang, H.; Qin, J.G.; Chen, L.; Li, E. Partial replacement of soybean meal with fermented cottonseed meal in a low fishmeal diet improves the growth, digestion and intestinal microbiota of juvenile white shrimp *Litopenaeus vannamei*. *Aquac. Rep.* **2022**, *27*, 101339. [CrossRef]

*fishes*

MDPI

*Article*

# Effects of Replacing Fish Meal with Stickwater Hydrolysate and Meal on the Growth, Serum Biochemical Indexes, and Muscle Quality of Yellow Catfish (*Tachysurus fulvidraco*)

**Shuang Zheng [1,2], Yong Shi [1,2], Junzhi Zhang [1,2], Jihong Dai [1,2], Yi Hu [1,2,*] and Lei Zhong [1,2,*]**

1 College of Fisheries, Hunan Agricultural University, Changsha 410128, China; zhengs@stu.hunau.edu.cn (S.Z.); shiyong@stu.hunau.edu.cn (Y.S.); zhjun123@hunau.edu.cn (J.Z.); 429022456@hunau.edu.cn (J.D.)

2 Hunan Engineering Research Center for Utilization of Characteristics of Aquatic Resources, Hunan Agricultural University, Changsha 410128, China

* Correspondence: huyi0231@hunau.edu.cn (Y.H.); zhonglei@hunau.edu.cn (L.Z.); Tel.: +86-1387-591-0605 (Y.H.); +86-1354-853-3428 (L.Z.)

**Abstract:** An eight-week feeding experiment was conducted to evaluate the effects of replacing fish meal with stickwater hydrolysate (SWH) or stickwater hydrolysate meal (SWM) on the growth, serum biochemical parameters, intestinal digestive enzyme activity, and muscle quality of yellow catfish (*Tachysurus fulvidraco*). The control diet (CON) contained 30% fish meal and the remaining five diets were substituted for fish meal with 2.5% (SWM2.5), 5% (SWM5) SWM, and 5% (SWH5), 10% (SWH10), and 15% (SWH15) SWH, respectively. The results showed that there were no significant differences in weight gain rate, feed conversion rate, survival rate, hepatosomatic index, and viscerosomatic index among the groups. The substitution of fish meal with SWH significantly augmented the serum triglyceride and total cholesterol levels, whereas urea nitrogen content exhibited a reduction proportional to the replacement ratio. The incorporation of SWH led to a notable rise in glutamate-pyruvate transaminase activity, albeit with a gradual decline as the substitution ratio escalated. Relative to the CON group, the SWH5 group displayed a significant reduction in serum superoxide dismutase activity and a significant elevation in serum catalase activity. The substitution of fish meal with SWM yielded noticeable increments in the activities of complement 3, immunoglobulin M, and alkaline phosphatase. Neither SWH nor SWM exerted a substantial influence on intestinal amylase activity. Regarding muscle characteristics, neither SWH nor SWM showed a marked effect on hardness and springiness; however, adhesiveness, cohesiveness, gumminess, and chewiness properties exhibited enhancement as the proportion of fish meal replacement increased. In conclusion, within this experimental context, substituting fish meal with SWH and SWM did not adversely impact the growth and meat quality of yellow catfish. Specifically, replacing 15% fish meal with stickwater hydrolysate and 5% fish meal with stickwater hydrolysate meal contributed to an enhanced immune capacity in yellow catfish to a certain extent.

**Keywords:** *Tachysurus fulvidraco*; fish meal substitution; stickwater hydrolysate (meal); growth; serum biochemistry; serum immunity

**Key Contribution:** Substituting fish meal with SWH and SWM did not adversely impact the growth performance and contributed to an enhanced immune capacity in yellow catfish.

**Citation:** Zheng, S.; Shi, Y.; Zhang, J.; Dai, J.; Hu, Y.; Zhong, L. Effects of Replacing Fish Meal with Stickwater Hydrolysate and Meal on the Growth, Serum Biochemical Indexes, and Muscle Quality of Yellow Catfish (*Tachysurus fulvidraco*). *Fishes* **2023**, *8*, 566. https://doi.org/10.3390/fishes8120566

Academic Editors: Qiyou Xu, Jianhua Ming, Fei Song, Changle Qi, Chuanpeng Zhou and Seyed Hossein Hoseinifar

Received: 9 October 2023
Revised: 12 November 2023
Accepted: 18 November 2023
Published: 21 November 2023



## 1. Introduction

Over the last few years, the rapid development and expansion of aquaculture have highlighted a growing concern regarding the shortage of raw materials for aquatic feed, particularly protein feedstuffs [1]. Recently, protein feedstuffs, exemplified by fish meal, are currently experiencing global scarcity, leading to an imbalance between supply and demand that subsequently drives up fish meal prices and, consequently, escalates aquaculture production costs [2]. These factors seriously restrict the sustainable development of aquaculture, and reducing the content of fish meal in compound feed has become a hot spot in current research. Fish meal has always been an indispensable high-quality protein source in aquatic feed due to its high content of essential amino acids and fatty acids, good palatability, and less anti-nutritional factors [3]. However, the majority of plant protein sources that contain antinutritional factors, such as free gossypol, phytic acid, and soybean antigen proteins, have the potential to stunt growth, reduce digestive enzyme activity, and decrease immunity of aquatic animals [4–6]. Moreover, plant-derived protein sources exhibit markedly lower contents of minor nitrogenous compounds, such as peptides, free amino acids, and taurine in comparison to animal-derived counterparts. Several of these compounds, including antigenic proteins, gossypol, and its derivatives, phytic acid, and tannins, hold the capacity to exert adverse impacts on fish growth and overall health [7]. Therefore, the exploration of innovative animal protein sources as substitutes for fish meal stands as a paramount research focus in contemporary aquaculture [8].

In industrial fish meal production lines, stickwater is extracted from the liquid produced during the fish meal pressing process. Stickwater hydrolysate (SWH) is obtained by adding hydrolytic enzymes, such as pineapple proteinase or a solution of pa-paya protease, to stickwater at temperatures below 50–55 °C for 3–5 h. [9]. Owing to the substantial presence of suspended particulates, encompassing proteins, peptides, amino acids, biogenic amines, and trimethylamine oxides, among others, stickwater also encompasses minor quantities of oil constituents and trace elements, all of which bestow an indispensable impact upon the overall nutritional profile of fish [10]. Research has demonstrated that the addition of stickwater to the feed improves the growth performance and feed utilization of juvenile snakehead (*Ophiocephalus argus*) [11]. Over the years, plenty of studies have demonstrated that hydrolyzed fish proteins obtained by protease-mediated hydrolysis not only induce fish feeding but also promote fish growth [10–12].

Yellow catfish (*Tachysurus fulvidraco*), is one of China's important economic aquaculture fish, and the production of it has been growing steadily, reaching 587,800 tonnes in 2022 [13]. The proportion of fish meal in its compound feed is as high as 28–35%, which is much higher than that of many other freshwater cultured fish [14]. Therefore, this study aimed to evaluate the effects of replacing fish meal with stickwater hydrolysate and its meal on the growth, serum biochemical, immune indexes, intestinal digestive enzyme activity, and muscle quality of yellow catfish. This provides a theoretical basis for improving the application of enzymatic hydrolysis of stickwater hydrolysate and stickwater hydrolysate meal in fish feed and effectively reducing the cost of aquaculture.

## 2. Materials and Methods

### 2.1. Preparation of the Experimental Diet Stickwater Hydrolysate (Meal)

The SWH used in this study was supplied by Zhejiang Fengfu Marine Organism Products Co. Ltd. (Zhoushan, China). The chemical composition and peptide content of fish meal (FM), stickwater hydrolysate (SWH), and stickwater hydrolysate (SWM) are shown in Table 1. The free amino acid composition of FM, SWH, and SWM is shown in Table 2. It can be seen from Tables 1 and 2 that the peptide content, acid-soluble protein, and free amino acid content of SWH and SWM are significantly higher than FM.

**Table 1.** Nutrient levels of FM, SWH, and SWM (DM basis).

| Items | FM | SWH | SWM |
|---|---|---|---|
| Moisture% | 8.27 | 42.10 | 2.69 |
| Crude protein% | 72.48 | 70.64 | 75.54 |
| Crude fat% | 9.72 | 4.43 | 1.13 |
| Ash% | 12.37 | 14.72 | 15.01 |
| ASP% [1] | 4.08 | 65.47 | 71.09 |

Table 1. *Cont.*

| Items | FM | SWH | SWM |
|---|---|---|---|
| | Peptide molecular weight distribution/u [2] | | |
| <180 Da | 1.97 (48.31) | 15.76 (24.06) | 16.27 (22.89) |
| 500–180 Da | 0.81 (19.78) | 20.69 (31.61) | 29.98 (42.17) |
| 1000–500 Da | 0.16 (3.86) | 9.4 (14.33) | 9.66 (13.59) |
| 2000–1000 Da | 0.19 (4.62) | 8.32 (12.73) | 7.41 (10.43) |
| 3000–2000 Da | 0.16 (3.97) | 4.04 (6.19) | 2.95 (4.15) |
| 5000–3000 Da | 0.24 (5.94) | 3.81 (5.82) | 2.42 (3.40) |
| 10,000–5000 Da | 0.25 (6.13) | 2.73 (4.18) | 1.54 (2.17) |
| >10,000 Da | 0.30 (7.39) | 0.72 (1.08) | 0.85 (1.20) |

[1] The content of acid soluble protein (ASP) was analyzed by the Kjeldahl method (GB/T22729-2008) in the Analysis and Testing Center, Hunan Agricultural University (Changsha, China). [2] The peptide molecular weight distribution was measured by HPLC (GB/T22729-2008) in the Analysis and Testing Center, Hunan Agricultural University (Changsha, China). The values outside brackets were the contents of acid soluble protein, and the values inside brackets were the percentages of peptide content.

**Table 2.** Composition of free amino acids of FM, SWH, and SWM (g/100 g, dry matter).

| Items | FM | SWH | SWM |
|---|---|---|---|
| Valine | 0.01 | 0.13 | 0.60 |
| Methionine | 0.02 | 0.06 | 0.18 |
| Isoleucine | 0.01 | 0.72 | 0.91 |
| Leucine | 0.02 | 1.43 | 1.21 |
| Phenylalanine | 0.01 | 0.84 | 0.79 |
| Histidine | 0.01 | 0.41 | 0.42 |
| Lysine | 0.03 | 0.78 | 0.75 |
| Arginine | 0.01 | 0.26 | 0.65 |
| Threonine | 0.00 | 0.42 | 0.64 |
| Tryptophan | 0.01 | 1.82 | 0.98 |
| Aspartic acid | 0.01 | 0.08 | 0.19 |
| Serine | 0.01 | 0.06 | 0.15 |
| Glutamic acid | 0.00 | 0.02 | 0.01 |
| Glycine | 0.24 | 0.51 | 0.47 |
| Alanine | 0.02 | 0.86 | 0.83 |
| Cystine | ND [1] | ND | 0.12 |
| Tyrosine | 0.01 | 0.12 | 0.32 |
| Proline | 0.02 | 0.27 | 0.26 |
| Total amino acid | 0.44 | 8.79 | 9.48 |

[1] "ND" indicates that it cannot be measured.

This experiment consisted of a total of six treatment groups, designated as the control group, two groups in which SWM replaced 2.5% and 5.0% of fish meal, and three groups in which SWH replaced 5%, 10%, and 15% of fish meal. These groups were labeled as CON, SWM2.5, SWM5, SWH5, SWH10, and SWH15. Before feed preparation, the feed raw materials were crushed, screened through a 60-mesh sieve, and weighed accurately according to the formula. The mixture of raw materials was blended from small to large proportions, and the trace components were evenly mixed using the stepwise expansion method. All dry matter raw materials were mixed first, followed by the oil source and water, which were mixed separately and thoroughly blended. The mixed raw materials were then processed using a single-screw extrusion-type feed puffing machine to produce pellet feed with particle size of 1.5 mm and 2.0 mm, with a ratio of 3:7. The pellet feed was air-dried in a cool place and stored for later use. The composition and nutrient levels of basic diet are shown in Table 3. The free amino acid and total amino acid composition of experimental diet are shown in Tables 4 and 5.

**Table 3.** Composition and nutrient levels of experimental diet (air-dry basis), percentages.

| Ingredients | CON | SWM2.5 | SWM5 | SWH5 | SWH10 | SWH15 |
|---|---|---|---|---|---|---|
| Fish meal | 30.00 | 27.50 | 25.00 | 25.00 | 20.00 | 15.00 |
| Stickwater hydrolysate meal | | 2.23 | 4.46 | | | |
| Stickwater hydrolysate | | | | 5.13 | 10.26 | 15.39 |
| Soybean meal | 18.00 | 18.00 | 18.00 | 18.00 | 18.00 | 18.00 |
| Cottonseed meal | 4.00 | 4.00 | 4.00 | 4.00 | 4.00 | 4.00 |
| Rapeseed meal | 4.00 | 4.00 | 4.00 | 4.00 | 4.00 | 4.00 |
| Beer yeast | 6.00 | 6.00 | 6.00 | 6.00 | 6.00 | 6.00 |
| Corn gluten meal | 5.00 | 5.00 | 5.00 | 5.00 | 5.00 | 5.00 |
| Fish oil | 0.00 | 0.24 | 0.49 | 0.47 | 0.93 | 1.40 |
| High-protein flour | 24.00 | 24.00 | 24.00 | 24.00 | 24.00 | 24.00 |
| $Ca(H_2PO_4)_2$ | 1.50 | 1.50 | 1.50 | 1.50 | 1.50 | 1.50 |
| Soybean oil | 3.40 | 3.40 | 3.40 | 3.40 | 3.40 | 3.40 |
| Premix [1] | 1.20 | 1.20 | 1.20 | 1.20 | 1.20 | 1.20 |
| Choline chloride | 0.50 | 0.50 | 0.50 | 0.50 | 0.50 | 0.50 |
| Antioxidants [2] | 0.01 | 0.01 | 0.01 | 0.01 | 0.01 | 0.01 |
| Mould inhibitor [3] | 0.03 | 0.03 | 0.03 | 0.03 | 0.03 | 0.03 |
| Wheat middling | 2.36 | 2.39 | 2.41 | 1.76 | 1.16 | 0.56 |
| Nutrient composition [4] | | | | | | |
| Crude protein | 38.85 | 38.74 | 38.68 | 38.81 | 38.64 | 38.75 |
| Crude fat | 8.09 | 8.15 | 8.24 | 8.14 | 8.08 | 8.11 |
| Crude ash | 11.15 | 11.23 | 11.19 | 11.32 | 11.27 | 11.21 |

[1] Premix provides KCl 200 mg, KI (1%) 60 mg, $CoCl_2 \cdot 6H_2O$ (1%) 50 mg, $CuSO_4 \cdot 5H_2O$ 30 mg, $FeSO_4 \cdot H_2O$ 400 mg, $ZnSO_4 \cdot H_2O$ 400 mg, $MnSO_4 \cdot H_2O$ 150 mg, $Na_2SeO_3 \cdot 5H_2O$ (1%) 65 mg, $MgSO_4 \cdot H_2O$ 2000 mg, Zeolite Powder 3645.85 mg, $VB_1$ 12 mg, Riboflavin 12 mg, $VB_6$ 8 mg, $VB_{12}$ 0.05 mg, $VK_3$ 8 mg, Inositol 100 mg, Pantothenic Acid 40 mg, Niacin Acid 50 mg, Folic Acid 5 mg, Biotin 0.8 mg, VA 25 mg, VD 35 mg, VE 50 mg, VC 100 mg, Ethoxyquin 150 mg, and Wheat Flour 2434.15 mg per kilogram of diets; [2] The main component of Antioxidant is Ethoxyquin; [3] The main component of Mould Inhibitor is Calcium Propionate; [4] Nutrient compositions were calculated values.

**Table 4.** Free amino acid compositions of experimental diet (air-dry basis, g/kg).

| Items | CON | SWM2.5 | SWM5 | SWH5 | SWH10 | SWH15 |
|---|---|---|---|---|---|---|
| | | Essential amino acid | | | | |
| Valine | 0.26 | 0.46 | 0.90 | 0.57 | 0.91 | 1.12 |
| Methionine | 0.08 | 0.15 | 0.27 | 0.13 | 0.29 | 0.36 |
| Isoleucine | 0.23 | 0.37 | 0.62 | 0.38 | 0.56 | 0.74 |
| Leucine | 0.47 | 0.60 | 1.07 | 0.68 | 1.22 | 1.39 |
| Phenylalanine | 0.21 | 0.29 | 0.51 | 0.37 | 0.59 | 0.76 |
| Histidine | 1.46 | 1.51 | 2.24 | 1.49 | 2.46 | 2.67 |
| Lysine | 0.37 | 0.54 | 0.92 | 0.54 | 0.82 | 1.08 |
| Arginine | 0.97 | 1.12 | 1.93 | 1.67 | 2.19 | 2.85 |
| Threonine | 0.18 | 0.22 | 0.46 | 0.29 | 0.41 | 0.52 |
| Tryptophan | 0.19 | 0.21 | 0.30 | 0.22 | 0.27 | 0.34 |
| | | Non-essential amino acid | | | | |
| Aspartic acid | 0.24 | 0.31 | 0.50 | 0.32 | 0.43 | 0.67 |
| Serine | 0.18 | 0.24 | 0.49 | 0.34 | 0.52 | 0.79 |
| Glutamic acid | 0.55 | 0.71 | 1.12 | 0.83 | 1.25 | 1.69 |
| Glycine | 0.25 | 0.76 | 1.20 | 0.62 | 0.88 | 1.42 |
| Alanine | 1.21 | 1.46 | 2.27 | 1.54 | 1.87 | 2.32 |
| Cystine | 0.02 | 0.03 | 0.10 | 0.03 | 0.09 | 0.14 |
| Tyrosine | 0.22 | 0.33 | 0.69 | 0.36 | 0.63 | 0.72 |
| Proline | 0.39 | 0.45 | 0.68 | 0.44 | 0.53 | 0.71 |
| Total amino acid | 7.48 | 9.76 | 16.27 | 10.82 | 15.92 | 20.29 |

**Table 5.** Total amino acid compositions of experimental diet (air-dry basis, g/kg).

| Items | CON | SWM2.5 | SWM5 | SWH5 | SWH10 | SWH15 |
|---|---|---|---|---|---|---|
| | | Essential amino acid | | | | |
| Valine | 2.11 | 2.06 | 2.01 | 1.98 | 2.12 | 2.06 |
| Methionine | 0.83 | 0.74 | 0.79 | 0.68 | 0.88 | 0.86 |
| Isoleucine | 1.89 | 1.78 | 1.75 | 1.74 | 1.83 | 1.81 |
| Leucine | 3.50 | 3.39 | 3.32 | 3.40 | 3.51 | 3.46 |
| Phenylalanine | 1.96 | 1.91 | 1.88 | 1.87 | 1.99 | 1.91 |
| Histidine | 1.49 | 1.36 | 1.42 | 1.55 | 1.52 | 1.58 |
| Lysine | 2.56 | 2.33 | 2.38 | 2.37 | 2.46 | 2.43 |
| Arginine | 2.68 | 2.73 | 2.70 | 2.69 | 2.62 | 2.72 |
| Threonine | 1.75 | 1.62 | 1.64 | 1.68 | 1.76 | 1.69 |
| Tryptophan | ND [1] | ND | ND | ND | ND | ND |
| | | Non-essential amino acid | | | | |
| Aspartic acid | 3.98 | 3.94 | 4.02 | 3.85 | 4.01 | 3.94 |
| Serine | 1.93 | 1.91 | 1.88 | 1.87 | 1.89 | 1.81 |
| Glutamic acid | 7.43 | 7.74 | 7.92 | 7.44 | 7.58 | 7.55 |
| Glycine | 2.25 | 2.57 | 2.64 | 2.61 | 2.40 | 2.32 |
| Alanine | 2.62 | 2.55 | 2.76 | 2.68 | 2.66 | 2.61 |
| Cystine | 0.12 | 0.16 | 0.21 | 0.19 | 0.20 | 0.14 |
| Tyrosine | 1.14 | 1.23 | 1.19 | 1.21 | 1.17 | 1.30 |
| Proline | 1.94 | 2.02 | 2.11 | 2.20 | 2.13 | 1.95 |
| Total amino acid | 40.18 | 40.04 | 40.62 | 40.01 | 40.73 | 40.14 |

[1] "ND" indicates that it cannot be measured.

## 2.2. Experimental Design

Juvenile yellow catfish were purchased from a commercial farm and reared in cages at Chetianjiang Reservoir, Loudi, China. Fish were acclimatized to 2.0 m × 1.5 m × 1.5 m floating cages. Nine-hundred juvenile yellow catfish (initial body weight of 9.99 ± 0.02 g) were divided equally and randomly into six groups. Each group was fed in triplicate for a period of 56 days. The experimental yellow catfish were fed with 1.5 mm grain size feed in the early stage and 2.0 mm grain size feed in the late stage for better palatability. The fish were fed by the experimental diet twice per day according to 3–5% of body weight. The physicochemical parameters of the water were kept at optimal conditions during the culture period: temperature (27.3 ± 3.6 °C), pH (7.1 ± 0.3), dissolved oxygen (6.65 ± 0.25 mg/L), ammonia (0.01 ± 0.005 mg/L), and nitrite (0.01 ± 0.004 mg/L).

## 2.3. Sample Collection and Determination Method

### 2.3.1. Growth Parameters

At the end of the breeding trial, the yellow catfish were fasted for 24 h and weighed to determine growth performance. Total mantissa, total body weight, and total feeding volume of experimental fish in each cage were recorded. Five fish were randomly selected from each cage, body length, body weight, visceral mass, and liver mass were measured, and growth indicators were calculated.

The growth performance was evaluated in terms of Survival Rate (SR), Weight Gain Rate (WGR), Feed Conversion Ratio (FCR), Condition Factor (CF), Hepatosomatic Index (HSI), and Viseromatic index (VSI), which were calculated using the following formula.

$$\text{Survival rate (SR, \%)} = \text{Nf}/\text{Ni} \times 100$$

$$\text{Weight gain rate (WGR, \%)} = (\text{Wt} - \text{Wo})/\text{Wo} \times 100$$

$$\text{Feed conversion rate (FCR)} = \text{total amount of the feed consumed (g)}/(\text{Wt} - \text{Wo})$$

$$\text{Condition factor (CF, g/cm}^3\text{)} = \text{Wt} \times 100/(\text{body length})^3$$

$$\text{Hepatosomatic index (HSI, \%)} = \text{liver weight (g)}/\text{Wt} \times 100$$

$$\text{Viserosomatic index (VSI, \%)} = \text{visceral weight (g)}/\text{Wt} \times 100$$

where Ni and Nf are the initial and final numbers of fish, and Wo (g) and Wt (g) are the initial and final weights.

2.3.2. Serum Biochemical, Antioxidant, and Immune Parameters

Five yellow catfish were randomly selected from each cage, and blood was collected from the tail vein with a sterile syringe, placed in a 1.5 mL enzyme-free tube, stood at 4 °C for 24 h, was centrifuged at 3000× $g$ for 10 min, and the upper layer of serum was collected and stored at −80 °C until use.

The concentrations of Triglycerides (TG), Total Cholesterol (TCHO), Urea Nitroge (UN), Glutamic Oxalacetic Transaminase (GOT), Glutamate Pyruvic Transaminase (GPT), Superoxide Dismutase (SOD), Malondialdehyde (MDA), Complement 3 (C3), Complement 4 (C4), Immunoglobulin M (IgM), Alkaline Phosphatase (AKP), and Catalase (CAT) in serum were determined by the commercial kit from Nanjing Jianxian Biotechnology Co., LTD (Nanjing, China). The determination method refers to the instruction of the kit.

2.3.3. Digestive Enzyme Activity Analysis

At the end of the culture experiment, three yellow catfish were randomly selected from each cage after 24 h of fasting, and the intestinal tissues were rapidly isolated, washed with normal saline, packed in 1.5 mL enzyme-free centrifuge tubes, and stored at −80 °C until use. The activities of trypsin and amylase in the intestine were determined by the commercial kit from Nanjing Jianxian Biotechnology Co., LTD (Nanjing, China).

2.3.4. Muscle Texture Determination

Three yellow catfish were randomly selected from each cage, and the TPA (TMS-PRO, FTC, Sterling, VA, USA) was used to measure the texture parameters of muscle hardness, adhesiveness, cohesiveness, springiness, gumminess, and chewiness. The test conditions consisted of 2 consecutive compressional sessions at a constant rate of 30 mm/min with a deformation of 60% of the original length and an initial force of 0.1 N.

### *2.4. Statistical Analysis*

The experimental data were analyzed by one-way analysis of variance (ANOVA), and Duncan's multiple-range test was used to compare the difference between the means. All results are expressed as means ± standard deviation of the means, and all statistical analyses were performed using SPSS 19.0. Differences were considered significant at $p$-value < 0.05.

## 3. Results

### *3.1. Growth Performance*

There were no significant differences in final weight, WGR, FCR, SR, HSI, and VSI among all groups ($p > 0.05$). However, the WGR and VSI of experimental yellow catfish were reduced by 5% SWM and 5% SWH instead of fish meal (Table 6).

### *3.2. Serum Biochemical Indexes*

The levels of TG, TCHO, and GPT in the SWM2.5 and SWM5 groups were significantly increased ($p < 0.05$), and the levels of UN and GOT were not significantly changed ($p > 0.05$). The TG and TCHO in the SWH5, SWH10, and SWH15 groups were significantly increased ($p < 0.05$), and the UN content showed a downward trend with the increase in replacement ratio ($p < 0.05$). The SWH10 group exhibited a notable rise in GOT activity, while the introduction of SWH as a replacement for fish meal resulted in a substantial elevation in GPT activity ($p < 0.05$) (Table 7).

**Table 6.** Growth performance of experimental yellow catfish.

| Index | CON | SWM2.5 | SWM5 | SWH5 | SWH10 | SWH15 |
|---|---|---|---|---|---|---|
| Initial weight (g) | 9.97 ± 0.01 | 10 ± 0.01 | 10 ± 0.02 | 9.98 ± 0.01 | 10.01 ± 0.01 | 9.99 ± 0.001 |
| Final weight (g) | 50.37 ± 0.77 | 50.90 ± 0.51 | 47.92 ± 4.88 | 47.7 ± 2.68 | 50.77 ± 1.35 | 50.93 ± 0.66 |
| WGR (%) | 405.24 ± 8.22 | 403.52 ± 6.25 | 379.3 ± 49.48 | 377.84 ± 27.03 | 407.38 ± 13.15 | 409.89 ± 6.67 |
| SGR (%) | 2.7 ± 0.03 | 2.69 ± 0.06 | 2.63 ± 0.28 | 2.6 ± 0.09 | 2.71 ± 0.04 | 2.71 ± 0.02 |
| SR (%) | 100.00 | 100.00 | 100.00 | 100.00 | 100.00 | 98.00 |
| FCR | 1.48 ± 0.03 | 1.50 ± 0.07 | 1.63 ± 0.2 | 1.6 ± 0.11 | 1.47 ± 0.05 | 1.45 ± 0.02 |
| HIS (%) | 2.07 ± 0.09 | 2.14 ± 0.09 | 1.99 ± 0.21 | 2.11 ± 0.14 | 2.16 ± 0.08 | 1.96 ± 0.09 |
| VSI (%) | 18.12 ± 0.4 | 16.96 ± 0.86 | 16.02 ± 1.64 | 15.35 ± 0.7 | 16.22 ± 0.66 | 15.44 ± 0.63 |
| CF | 1.83 ± 0.03 | 1.97 ± 0.04 | 1.89 ± 0.08 | 1.89 ± 0.11 | 1.86 ± 0.04 | 1.98 ± 0.1 |

**Table 7.** Serum biochemical indexes of experimental yellow catfish.

| Index | CON | SWM2.5 | SWM5 | SWH5 | SWH10 | SWH15 |
|---|---|---|---|---|---|---|
| TG (mmol/L) | $2.92 \pm 0.16^{a}$ | $4.04 \pm 0.51^{b}$ | $4.57 \pm 0.23^{b}$ | $3.47 \pm 0.29^{a}$ | $6.28 \pm 0.6^{b}$ | $6.81 \pm 1.21^{b}$ |
| TCHO (mmol/L) | $5.58 \pm 0.11^{a}$ | $6.8 \pm 0.2^{b}$ | $9.36 \pm 0.09^{c}$ | $7.92 \pm 0.35^{b}$ | $6.7 \pm 0.92^{ab}$ | $7.5 \pm 0.22^{b}$ |
| UN (mmol/L) | $5.09 \pm 0.16^{b}$ | $4.59 \pm 0.04^{a}$ | $5.27 \pm 0.06^{b}$ | $4.53 \pm 0.08^{b}$ | $4.65 \pm 0.16^{b}$ | $3.36 \pm 0.14^{c}$ |
| GOT (U/L) | $23.81 \pm 1.59^{a}$ | $24.68 \pm 0.74^{a}$ | $22.7 \pm 0.58^{a}$ | $23.89 \pm 0.55^{a}$ | $32.2 \pm 2.37^{b}$ | $27.81 \pm 0.34^{ab}$ |
| GPT (U/L) | $1.19 \pm 0.05^{a}$ | $2.34 \pm 0.15^{b}$ | $2.22 \pm 0.14^{b}$ | $3.24 \pm 0.17^{c}$ | $2.16 \pm 0.35^{b}$ | $1.25 \pm 0.08^{a}$ |

Values with different superscripts in the same row are significantly different ($p < 0.05$).

### 3.3. Serum Antioxident Indexes

The SOD activity in SWM5 group was significantly lower than that in the CON group ($p < 0.05$). The serum CAT activity in the SWM2.5 and SWM5 groups increased with the increase in replacement ratio, and the activity in the SWM5 group was significantly higher than that in the CON group ($p < 0.05$). MDA content did not change significantly ($p > 0.05$), but its content showed an increasing trend with the increase in substitution ratio (Table 8).

**Table 8.** Serum antioxidant indexes of experimental yellow catfish.

| Index | CON | SWM2.5 | SWM5 | SWH5 | SWH10 | SWH15 |
|---|---|---|---|---|---|---|
| SOD (U/mL) | $66.65 \pm 2.8^{b}$ | $63.39 \pm 2.12^{b}$ | $45.01 \pm 2.09^{a}$ | $64.79 \pm 1.42^{b}$ | $73.68 \pm 2.84^{c}$ | $53.16 \pm 1.86^{a}$ |
| CAT (U/mL) | $1.84 \pm 0.09^{a}$ | $2.42 \pm 0.37^{ab}$ | $3.42 \pm 0.11^{b}$ | $2.68 \pm 0.19^{b}$ | $2.8 \pm 0.18^{b}$ | $2.63 \pm 0.3^{b}$ |
| MDA (mmol/mL) | 8.77 ± 0.07 | 8.77 ± 0.27 | 8.84 ± 0.22 | 8.75 ± 0.24 | 9.06 ± 0.67 | 9.27 ± 0.44 |

Values with different superscripts in the same row are significantly different ($p < 0.05$).

### 3.4. Serum Immune Indexes

Substitution of fish meal by SWM significantly increased serum C3 and IGM activities ($p < 0.05$). The C3 and IGM activities in the SWH5 group were significantly lower than that in the CON group ($p < 0.05$), but its content showed an increasing trend with the increase in substitution ratio. Moreover, the IGM activity was significantly increased in the SWH15 group ($p < 0.05$). The AKP activity was significantly higher in the SWM2.5 and SWH10 groups than in the CON group ($p < 0.05$), but its activity decreased when the substitution ratio increased (Table 9).

### 3.5. Digestive Enzyme Activities in the Intestine

There was no statistically significant influence ($p > 0.05$) on the Amy activity among the different diets ($p > 0.05$). However, the intestinal Amy activity in the SWH10 group had the highest difference. The substitution of 2.5% and 5% of fish meal with SWM or SWH demonstrated no significant impact on the intestinal Try activity of yellow catfish

($p > 0.05$). However, when 15% of fish meal was replaced by SWH, a notable enhancement in intestinal Try activity was observed ($p < 0.05$). (Table 10).

**Table 9.** Serum immune indexes of experimental yellow catfish.

| Index | CON | SWM2.5 | SWM5 | SWH5 | SWH10 | SWH15 |
|---|---|---|---|---|---|---|
| C3 (g/L) | 0.4 ± 0.04 [a] | 0.58 ± 0.07 [ab] | 0.77 ± 0.15 [b] | 0.24 ± 0.003 [b] | 0.36 ± 0.03 [a] | 0.43 ± 0.03 [a] |
| IGM (g/L) | 0.7 ± 0.06 [a] | 0.91 ± 0.07 [ab] | 1.07 ± 0.14 [b] | 0.54 ± 0.03 [a] | 0.76 ± 0.05 [ab] | 1.04 ± 0.33 [b] |
| AKP (King's unit/100 mL) | 4.12 ± 0.23 [a] | 4.59 ± 0.02 [b] | 4.26 ± 0.13 [ab] | 4.35 ± 0.19 [ab] | 4.56 ± 0.18 [b] | 4.08 ± 0.28 [a] |

Values with different superscripts in the same row are significantly different ($p < 0.05$).

**Table 10.** Digestive enzyme activities in the intestine of experimental yellow catfish.

| Index | CON | SWM2.5 | SWM5 | SWH5 | SWH10 | SWH15 |
|---|---|---|---|---|---|---|
| Amy (U/mg) | 1.87 ± 0.1 | 1.7 ± 0.34 | 1.27 ± 0.12 | 1.46 ± 0.11 | 2.02 ± 0.21 | 1.76 ± 0.27 |
| Try (U/mg) | 663.56 ± 35.46 [a] | 672.95 ± 30.62 [a] | 589.69 ± 37.69 [a] | 610.34 ± 34.77 [a] | 661.36 ± 33.3 [a] | 463.88 ± 19.73 [b] |

Values with different superscripts in the same row are significantly different ($p < 0.05$).

### 3.6. Muscle Textural Indexes

The hardness, adhesiveness, and chewiness of muscle in the SWM2.5 and SWM5 groups did not change significantly ($p > 0.05$). The adhesiveness and cohesiveness of the SWM5 group was significantly higher than the CON group ($p < 0.05$). The SWM2.5 group had the highest gumminess. There was no significant difference in hardness and springiness among the SWH5, SWH10, and SWH15 groups ($p > 0.05$), and SWH5 group had the highest hardness and springiness (Table 11).

**Table 11.** Muscle textural indexes of experimental yellow catfish.

| Index | CON | SWM2.5 | SWM5 | SWH5 | SWH10 | SWH15 |
|---|---|---|---|---|---|---|
| Hardness (N) | 5.08 ± 0.33 | 5.13 ± 0.39 | 4.75 ± 0.23 | 5.5 ± 0.28 | 5.24 ± 0.33 | 5.35 ± 0.36 |
| Adhesiveness (N.mm) | 0.03 ± 0.002 [a] | 0.02 ± 0.002 [a] | 0.06 ± 0.002 [b] | 0.06 ± 0.002 [b] | 0.07 ± 0.003 [b] | 0.07 ± 0.001 [b] |
| Cohesiveness (Ratio) | 0.49 ± 0.02 [a] | 0.57 ± 0.02 [b] | 0.68 ± 0.01 [c] | 0.65 ± 0.01 [c] | 0.63 ± 0.01 [bc] | 0.6 ± 0.01 [b] |
| Springinesse (mm) | 1.71 ± 0.07 | 1.76 ± 0.07 | 1.82 ± 0.06 | 1.8 ± 0.04 | 1.75 ± 0.03 | 1.7 ± 0.06 |
| Gumminess (N) | 3.54 ± 0.26 [a] | 4.48 ± 0.22 [b] | 3.63 ± 0.13 [a] | 3.88 ± 0.19 [b] | 4.02 ± 0.15 [b] | 3.28 ± 0.16 [a] |
| Chewiness (mJ) | 5.29 ± 0.38 [a] | 6.1 ± 0.48 [a] | 5.84 ± 0.27 [a] | 6.59 ± 0.28 [b] | 6.82 ± 0.34 [b] | 5.4 ± 0.32 [a] |

Values with different superscripts in the same row are significantly different ($p < 0.05$).

### 3.7. Body Composition

There was no significant difference in moisture content, crude fat, and crude ash of experimental yellow catfish in the SWM2.5 and SWM5 groups ($p > 0.05$). Substitution of fish meal by SWH resulted in significantly higher crude fat than that of the CON group, and the substitution group showed an upward trend in crude fat (Table 12).

**Table 12.** Body composition of experimental yellow catfish.

| Index | CON | SWM2.5 | SWM5 | SWH5 | SWH10 | SWH15 |
|---|---|---|---|---|---|---|
| Moisture content | 71.22 ± 0.62 | 71.59 ± 0.56 | 71.61 ± 0.52 | 70.97 ± 1.02 | 70.34 ± 0.06 | 70.71 ± 0.42 |
| Crude fat | 28.5 ± 0.47 | 28.2 ± 0.45 | 29.7 ± 0.52 | 30.7 ± 0.32 | 31.1 ± 0.47 | 31.8 ± 0.52 |
| Crude ash | 13.58 ± 0.17 | 14.05 ± 0.11 | 14.14 ± 0.3 | 13.96 ± 0.12 | 14.11 ± 0.1 | 14.03 ± 0.5 |

## 4. Discussion

SWH and SWM represent new sources of animal protein which are enzymatic hydrolysis products of the industrial production of fish meal by-products. This not only reduces the amount of fish meal but also enhances its utilization rate. A previous study found that replacing 60% of Peruvian super steam fish meal with stickwater hydrolyzsate combined with plant protein diet did not affect feeding, feed efficiency, or growth performance of pearl gentiana grouper (*Epinephelus lanceolatus* ♂ × *E. fuscoguttatus* ♀). [15]. In this study, the substitution of 2.5% and 5% of fish meal with SWM, and the replacement of 5%, 10%, and 15% of fish meal with SWH, had no significant effect on the weight gain rate, specific growth rate, survival rate, and feed conversion rate of yellow catfish, which was similar to the results of the study of Atlantic salmon (*Salmo salar* L.) [16]. This may be attributed to SWH and SWM being rich in water-soluble proteins, small peptides, and free amino acids. These ingredients are different from other animal and plant protein sources of unique nutrients, among which taurine [17,18] and small peptides [19] have a certain growth-promoting effect on fish. However, excessive substitution of fish meal with SWM can result in growth inhibition in certain fish species, such as red sea bream (*Pagrus major*) [20] and sea bass larvae (*Dicentrarchus labrax*) [21]. The results of our experiment also revealed that the growth performance of yellow catfish decreased when 5% of fish meal was replaced with SWM, as compared to the CON group. The primary reason for this may be the excessive presence of small peptides and free amino acids in the diet due to high levels of SWH and SWM replacing fish meal, resulting in amino acid saturation and competition in amino acid transport mechanisms [22], leading to an imbalance in amino acid absorption in the bodies of yellow catfish.

Another aspect that can illustrate the growth performance of fish is the activity of intestinal digestive enzymes. Intestinal digestive enzymes degrade macromolecular nutrients in feed into small molecular nutrients, such as amino acids and glucose. Their activity levels directly affect the fish's ability to digest and absorb nutrients [23]. Shi et al. demonstrated that replacing 10% of fish meal with stickwater hydrolysate in the compound diet of rice field eel (*Monopterus albus*) significantly improved the digestive enzyme activities of the fish [24]. In this study, substituting fish meal with SWM did not affect the amylase activity of yellow catfish, but it exhibited a decreasing trend and had no impact on trypsin activity. The substitution of fish meal with SWH did not affect the amylase activity of Pelteobagrus fulvidraco, with the highest activity observed in the SWH10 group. However, when fish meal was replaced by 15% SWH, trypsin activity was significantly lower than that in the CON group. The main reason for this is that SWH contains biogenic amines, which are products of protein degradation. The high concentration of biogenic amine will have certain negative effects on the biological organism [25], and excess free amino acids can reduce the organism's ability to consume and utilize protein [26,27].

Serum physiological indicators can reflect the health and physiological status of fish. Serum cholesterol and triglycerides play a crucial part in human health, and to a certain extent reflect the ability of liver fat metabolism and lipid absorption [28]. The results of this study showed that SWM and SWH significantly increased the serum total cholesterol and triglyceride levels of yellow catfish when replacing fish meal, indicating that SWM and SWH had an inhibitory effect on liver fat metabolism. This finding is consistent with the results reported by Wu et al. in grass carp (*Ctenopharyngodon idella*) [29]. In this study, serum urea nitrogen decreased after fish meal replacement with SWH and SWM. This may be attributed to the increase in small molecular substances in enzymatic hydrolysates, which promoted protein decomposition and accelerated nitrogen excretion, ultimately reducing urea nitrogen production [30,31]. Transaminases related to protein metabolism in aquatic animals mainly exist in the liver cells of the body. These enzymes metabolize and transform proteins through transamination and deamination processes. When the body's liver cells experience inflammation and toxic reactions, damaged liver cells release a significant amount of liver transaminase into the bloodstream, resulting in elevated serum transaminase levels [32,33]. This study demonstrated that the activities of glutamate pyruvic transaminase (GPT) and glutamic oxalacetic transaminase (GOT) in the serum of yellow catfish increased with an increase in SWH replacement rate, suggesting that a high replacement rate may cause some degree of liver damage in yellow catfish. Serum GOT and GPT activities increased with higher levels of dietary SWH and SWM supplementation, consistent with the findings in turbot [34] and yellow catfish [35]. It is possible that an increase in the substitution ratio leads to an increase in biogenic amines in the diet, causing damage to the fish's liver, subsequently resulting in higher transaminase levels in the bloodstream, which negatively impacts the animals' health [36].

Complement is an essential component of antimicrobial defense and is mainly composed of complement 3 (C3) and complement 4 [37]. Alkaline phosphatase (AKP), a phosphohydrolase, is a crucial nonspecific immune marker and an evaluation indicator reflecting the health status of aquatic animals [38]. Various studies have demonstrated that fish have developed specific immune response systems, including cellular and humoral immunity [39]. Immunoglobulins play a significant role in the humoral immunity of fish. Immunoglobulin M (IgM) is one of the most important immunoglobulins in fish and is the first antibody produced when the body is stimulated by antigens [40]. This study revealed that as the proportion of SWH and SWM replacing fish meal increased, the levels of C3, AKP, and IgM in the serum of yellow catfish exhibited an upward trend. In a study involving largemouth bass (*Micropterus salmoides*), it was observed that adding an appropriate amount of small peptides to the diet could enhance its immune capacity [41]. Small peptides generally refer to oligopeptides containing between two and three amino acids, which can be completely absorbed and utilized by the body, playing a pivotal role in amino acid digestion, absorption, and metabolism [42]. These results suggest that substituting a portion of the fish meal with SWM and SWH in the diet can enhance the immunity of yellow catfish, which may be attributed to the presence of small peptides in SWH and SWM.

In addition, the overall antioxidant capacity of the organism's defense system of fish is closely related to the degree of health, including the enzymatic antioxidant system and non-enzymatic antioxidant system [43]. The enzymatic antioxidant system relies on the role of various antioxidant enzymes in the body, while the non-enzymatic antioxidant system relies on the role of metalloproteins, amino acids, and vitamins [44]. Superoxide dismutase (SOD) is a vital part of the enzymatic antioxidant system in animals, and its activity reflects the ability to scavenge reactive oxygen free radicals in animals. SOD can convert harmful superoxide free radicals into $H_2O_2$ through a reaction [45]. Catalase (CAT) is an antioxidant enzyme present in nearly all organisms. Its primary role is to catalyze the decomposition of hydrogen peroxide into water and oxygen, thereby removing $H_2O_2$ from the body to prevent cell damage caused by $H_2O_2$. As one of the key enzymes in the biological defense system, CAT provides an antioxidant defense mechanism for the body [46]. The results indicate that replacing fish meal with 10% SWH significantly improved the antioxidant capacity of yellow catfish. Malondialdehyde (MDA) is a product of lipid peroxidation, and its content indirectly reflects the content of reactive oxygen species and the degree of lipid peroxidation in tissues and cells [47]. The findings from this experiment revealed that the replacement of fish meal with SWH and SWM increased the serum MDA content of yellow catfish as the replacement ratio increased. Xu et al. found that substituting fermented fish soluble pulp for fish meal could significantly reduce MDA content in the serum of juvenile turbot [16]. This may be attributed to the presence of a certain amount of MDA in SWH and SWM. In Cao et al.'s study, it was observed that fish stickwater has the characteristics of high moisture, high fat, and high unsaturation, making its quality more susceptible to destruction due to fat oxidation. MDA is the end-product of lipid oxidation, and the degree of rancidity in fish soluble pulp affects its MDA content [48].

The texture parameters of fish muscle are important indicators for evaluating the taste of fish. These parameters depend on muscle hardness, viscosity, cohesiveness, springiness, gumminess, chewiness, and other factors [49]. Consumers typically prefer meat that is firm and elastic, making muscle texture a key consideration for consumers. In general, the greater the hardness and viscosity of the fish, the better its taste. Hardness is a crucial texture parameter that reflects the internal bonding force within the sample [50]. It has been demonstrated that greater viscosity results in meat that is chewier and crisper, mimicking the mouthfeel of natural muscles [51]. The results indicated that replacing fish meal with SWM had no significant impact on the hardness, viscosity, and chewiness of the meat.

However, with an increase in the replacement ratio, the meat's hardness and chewiness decreased while viscosity increased. The replacement of fish meal with SWH had no significant effect on the hardness and viscosity of yellow catfish meat, with the highest values for hardness and viscosity observed at a replacement ratio of 5%. Substituting 2.5% and 5% of fish meal with SWM, as well as replacing 5%, 10%, and 15% of fish meal with SWH, had no significant effect on the moisture content, crude fat, and crude ash of yellow catfish. This suggests that replacing fish meal with SWH and SWM does not adversely affect the muscle quality of yellow catfish.

## 5. Conclusions

In conclusion, the substitution of fish meal by 2.5% and 5% SWM, and 5%, 10%, and 15% SWH did not affect the growth performance of yellow catfish. Specifically, the substitution of fish meal by 5% SWM and 15% SWH contributed to an enhanced immune capacity in yellow catfish to a certain extent. However, the substitution of fish meal by 5% SWM and 15% SWH will inhibit the antioxidant capacity of fish, and also cause damage to the liver of fish.

**Author Contributions:** Conceptualization, S.Z. and Y.S.; methodology, J.Z.; software, S.Z.; validation, J.D.; formal analysis, S.Z. and Y.S.; investigation, L.Z.; resources, L.Z.; data curation, S.Z. and Y.S.; writing—original draft preparation, S.Z.; writing—review and editing, Y.S. and Y.H.; visualization, S.Z.; supervision, Y.H.; project administration, Y.H.; funding acquisition, Y.H. All authors have read and agreed to the published version of the manuscript.

**Funding:** This research was funded by the National Natural Science Foundation of China (No. 32172986), and National Key R&D Program of China (2019YFD0900200).

**Institutional Review Board Statement:** The animal subjects used in the experiment were fish. All the fish were handled in accordance with EU regulations on the protection of laboratory animals and approved by Animal Welfare of Hunan Agricultural University (protocol code 2022m016, 16 May 2022).

**Informed Consent Statement:** Informed consent was obtained from all subjects involved in the study.

**Data Availability Statement:** The data that support the findings of this study are available upon reasonable request to the corresponding author.

**Conflicts of Interest:** The authors declare no conflict of interest.

## References

1. Naylor, R.L.; Hardy, R.W.; Bureau, D.P.; Chiu, A.; Elliott, M.; Farrell, A.P.; Forster, I.; Gatlin, D.M.; Goldburg, R.J.; Hua, K.; et al. Feeding aquaculture in an era of finite resources. *Proc. Natl. Acad. Sci. USA* **2009**, *106*, 15103–15110. [CrossRef] [PubMed]
2. Rajabdeen, J.; Vanjiappan, R.; Rajamohamed, K.; Kondusamy, A.; Moturi, M.; Jagabattulla, S.D. Fish meal availability in the scenarios of climate change: Inevitability of fish meal replacement in aquafeeds and approaches for the utilization of plant protein sources. *Aquac. Res.* **2019**, *50*, 3493–3506.
3. Zhou, Q.C.; Mai, K.S.; LIU, Y.J.; Tan, B.P. Advances in animal and plant protien sources in place of fish meal. *J. Fish. China* **2005**, *29*, 404–410.
4. Deng, J.M.; Mai, K.S.; Ai, Q.H.; Zhang, W.B.; Wang, X.J.; Xu, W.; Liufu, Z.G. Effects of replacing fish meal with soy protein concentrate on feed intake and growth of juvenile Japanese flounder, *Paralichthys olivaceus*. *Aquaculture* **2006**, *258*, 503–513. [CrossRef]
5. Wang, J.X.; Zhang, H.T.; Yang, Q.H.; Tan, B.P.; Dong, X.H.; Chi, S.Y.; Liu, H.Y.; Zhang, S. Effects of replacing soybean meal with cottonseed meal on growth, feed utilization and non-specific immune enzyme activities for juvenile white shrimp, *Litopenaeus vannamei*. *Aquac. Rep.* **2022**, *16*, 100255. [CrossRef]
6. Ye, J.D.; Liu, X.H.; Wang, Z.J.; Wang, K. Effect of partial fish meal replacement by soybean meal on the growth performance and biochemical indices of juvenile Japanese flounder *Paralichthys olivaceus*. *Aquac. Int.* **2011**, *19*, 143–153. [CrossRef]
7. Burrells, C.; Williams, P.D.; Forno, P.F. Dietary nucleotides: A novel supplement in fish feeds. *Aquaculture* **2001**, *199*, 159–169. [CrossRef]
8. Mo, W.Y.; Man, Y.B.; Wong, M.H. Use of food waste, fish waste and food processing waste for China's aquaculture industry: Needs and challenge. *Sci. Total Environ.* **2018**, *613–614*, 635–643. [CrossRef]
9. Wu, D.W.; Zhou, L.Y.; Gao, M.M.; Wang, M.Y.; Wang, B.; He, J.; Luo, Q.G.; Ye, Y.T.; Cai, C.F.; Wu, P.; et al. Effects of stickwater hydrolysates on growth performance for yellow catfish (*Pelteobagrus fulvidraco*). *Aquaculture* **2018**, *488*, 161–173. [CrossRef]

10. Chaklader, M.R.; Fotedar, R.; Howieson, J.; Siddik, M.A.B.; Foysal, M.J. The ameliorative effects of various fish protein hydrolysates in poultry by-product meal based diets on muscle quality, serum biochemistry and immunity in juvenile barramundi, *Lates calcarifer*. *Fish Shellfish. Immunol.* **2020**, *104*, 567–578. [CrossRef]
11. Yun, B.; Ai, Q.H.; Xue, H.; Qian, X.Q. Effects of Dietary Squid Soluble Fractions on Growth Performance and Feed Utilization in Juvenile Snakehead (*Ophiocephalus argus*) Fed Practical Diets. *Isr. J. Aquac.-Bamidgeh* **2014**, *66*, 1–8.
12. Kotzamanis, Y.P.; Gisbert, E.; Gatesoupe, F.J.; Infante, J.Z.; Cahu, C. Effects of different dietary levels of fish protein hydrolysates on growth, digestive enzymes, gut microbiota, and resistance to Vibrio anguillarum in European sea bass (*Dicentrarchus labrax*) larvae. *Comp. Biochem. Physiol. Part A Mol. Integr. Physiol.* **2007**, *147*, 205–214. [CrossRef]
13. Fisheries and Fisheries Administration Bureau of the Ministry of Agriculture and Rural Affairs of the People's Republic of China, National Fisheres Technology Extension Center, China Society of Fisheries. In *China Fishery Statistical Yearbook*; Agriculture Press: Beijing, China, 2022.
14. Mahdabi, M.; Hosseini, S. A Comparative Study on Some Functional and Antioxidant Properties of Kilka Meat, fish meal, and Stickwater Protein Hydrolysates. *J. Aquat. Food Prod. Technol.* **2018**, *27*, 844–858. [CrossRef]
15. Zhang, Y.X.; Zhang, L.; Huang, L.Y.; Dong, Z.Y.; Lu, Q.; Zou, Y.; Tang, F.; Zhao, S.B.; Storebakken, T. Evaluation of conventional or hydrolyzed stickwater from food-grade skipjack tuna by-product in diet for hybrid grouper (*Epinephelus fuscoguttatus* ♀ × *Epinephelus lanceolatus* ♂). *Aquaculture* **2022**, *548*, 737714. [CrossRef]
16. Kousoulaki, K.; Albrektsen, S.; Langmyhr, E.; Olsen, H.J.; Campbell, P.; Aksnes, A. The water soluble fraction in fish meal (stickwater) stimulates growth in Atlantic salmon (*Salmo salar* L.) given high plant protein diets. *Aquaculture* **2009**, *289*, 74–83. [CrossRef]
17. Kim, S.K.; Takeuchi, T.; Yokoyama, M.; Murata, Y.; Kaneniwa, M.; Sakakura, Y. Effect of dietary taurine levels on growth and feeding behavior of juvenile Japanese flounder *Paralichthys olivaceus*. *Aquaculture* **2005**, *250*, 765–774. [CrossRef]
18. Li, M.; Lai, H.; Li, Q.; Gong, S.Y.; Wang, R.X. Effects of dietary taurine on growth, immunity and hyperammonemia in juvenile yellow catfish *Pelteobagrus fulvidraco* fed all-plant protein diets. *Aquaculture* **2016**, *450*, 349–533. [CrossRef]
19. Nguyen, H.T.M.; Pérez-Gálvez, R.; Bergé, J.P. Effect of diets containing tuna head hydrolysates on the survival and growth of shrimp *Penaeus vannamei*. *Aquaculture* **2012**, *324–325*, 127–134. [CrossRef]
20. Kader, M.A.; Koshio, M. Effect of composite mixture of seafood by-products and soybean proteins in replacement of fish meal on the performance of red sea bream, *Pagrus major*. *Aquaculture* **2012**, *368–369*, 95–102. [CrossRef]
21. Cahu, C.; Rønnestad, I.; Grangier, V.; Zambonino Infante, J.L. Expression and activities of pancreatic enzymes in developing sea bass larvae (*Dicentrarchus labrax*) in relation to intact and hydrolyzed dietary protein; involvement of cholecystokinin. *Aquaculture* **2004**, *238*, 295–308. [CrossRef]
22. Wei, Y.L.; Liu, J.S.; Duan, M.; Ma, Q.; Xu, H.G.; Liang, M.Q. The expressions of two myostatin paralogs in turbot *Scophthalmus maximus* fasted for 2 weeks and fed diets containing graded levels of fish protein hydrolysate. *Aquac. Res.* **2022**, *53*, 5720–5730. [CrossRef]
23. Hidalgo, M.; Urea, E.; Sanz, A. Comparative study of digestive enzymes in fish with different nutritional habits. Proteolytic and amylase activities. *Aquaculture* **1999**, *170*, 267–283. [CrossRef]
24. Shi, Y.; Zhong, L.; Ma, X.; Liu, Y.; Tang, T.; Hu, Y. Effect of replacing fish meal with stickwater hydrolysate on the growth, serum biochemical indexes, immune indexes, intestinal histology and microbiota of rice field eel (*Monopterus albus*). *Aquac. Rep.* **2019**, *15*, 100223. [CrossRef]
25. Cinquina, A.L.; Calı, A.; Longo, F.; Santis, L.D.; Severoni, A.; Abballe, F. Determination of biogenic amines in fish tissues by ion-exchange chromatography with conductivity detection. *J. Chromatogr. A* **2004**, *1032*, 73–77. [CrossRef] [PubMed]
26. Cudennec, B.; Fouchereau-Peron, M.; Ferry, F.; Duclos, E.; Ravallec, R. In vitro and in vivo evidence for a satiating effect of fish protein hydrolysate obtained from blue whiting (*Micromesistius poutassou*) muscle. *J. Funct. Foods* **2012**, *4*, 271–277. [CrossRef]
27. Hevrøy, E.M.; Espe, M.; Waagbø, R.; Sandnes, K.; Ruud, M.; Hemre, G.I. Nutrient utilization in Atlantic salmon (*Salmo salar* L.) fed increased levels of fish protein hydrolysate during a period of fast growth. *Aquac. Nutr.* **2005**, *11*, 301–313. [CrossRef]
28. Li, F.L.; Tang, Y.; Wei, L.X.; Yang, M.X.; Lu, Z.J.; Shi, F.; Zhan, F.B.; Li, Y.N.; Liao, W.C.; Lin, L.; et al. Alginate oligosaccharide modulates immune response, fat metabolism, and the gut bacterial community in grass carp (*Ctenopharyngodon idellus*). *Fish Shellfish. Immunol.* **2022**, *130*, 103–113. [CrossRef]
29. Wu, D.W.; Ye, Y.T.; Cai, C.F.; Xu, J.Y.; Zhang, L.L.; Chen, K.Q.; Huang, Y.W.; Xu, D.H.; Peng, K.; Luo, Q.G.; et al. Effects of fish meal replacement by stickwater meal on growth and health of grass carp (*Ctenopharyngodon idellus*). *Chin. J. Anim. Nutr.* **2015**, *27*, 2094–2105.
30. Zimmer, A.M.; Wright, P.A.; Wood, C.M. Ammonia and urea handling by early life stages of fishes. *J. Exp. Biol.* **2017**, *220*, 3843–3855. [CrossRef]
31. Peres, H.; Oliva-Teles, A. The effect of dietary protein replacement by crystalline amino acid on growth and nitrogen utilization of turbot *Scophthalmus maximus* juveniles. *Aquaculture* **2005**, *250*, 755–764. [CrossRef]
32. Song, Z.D.; Li, H.Y.; Wang, J.Y.; Li, P.Y.; Sun, Y.Z.; Zhang, L.M. Effects of fish meal replacement with soy protein hydrolysates on growth performance, blood biochemistry, gastrointestinal digestion and muscle composition of juvenile starry flounder (*Platichthys stellatus*). *Aquaculture* **2014**, *426–427*, 96–104. [CrossRef]

33. Wang, L.N.; Liu, W.B.; Lu, K.L.; Xu, W.N.; Cai, D.S.; Zhang, C.N.; Qian, Y. Effects of dietary carbohydrate/lipid ratios on non-specific immune responses, oxidative status and liver histology of juvenile yellow catfish *Pelteobagrus fulvidraco*. *Aquaculture* **2014**, *426–427*, 41–48. [CrossRef]
34. Zhang, L.L.; Liang, M.Q.; Xu, H.G.; Liu, X.; Zheng, K.K. Effects of dietary krill hydrolysates on growth performance, body composition and related enzyme activities of juvenlie turbot (*Scophthalmus maximus* L.). *Acta Hydrobiol. Sin.* **2017**, *41*, 497–505.
35. Zhou, Q.C.; Jin, M.; Elmada, Z.C.; Liang, X.P.; Mai, K.S. Growth, immune response and resistance to Aeromonas hydrophila of juvenile yellow catfish, *Pelteobagrus fulvidraco*, fed diets with different arginine levels. *Aquaculture* **2015**, *437*, 84–91. [CrossRef]
36. Tapia-Salazar, M.; Cruz-Suárez, L.E.; Ricque-Marie, D.; Pike, I.H.; Smith, T.K.; Harris, A.; Nygård, E.; Opstvedt, J. Effect of fish meal made from stale versus fresh herring and of added crystalline biogenic amines on growth and survival of blue shrimp *Litopenaeus stylirostris* fed practical diets. *Aquaculture* **2004**, *242*, 437–453. [CrossRef]
37. Kantserova, N.P.; Ushakova, N.V.; Lysenko, L.A.; Nemova, N.N. Calcium-Dependent Proteinases of Some Invertebrates and Fish. *J. Evol. Biochem. Physiol.* **2010**, *46*, 585–591. [CrossRef]
38. Guerra-Olvera, F.M.; Viana, M.T. Effect of dietary cholesterol content on growth and its accumulation in liver and muscle tissues of juvenile yellowtail kingfish (*Seriola lalandi*). *Cienc. Mar.* **2015**, *41*, 143–153. [CrossRef]
39. Van-Muiswinkel, W.B. A history of fish immunology and vaccination I. The early days. *Fish Shellfish. Immunol.* **2008**, *25*, 397–408. [CrossRef]
40. Pilström, L.; Bengtén, E. Immunoglobulin in fish-genes, expression and structure. *Fish Shellfish. Immunol.* **1996**, *6*, 243–262. [CrossRef]
41. Li, X.; Wei, X.; Guo, X.; Mi, S.C.; Hua, X.M.; Li, N.Y.; Yao, J.T. Enhanced growth performance, muscle quality and liver health of largemouth bass (*Micropterus salmoides*) were related to dietary small peptides supplementation. *Aquac. Nutr.* **2020**, *26*, 2169–2177. [CrossRef]
42. Tiziano, V.; Genciana, T.; Konrad, D.; Marco, S. Peptide transport and animal growth: The fish paradigm. *Biol. Lett.* **2011**, *7*, 597–600.
43. Helland, S.J.; Grisdale-Helland, B. Replacement of fish meal with wheat gluten in diets for Atlantic halibut (*Hippoglossus hippoglossus*): Effect on whole-body amino acid concentrations. *Aquaculture* **2006**, *261*, 1363–1370. [CrossRef]
44. Shiells, R.; Falk, G. Retinal on-bipolar cells contain a nitric oxide-sensitive guanylate cyclase. *Neuroreport* **1992**, *3*, 845–848. [CrossRef]
45. Mccord, M.J. The evolution of free radicals and oxidative stress. *Am. J. Med.* **2000**, *108*, 652–659. [CrossRef] [PubMed]
46. Ramalingam, M.; Kim, S.J. Reactive oxygen/nitrogen species and their functional correlations in neurodegenerative diseases. *J. Neural Transm.* **2012**, *119*, 891–910. [CrossRef] [PubMed]
47. Kanner, J.; Lapidot, T. The stomach as a bioreactor: Dietary lipid peroxidation in the gastric fluid and the effects of plant-derived antioxidants. *Free. Radic. Biol. Med.* **2001**, *31*, 1388–1395. [CrossRef]
48. Wu, T.H.; Bechtel, P.J.; Bower, C.K. Effects of Delayed Processing of Pink Salmon (*Oncorhynchus gorbuscha*) By-products on fish meal Quality. *J. Aquat. Food Prod. Technol.* **2009**, *18*, 345–359. [CrossRef]
49. Cheng, J.H.; Sun, D.W.; Han, Z.; Zeng, X.A. Texture and Structure Measurements and Analyses for Evaluation of Fish and Fillet Freshness Quality: A Review. *Compr. Rev. Food Sci. Food Saf.* **2014**, *13*, 52–61. [CrossRef]
50. Li, H.H.; Pan, Y.X.; Liu, L.; Li, Y.L.; Huang, X.Q.; Zhong, Y.W.; Tang, T.; Zhang, J.S.; Chu, W.Y.; Shen, Y.D. Effects of high-fat diet on muscle textural properties, antioxidant status and autophagy of Chinese soft-shelled turtle (*Pelodiscus sinensis*). *Aquaculture* **2019**, *511*, 734228. [CrossRef]
51. Mitchell, J. Food texture and viscosity: Concept and measurement. *Int. J. Food Sci. Technol.* **2003**, *38*, 839–840. [CrossRef]

*fishes*

MDPI

*Article*

# Tryptophan Reduces Intracohort Cannibalism Behavior in Tropical Gar (*Atractosteus tropicus*) Larvae

**César Antonio Sepúlveda-Quiroz [1,2], Graciela María Pérez-Jiménez [1], Gloria Gertrudis Asencio-Alcudia [1,2], Omar Mendoza-Porras [3], Luis Daniel Jiménez-Martínez [4], Mario A. Galaviz-Espinoza [5], Dariel Tovar-Ramirez [6], Rafael Martinez-Garcia [1,*], Carina Shianya Alvarez-Villagomez [1,*] and Carlos Alfonso Alvarez-Gonzalez [1]**

1 Laboratorio de Fisiología en Recursos Acuáticos, División Académica de Ciencias Biológicas, Universidad Juárez Autónoma de Tabasco, Carretera Villahermosa-Cárdenas Km. 0.5, Villahermosa 86039, Tabasco, Mexico; casq15@gmail.com (C.A.S.-Q.); gjimenez9@outlook.com (G.M.P.-J.); yoya_asencio@live.com.mx (G.G.A.-A.); alvarez_alfonso@hotmail.com (C.A.A.-G.)
2 Tecnológico Nacional de México Campus Villahermosa, Km. 3.5 Carretera Villahermosa–Frontera, Cd. Industrial, Villahermosa 86010, Tabasco, Mexico
3 CSIRO Livestock and Aquaculture, Queensland Bioscience Precinct, 306 Carmody Rd, St Lucia 4067, Australia; omar.mendozaporras@csiro.au
4 División Académica Multidisciplinaria de Jalpa de Méndez, Universidad Juárez Autónoma de Tabasco (DAMJM-UJAT), Carretera Estatal Libre Villahermosa-Comalcalco Km. 27+000 s/n Ranchería Ribera Alta, Jalpa de Méndez 86205, Tabasco, Mexico; luisd1984@hotmail.com
5 Facultad de Ciencias Marinas, Universidad Autónoma de Baja California (UABC), Ensenada 22860, Baja California, Mexico; mgalaviz@uabc.edu.mx
6 Laboratorio de Fisiología Comparada y Genómica Funcional, Centro de Investigaciones Biológicas del Noroeste, La Paz 23096, Baja California Sur, Mexico; dtovar04@cibnor.mx
* Correspondence: biologomartinez@hotmail.com (R.M.-G.); gushi_2@hotmail.com (C.S.A.-V.)

**Citation:** Sepúlveda-Quiroz, C.A.; Pérez-Jiménez, G.M.; Asencio-Alcudia, G.G.; Mendoza-Porras, O.; Jiménez-Martínez, L.D.; Galaviz-Espinoza, M.A.; Tovar-Ramirez, D.; Martinez-Garcia, R.; Alvarez-Villagomez, C.S.; Alvarez-Gonzalez, C.A. Tryptophan Reduces Intracohort Cannibalism Behavior in Tropical Gar (*Atractosteus tropicus*) Larvae. *Fishes* **2024**, *9*, 40. https://doi.org/10.3390/fishes9010040

Academic Editors: Qiyou Xu, Jianhua Ming, Fei Song, Changle Qi and Chuanpeng Zhou

Received: 15 December 2023
Revised: 13 January 2024
Accepted: 18 January 2024
Published: 21 January 2024



**Abstract:** The intracohort cannibalism present in tropical gar larvae (*A. tropicus*) generates great problems in its culture, as in other fish species around the world. The addition of tryptophan (Trp) (10, 20, and 30 g/kg) and a control diet (CD) without Trp were evaluated in *A. tropicus* larvae regarding growth, survival, cannibalism, behavior, digestive enzymatic activity, and genes related to aggressiveness and/or cannibalism in two stages: 0–13 days after hatching (DAH); and only cannibals (14–24 DAH). In the first stage, no differences were observed in growth parameters; cannibalism was lower with the use of Trp, with the lowest percentage being the 10 g/kg Trp treatment (56.75 ± 2.47%) compared to CD (64.75 ± 1.76%). In the second stage, survival was greater in 10 g/kg Trp (75.00 ± 7.07%) than in CD (23.33 ± 5.77%). Thus, cannibalism was lower with 10 g/kg Trp (20.0 ± 10.0%) compared to CD (76.66 ± 5.77%). Cannibal larvae fed with 10 g/kg Trp had a greater enzymatic activity in acid and alkaline proteases and leucine aminopeptidase, as well as the overexpression of *avpi1*, *crh*, and *htr1a* and the subexpression of *tph1*, *th*, *sstr1*, and *hdc* ($p < 0.05$). No aggressive behaviors were recorded in the larvae fed with the 10 g/kg Trp treatment, unlike those fed with CD. The use of 10 g/kg Trp improves survival and reduces cannibalism in *A. tropicus* larvae.

**Keywords:** intracohort cannibalism; mitigation; survival; fish larvae; behavior

**Key Contribution:** The use of 10 g/Kg of Trp in the diet improves survival and reduces cannibalism in *A. tropicus* larvae. Cannibal larvae of *A. tropicus* show a greater growth due to the "jumper" effect, unlike non-cannibal larvae.

## 1. Introduction

In fish, around 390 species have been documented that show some type of cannibalism, a behavior that occurs most of the time throughout their life or exclusively in the early stages of development (larvae and/or juvenile) [1]. Tropical gar (*Atractosteus tropicus*) shows cannibalism in the larvae stage (10 DAH (days after-hatching)) in captivity and in the wild [2,3]. This species is considered of importance; it has economic value due to its culture and fishing, as well as ecological and cultural value in Mexico and Central America [4,5]. However, the presence of cannibalism in the early stages of development limits the growth, survival, and profitability values of the production. For example, Frías-Quintana et al. [6] reported, for *A. tropicus* in laboratory conditions, survival and cannibalism values of 24% and 33%, respectively. In another study, Palma-Cancino et al. [7] used co-feeding (commercial diet and *Artemia nauplii*) on *A. tropicus* larvae; cannibalism was recorded with complete (11 DAH) and incomplete (21 DAH) ingestion, obtaining survival values ranging from 1 to 33%. The problem is magnified because the fish that are not consumed but are injured by the attacks die. When modifying the diet of *A. tropicus* larvae with a low concentration of polyunsaturated fatty acid, similar cannibalism values (40%) and survival values ranging from 15 to 30% continued to be found [8].

Because of this, different strategies have been evaluated in several species to mitigate cannibalism in fish, one of them being the incorporation of Tryptophan (Trp) in the diet. Trp is an essential amino acid, which is the precursor of 5-hydroxytryptamine (5-HT) (serotonin) [9], a neurotransmitter responsible for controlling appetite [10], reproduction [11], and physiological processes related to immunity and intestinal homeostasis [12]. It has been observed that the administration of Trp in the fish *Aequidens pulcher* and *Apteronotus leptorhynchus* reduced their aggressiveness [13,14]. Meanwhile, the administration of the inhibitor 5-HT (p-chlorophenylalanine) increased aggressiveness in *Cichlasoma meeki* [15]. Likewise, the incorporation of Trp at concentrations of 2, 4, and 6 ppm and 2 and 3% in the diet induced a significant decrease of 50% in the cannibalism of Pabda (*Ompok bimaculatus*) larvae [16,17].

Therefore, the objective of this work is to evaluate the effect of Trp administration in the diet of *A. tropicus* larvae, as a possible mitigant of cannibalistic behavior. The rate of cannibalism, survival, growth, enzymatic activity, and the expression of genes related to cannibalism and behavior are analyzed.

## 2. Materials and Methods

### 2.1. Biological Material

The *A. tropicus* larvae were obtained for this experiment by inducing a female (3.4 kg, 91 cm) to spawn using the hormone LHRHa (Sigma-Aldrich, Taufkirchen, Germany) (30 μg/kg fish$^{-1}$) applied intramuscularly. Subsequently, the female and three males (1.6 kg, 35 cm, no hormone induction) were placed in a circular tank (2000 L) with raffia thread to simulate the natural spawning site. The above was conducted with the broodstock batch from the Laboratorio de Fisiología en Recursos Acuáticos (LAFIRA) from Division Academica de Ciencias Biologicas of the Universidad Juarez Autonoma de Tabasco.

### 2.2. Experimental Design

The present study consisted of two stages. In the first stage, three concentrations of Trp (10, 20, and 30 g/kg) and a control diet (CD; no Trp) were evaluated; this stage lasted from day 3 to 13 DAH of larvae. Larvae (n = 200, 0.018 ± 0.001 g; 1.28 ± 0.09 cm) were placed in circular-shaped 70 L tanks (2.8 larvae/L). Each treatment was carried out in triplicate. The feeding regime was as follows: after yolk sac absorption (3 DAH), co-feeding was supplied; treatment diet and Artemia nauplii were administered for additional five days. At the end of the co-feeding, fish were fed only with the formulated feed, four times per day (8:00, 12:00, 16:00, and 20:00 h) ad libitum. In the second stage, only cannibal larvae were selected and fed with Trp (10 g/kg) and CD. Treatments were carried out in triplicate with 10 fish per replicate (0.14 larvae/L). This stage lasted from 14 to 24 DAH. Fish were weighted and sized at the beginning and at the end of each stage and cannibalism was quantified. For both stages, the tanks were connected to a recirculation system powered by a 0.5 HP water pump (Jacuzzi, JWPA5D-230A, Delavan, WI, USA) and a 1500 L reservoir for solid deposition and a biological filter. A partial daily exchange of 10% water was carried out by siphoning feces and uneaten feed. The water quality parameters were daily monitored temperature (27.36 ± 0.6 °C), dissolved oxygen (4.8 ± 0.4 mg/L, oximeter YSI 85; Yellow Springs, OH, USA), and pH (7.1 ± 0.3, HANNA HI 991001, Nusfalau, Romania).

### 2.3. Formulation and Preparation of the Experimental Diets

The formulation of the diets was carried out using the software MIXITWIN v.5.0. (Microsoft Windows, Washington, DC, USA), following Álvarez-González et al. [18]. The macronutrients were weighed and mixed, followed by the incorporation of micronutrients. In this step, we added tryptophan (Sigma-Aldrich, Taufkirchen, Germany reagent grade, ≥98%) in the different concentrations mentioned. This was followed by the addition of liquid ingredients. To achieve an adequate mixture, water was added 400 mL/kg per diet and mixed 15 min with each addition (a total of 60 min of mixing time per diet). The mixture was passed through a meat grinder (Torrey, M-22RI, Nuevo Leon, Mexico) and pellets were oven dried at 55 °C for 12 h (Coriat, HC-35-D, Ciudad de Mexico, Mexico). Finally, the pellets were manually ground and sieved to particles smaller than 0.5 mm (used during co-feeding) and larger than 0.7 mm (used after co-feeding). Diets were stored in hermetic plastic bags at −20 °C for later use. For all diets, the proximal components (moisture, ash, lipids, and protein) were analyzed according to AOAC [19] (Table 1), lipids (Table 2), and amino acids (Table 3).

**Table 1.** Composition of the experimental diets with different concentrations of Trp and CD.

| Ingredients | CD | Trp (g/kg) 10 | 20 | 30 |
|---|---|---|---|---|
| Fish meal [a] | 305.4 | 305.4 | 305.4 | 305.4 |
| Renderer meal [a] | 300.0 | 300.0 | 300.0 | 300.0 |
| Soy meal [a] | 150.0 | 150.0 | 150.0 | 150.0 |
| Corn starch [b] | 67.1 | 67.1 | 67.1 | 67.1 |
| Oil soy [c] | 116.5 | 116.5 | 116.5 | 116.5 |
| Cellulose [d] | 30.0 | 20.0 | 10.0 | 0.0 |
| Tryptophan [d] | 0.0 | 10.0 | 20.0 | 30.0 |
| Premix vit-min [e] | 15.0 | 15.0 | 15.0 | 15.0 |
| Grenetine [f] | 10.0 | 10.0 | 10.0 | 10.0 |
| Vit C [g] | 5.0 | 5.0 | 5.0 | 5.0 |
| Vit E [h] | 1.0 | 1.0 | 1.0 | 1.0 |
| Proximate composition (g/100 g dry matter), except energy | | | | |
| Energy (kJ/g) | 17.91 | 17.53 | 17.39 | 17.48 |
| Protein | 44.00 | 45.00 | 46.00 | 47.00 |
| Ether extract | 16.38 | 16.59 | 16.41 | 16.48 |
| Fibre | 1.05 | 1.12 | 1.07 | 1.01 |
| Ash | 13.43 | 13.23 | 13.02 | 12.93 |
| NFE [1] | 25.14 | 24.06 | 23.5 | 22.58 |

[a] Marine and agricultural proteins S.A. de C.V., Guadalajara, Jalisco; [b] IMSA Corn Industrializer S.A de C.V. Guadalajara, Jalisco, México; [c] Ragasa industries S.A. de C.V.; [d] Sigma-Aldrich Quimica S. de R.L. de C.V.; [e] Vitamin premix composition g, mg, or International Units per kg of diet: Vitamin A, 10,000,000 IU; Vitamin D3, 2,000,000 IU; Vitamin E, 100,000 IU; Vitamin K3, 4.0 g; Thiamine B1, 8.0 g; Riboflavin B2, 8.7 g; Pyridoxine B6, 7.3 g; Vitamin B12, 20.0 mg; Niacin, 50.0 g; Pantothenic acid, 22.2 g; Inositol, 0.15 mg; Nicotinic Acid, 0.16 mg; Folic Acid, 4.0 g; Biotin, 500 mg; Vitamin C, 10.0 g; Choline, 0.3 mg, Excipient q.s., 2 g; Manganese, 10 g; Magnesium, 4.5 g; Zinc, 1.6 g; Iron, 0.2 g; Copper, 0.2 g; Iodine, 0.5 g; Selenium, 40 mg; Cobalt, 60 mg; Excipient q.s., 1.5 g; [f] D'gari, food and diet products relámpago S.A. de C.V.; [g] ROVIMIX® STAY-C® 35-DSM, Guadalajara, México; [h] GELPHARMA, S.A. de C.V. NFE [1] = nitrogen-free extract: 100-(%protein-%ethereal extract-%ash-%fibre).

**Table 2.** Analysis of the total fatty acids in the experimental diets used for *A. tropicus* larvae.

| Fatty Acids (%) | CD | 10 | Trp (g/kg) 20 | 30 |
|---|---|---|---|---|
| C13:0 | 7.0 | 10.2 | 9.7 | 7.9 |
| C14:0 | 1.2 | 1.3 | 1.2 | 1.2 |
| C16:0 | 16.7 | 17.0 | 17.0 | 17.4 |
| C17:0 | ND | ND | ND | ND |
| C18:0 | 5.8 | 6.0 | 6.0 | 6.1 |
| C23:0 | ND | ND | ND | ND |
| **ΣSFA** | **30.7** | **34.4** | **33.9** | **32.7** |
| C16:1n7 | 2.0 | 2.1 | 2.1 | 2.2 |
| C18:1n9 | 21.0 | 20.4 | 20.5 | 21.1 |
| C18:1n7 | 1.6 | 1.8 | 1.8 | 1.8 |
| **ΣMUFAS** | **24.6** | **24.3** | **24.5** | **25.1** |
| C18:2n6 | 34.9 | 32.0 | 32.3 | 32.6 |
| C18:3n3 | 0.3 | 4.3 | 4.3 | 4.3 |
| C18:4n3 | 4.7 | ND | ND | ND |
| C20:3n3 | 0.5 | 0.6 | 0.7 | 0.7 |
| C20:4n6 | 0.3 | ND | ND | ND |
| C20:5n3 | 1.6 | 1.7 | 1.8 | 1.7 |
| C22:5n3 | 0.4 | 0.4 | 0.4 | 0.4 |
| C22:6n3 | 1.9 | 2.2 | 2.2 | 2.4 |
| **ΣPUFAS** | **44.6** | **41.3** | **41.6** | **42.2** |
| NID | **0.0** | 0.0 | 0.0 | 0.0 |
| | 100.0 | 100.0 | 100.0 | 100.0 |

**Table 3.** Analysis of the total amino acids in the experimental diets used for *A. tropicus* larvae.

| Amino Acid | CD | 10 | Trp (g/kg) 20 | 30 |
|---|---|---|---|---|
| | | Essential amino acids | | |
| HIS | 1.1 | 1.0 | 1.2 | 1.0 |
| ARG | 5.2 | 5.6 | 5.4 | 5.3 |
| THR | 1.7 | 1.6 | 1.7 | 1.7 |
| VAL | 1.7 | 1.6 | 1.6 | 1.6 |
| MET | 0.5 | 0.5 | 0.2 | 0.2 |
| LYS | 5.7 | 5.0 | 5.3 | 5.5 |
| ILE | 1.3 | 1.2 | 1.3 | 1.3 |
| LEU | 3.5 | 3.2 | 3.5 | 3.5 |
| PHE | 1.4 | 1.2 | 1.4 | 1.4 |
| subtotal | 22.1 | 20.9 | 21.5 | 21.7 |
| | | Non-essential amino acids | | |
| ASP | 2.4 | 2.2 | 2.4 | 2.4 |
| SER | 2.3 | 2.2 | 2.4 | 2.3 |
| GLU | 5.8 | 5.7 | 6.0 | 6.0 |
| GLY | 7.0 | 8.2 | 7.1 | 6.9 |
| ALA | 3.4 | 3.7 | 3.6 | 3.5 |
| TYR | 1.5 | 1.3 | 1.4 | 1.5 |
| subtotal | 22.3 | 23.4 | 22.9 | 22.7 |
| | | Others | | |
| TAU | 0.6 | 0.7 | 0.6 | 0.6 |
| Total | 45.0 | 45.0 | 45.0 | 45.0 |
| Tryptophan (mg/g) | 10.60 | 19.74 | 32.42 | 44.21 |

## 2.4. Growth Indexes and Feed Quality

Weight and length were measured as follows: At the beginning of the experiment (3 DAH) and at the end of the experiment (13 DAH) as well as in the second phase (14–24 DAH). The individual weight of each organism was determined by using an analytical balance (A&D Company, Limited mod.HR-250, Seoul, Republic of Korea). The total length was calculated by analyzing the photographs taken of the organisms, through a transparent container with a scale using the software ImageJ 1.51j8 (U.S. National Institutes of Health, Bethesda, MD, USA). The following biometrics were calculated as follows: Survival (S): (final fish number/initial fish number) × 100; feed intake (FI): total feed intake per experimental unit/number of rearing days; absolute weight gain (AWG): final weight (g)—initial weight (g); specific growth rate (SGR): [(ln final weight − ln initial weight)/days] × 100; feed conversion ratio (FCR): (feed intake, g dry matter)/(fish weight gain, g); condition factor (K): [(wet weight (g) × total length − 3 (cm)] × 100; and protein efficiency ratio (PER): fish live weight gain (g)/dry protein fed (g). Visually deformed organisms (scoliosis, crossbite, lower jaw reduction, and without eyes) were identified and counted by monitoring organisms with erratic swimming, high pigmentation, and visible malnutrition. These organisms were collected after death, and the deformity was confirmed visually using a stereomicroscope (Carl Zeiss mod. Stemi DV4, Göttingen, Germany). The percentage of deformed organisms was calculated using the following formula: Deformity (D) (fish with deformities/initial fish number) × 100. Also, the coefficient of variation (%) (CV) ((standard deviation of individual weight/mean individual weight) × 100) and the size heterogeneity (weight) (SH) ((final coefficient of variation/initial coefficient of variation)) were calculated.

## 2.5. Collection of Biological Samples

All procedures were performed according to the Official Mexican Norm (NOM-062-ZOO-1999) [20] of Animal Welfare and with the Declaration of Helsinki.

At the end of the experiment of the first stage, 15 larvae per treatment were euthanized with a cold temperature shock; larvae were in a tray and then placed in an ultra-freezer at −80 °C (Lexicon II ultra-low freezer, Singapore) for 3 min, and then dissected to determine the effect of Trp on growth, digestive enzymes, gene expression, and cannibalism. In the second stage, nine larvae per treatment were used, and the same euthanized protocol was used. For the analysis of the enzymatic activity, the larvae were preserved at −80 °C. Finally, to analyze the gene expression, the larvae were preserved in RNAlater buffer according to the manufacturer's instructions (Invitrogen, Waltham, MA, USA) at −80 °C. Whole larvae were used in all analyses.

## 2.6. Cannibalism

In both stages, cannibalism was measured by monitoring larvae 30 min before and 60 min after each feeding. The criteria to determine cannibalism was the following: attack by bites (a fish attacks another fish by bites without the attacker ingesting the prey), partial cannibalism (one fish partially eats another fish), and complete cannibalism (one fish entirely eats another fish). These three behaviors are classified as cannibalism. The formula (fish with cannibalistic behavior/initial fish number) × 100) was used to quantify cannibalism in each treatment. The attacking larvae (cannibal) and the attacked larvae (non-cannibal) were counted with the weight and size being registered. Five larvae were sampled for molecular analysis and five for enzymatic activity, as described earlier.

## 2.7. Digestive Enzyme Activity

For the quantification of digestive enzymes, we followed the Shuangyao et al. [21] protocol: The larvae were manually macerated inside 1.5 mL tubes on ice. For this, 100 mg of tissue were placed in a total volume of 0.5 mL and centrifuged at 12,000× $g$ at 4 °C for 15 min. The supernatant was recovered and stored at −80 °C in 30 µL aliquots. Soluble protein was determined using the Bradford assay [22]. For the quantification of acid proteases, 0.5% hemoglobin solubilized in 100 mM glycine-HCl pH 2 buffer was used as a substrate. Alkaline proteases were quantified using 0.5% casein solubilized in 50 mM Tris-HCl and 10 mM CaCl at pH 9 [23]. In both assays, the samples were incubated at 37 °C, and the reaction was stopped using 0.5 mL of 20% trichloroacetic acid and centrifuged at 16,000× *g* for five minutes. Absorbance was read at 280 nm. The extinction coefficient ($\varepsilon$) to calculate the activity of acid and alkaline proteases was 0.005 mL/µM cm. To quantify trypsin activity, 1 mM BAPNA (Nα-Benzoyl-DL-Arginine-P-nitroanilide) dissolved in 50 mM Tris-HCl was used as a substrate, in pH 8 at 37 °C. Trypsin was read at 410 nm using an $\varepsilon$ of 8800 mL/µM cm [24]. The Maroux et al. [25] method was used to determine the activity of leucine aminopeptidase, where 0.1 M leucine p-nitroanilide dissolved in DMSO with 50 mM sodium phosphate was the substrate, at pH 7.2, and incubated at 37 °C. Absorbance was measured at 410 nm with an $\varepsilon$ of 8800 mL/µM cm. Lipase activity was determined using β-naphthyl acetate (100 mM) dissolved in 50 mM Tris-HCl at as substrate, at pH 7.5, with sodium taurocholate (100 mM) at 37 °C. The reaction was stopped with 0.72 N TCA. Fast Blue (100 mM) and a 1:1 ethanol/ethyl acetate mixture was added, and the absorbance was quantified at 540 nm using $\varepsilon$ d 0.02 mL/µM cm [26].

The enzyme activity was determined using the following equations: units by mL (U/mL) = [$\Delta$abs × final reaction volume (mL)]/[$\varepsilon$ × time (min) × volume extracted (mL)] − 1; specific activity (U/mg protein) = U mL/mg of soluble protein; and the molar extinction coefficient ($\varepsilon$).

### 2.8. RNA Extraction and Quantitative Reverse Transcription PCR (RT-qPCR)

Total RNA was extracted from complete larvae samples using Trizol (Invitrogen, Waltham, MA, USA), according to the manufacturer's protocol. The concentration and purity of RNA samples were assessed by the ratio between the absorbance at 260 and 280 nm in a spectrophotometer (Jenway GenovaNano, Cole-Parmer, Staffordshire, UK). RNA (1 µg) was reverse-transcribed (RT) using the SuperScript II kit (Invitrogen, Waltham, MA, USA), with a final volume of 20 µL. RT reactions were performed in a thermocycler (Mastercycler nexus GSX1, Eppendorf AG, Hamburg, Germany). The standard RT program used was as follows: 5 min at 65 °C, 10 min at 25 °C, 50 min at 42 °C (cDNA strand extension), 15 min at 70 °C (reverse transcriptase inactivation), and finally 20 min at 37 °C. Somatostatin receptor 1 (*sstr1*), tyrosine hydroxylase (*th*), histidine decarboxylase (*hdc*), corticotropin-releasing hormone (*crh*), 5-hydroxytryptamine (serotonin) receptor 1A, G protein-coupled (*htr1a*), gonadotropin releasing hormone 1 (*gnrh1*), arginine vasopressin-induced 1 (*avpi1*), and tryptophan hydroxylase 1 (*tph1*) for *A. tropicus* were designed from the species transcriptome (NCBI Accession: PRJNA395289) [27]. These genes were selected due to their influence on aggressive behaviors and interactions with specific neurotransmitters [28,29] (Table 4). The RT-qPCR was performed in a CFX96 Real-Time System (BioRad, Hercules, CA, USA) using 5 µL of EvaGreen Supermix (BioRad), 0.5 µL primers mix, and 4.5 µL of cDNA for a final volume of 10 µL. The RT-qPCR program was used at: 50 °C for 2 min, 95 °C 10 s, followed by 40 cycles at 95 °C 15 s, and 62 °C 1 min. *β-actin* was used as the reference gene [30]. The relative gene expression was calculated as the fold-change compared to the control and using the $2^{-\Delta\Delta Ct}$ formula [31].

**Table 4.** Oligonucleotide design for the real-time polymerase chain reaction (qPCR) of aggressive genes in *A. tropicus* larvae.

| Protein | Gen | Primers (5′-3′) | Alignment Temperature (°C) |
|---|---|---|---|
| Somatostatin receptor 1 | *sstr1* | FW: CCTCAGCATTGACCGCTACA<br>RV: AATACCGCCATCCACTGACG | 60 |
| Tyrosine hydroxylase | *th* | FW: GGACCAGATGTACCAGCCAG<br>RV: GCAGTTCATCCCTCGCAGAT | 59 |
| Histidine decarboxylase | *hdc* | FW: GCATTTCGACTGCACTGCTT<br>RV: CTTCGGCTGAGTGGGATCTG | 59 |
| Corticotropin-releasing hormone | *crh* | FW: AACGTCAACAGGGCTTTCCA<br>RV: TCTTCCCGTCAGGTCTTCCA | 60 |
| 5-hydroxytryptamine (serotonin) receptor 1A, G protein-coupled | *htr1a* | FW: AAGCGCAGTGTGGAACCTAA<br>RV: GCTGTCGGGGTATTAGGCAG | 60 |
| Gonadotropin-releasing hormone 1 | *gnrh1* | FW: AGTCAGCACTGGTCATACGG<br>RV: CTCACCTCCTCCGCAATGTC | 59 |
| Dopamine receptor D1 | *drd1* | FW: TTTTTGGCCCTTTGGCTCATT<br>RV: AAGTTCAAAATGGAGGCTGTGG | 59 |
| Arginine vasopressin-induced 1 | *avpi1* | FW: AGGGAGGACCACTGAAGATGA<br>RV: CCAGCAGAGGACAAGTCTGC | 60 |
| Tryptophan hydroxylase 1 | *tph1* | FW: CCCCCGTATCGAGTTCACAG<br>RV: AGGGGCAGGTTCTTGAGGTA | 60 |

### 2.9. Effect of the Mitigants on the Ethology of Cannibal Larvae

We identified the effect of Trp (10 g/kg) on the cannibalistic behavior and/or ethology in the second stage experiment, comparing it to the larvae fed with the control diet (CD). Cannibal larvae were placed in 15 × 10 × 8 cm fish tanks. To reduce the effect of stress caused by the transference of the larvae to the experimental tanks, an acclimation time of 15 min was allowed. Subsequently, 15 min video recordings were made (Gopro Hero 7 Silver, Monterey, CA, USA). Two larvae per tank were evaluated under the following challenges: without shelter, with rocks, and with artificial vegetation. The videos were analyzed using Tracker 5.1.5 (Free Software Foundation, Inc., Franklin Street Boston, MA, USA) and BORIS 7.9.24 [32], which allow the identification and measurement of aggressive behavior and shelter preference (rocks and artificial vegetation). The same experiment was carried out on CD-fed larvae to determine the effects of Trp on cannibalism.

### 2.10. Statistics Analysis

Normality (Kolmogorov–Smirnov) and homoscedasticity (Bartlett) tests were performed. A one-way ANOVA was carried out for all the analysis, and in the case of finding differences, a posteriori test of unequal N HSD (Tukey) was used. A Student's *t*-test was used to compare among treatments where applicable. To analyze the gene expression data, non-parametric Kruskal–Wallis and Nemenyi posteriori tests were used. All tests were performed using the software Prism V. 9.0 with a significance value of 0.05.

## 3. Results

### 3.1. Growth Indexes and Survival

In the first stage, no significant differences were observed in the weight and total length between *A. tropicus* larvae fed with the Trp diets and CD (Figure 1). Likewise, no significant differences were observed in survival; however, the larvae fed with Trp presented a slight tendency of better survival (Table 5). There were no differences in AWG, SGR, FCR, PER, K, CV, and SH ($p < 0.05$). At the end of this stage, the distribution of the weights and sizes of the fish did not show significant differences among the treatments (Figure 2).

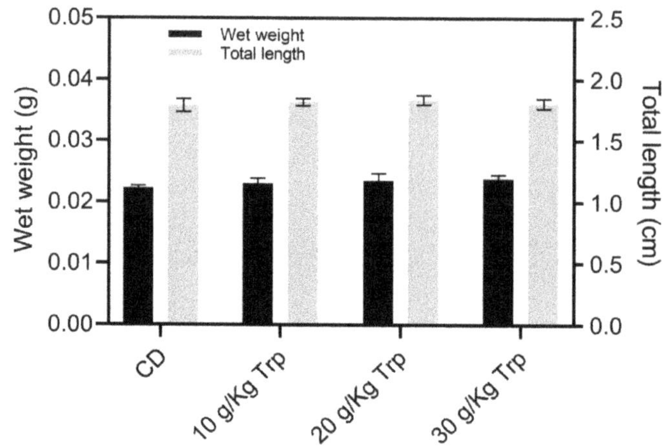


**Figure 1.** Growth in the weight (g) and total length (cm) of *A. tropicus* larvae fed with different Trp concentrations and the control diet (CD). Values are mean ± SD.

**Table 5.** Growth performance and feed utilization indexes of *A. tropicus* larve fed with different concentrations of Trp and the CD (mean ± standard deviation, SD).

| | | Trp (g/kg) | | |
|---|---|---|---|---|
| | **CD** | **10** | **20** | **30** |
| Initial weight (g) | 0.018 ± 0.001 | 0.018 ± 0.001 | 0.018 ± 0.001 | 0.018 ± 0.001 |
| Final weight (g) | 0.022 ± 0.0003 | 0.023 ± 0.001 | 0.023 ± 0.001 | 0.023 ± 0.001 |
| Initial total length (cm) | 1.28 ± 0.09 | 1.28 ± 0.09 | 1.28 ± 0.09 | 1.28 ± 0.09 |
| Final total length (cm) | 1.78 ± 0.05 | 1.81 ± 0.02 | 1.83 ± 0.03 | 1.80 ± 0.03 |
| S (%) | 28.25 ± 1.76 | 34.75 ± 5.30 | 28.66 ± 1.04 | 29.75 ± 1.76 |
| FI (g/d) | 0.031 ± 0.001 | 0.031 ± 0.003 | 0.031 ± 0.003 | 0.031 ± 0.001 |
| AWG (g/fish) | 0.004 ± 0.0003 | 0.005 ± 0.0008 | 0.005 ± 0.001 | 0.005 ± 0.00004 |
| SGR (%/d) | 5.07 ± 0.52 | 5.35 ± 0.29 | 5.53 ± 0.38 | 5.24 ± 0.39 |
| FCR | 7.23 ± 0.85 | 6.68 ± 0.53 | 5.77 ± 0.46 | 5.64 ± 0.02 |
| PER | 0.13 ± 0.01 | 0.15 ± 0.1 | 0.15 ± 0.04 | 0.15 ± 0.04 |
| K | 0.39 ± 0.02 | 0.39± 0.01 | 0.37± 0.01 | 0.38± 0.02 |
| CV (%) | 6.43 ± 2.98 | 8.63 ± 2.23 | 8.20 ± 3.05 | 6.95 ± 0.05 |
| SH | 0.71 ± 0.33 | 0.96 ± 0.25 | 0.91 ± 0.34 | 0.77 ± 0.006 |

Significant differences among the diets are indicated by different letters ($p < 0.05$). FI: feed intake; AWG: absolute weight gain; SGR: specific growth rate; S: survival; FCR: feed conversion rate; PER: protein efficiency rate; K: condition factor; CV: coefficient of variation; SH: size heterogeneity.

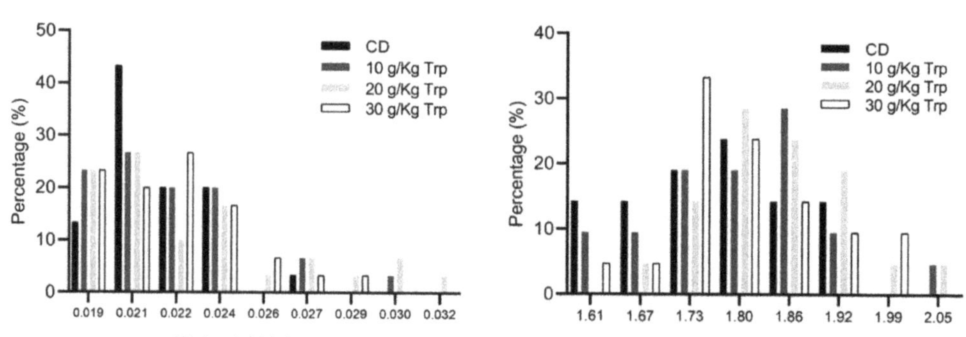


**Figure 2.** Wet weight (g) and total length (cm) class distribution by cannibalism effect in *A. tropicus* larvae fed with Trp and CD.

*3.2. Cannibalism and Deformities*

In all treatments where Trp was administered, the percentage of cannibalism was lower (56.75–64.25%) compared to that where CD was administered. The lowest percentage of cannibalism was observed for 10 g/kg Trp (56.75 ± 2.47%) and the highest value occurred in the larvae of the CD treatment (64.75 ± 1.76%), presenting a significant difference ($p < 0.05$). The percentage of fish with deformities was 5.75 ± 0.43%, and no differences were observed between treatments ($p > 0.05$) (Figure 3).

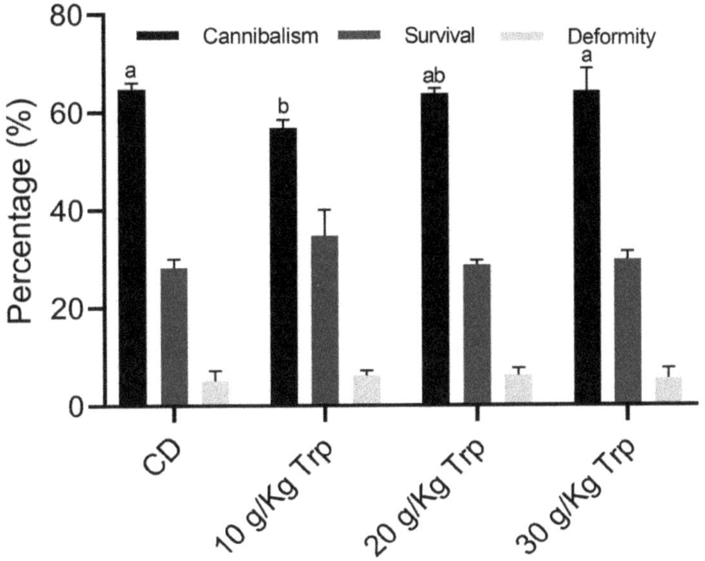


**Figure 3.** Cannibalism, survival, and deformity percentages in *A. tropicus* larvae fed with different Trp concentrations. Values are mean ± SD. Significant differences among the diets are indicated by different letters ($p < 0.05$).

The larvae identified as cannibals and non-cannibals had an average weight of 0.025 ± 0.002 g and 0.023 ± 0.003 g, respectively (*t*-test, $p = 0.007$). The larvae identified as cannibals had a total size of 1.83 ± 0.09 cm, and the larvae identified as non-cannibals a total size of 1.72 ± 0.10 cm (Figure 4) (t-test, $p < 0.0001$). The average difference recorded between the weight of a cannibal fish and its prey was 0.003 ± 0.003 g (16.12 ± 13.44%). In total size, the average difference between the cannibal fish and its prey was 0.15 ± 0.06 cm (8.96 ± 3.72%).

To continue to stage two, based on the results obtained, the 10 g/kg Trp diet was selected in which the lowest percentage of cannibalism and the highest survival occurred. The administration of this feed was continued only to larvae identified as cannibals for 10 more days.

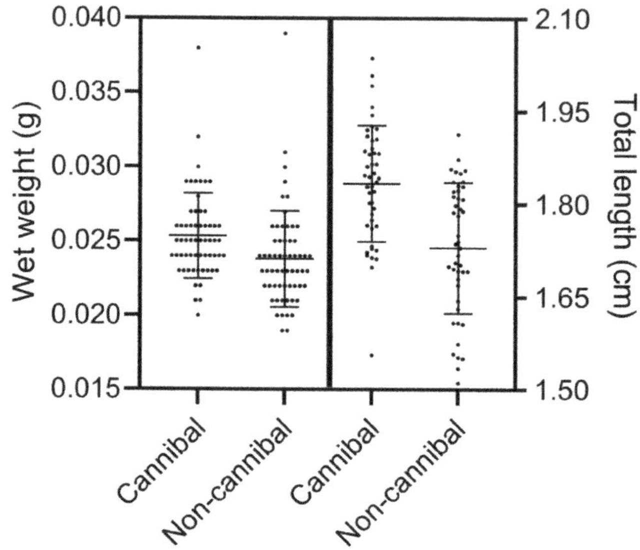


**Figure 4.** Wet weight (g) (n = 60) and total length (cm) (n = 45) of *A. tropicus* cannibal and non-cannibal larvae. Values are mean ± SD.

### 3.3. Growth Indexes and Survival of Cannibals

The cannibal larvae of the CD treatment obtained a greater final weight (0.179 ± 0.03 g), compared to those treated with 10 g/kg Trp (0.080 ± 0.01 g) (*t*-test, $p = 0.0224$). The same was observed for the total size (3.68 ± 0.34 cm in CD and 2.72 ± 0.18 cm in 10 g/kg Trp) (Figure 5) The highest survival was recorded in the 10 g/kg Trp group with 75.00 ± 7.07%, presenting significant differences compared to the CD group (30.00 ± 17. 32%) (*t*-test, $p = 0.0044$). The AWG (0.13 ± 0.03) (*t*-test, $p = 0.019$), SGR (14.62 ± 2.87) (*t*-test, $p = 0.029$), and PER (0.17 ± 0.01) (*t*-test, $p = 0.041$) were higher for CD. Also, the CD group showed a higher CV (55.64 ± 6.84) compared to the treatment of 10 g/kg Trp (26.74 ± 14.20) (*t*-test, $p = 0.015$) and SH (6.07 ± 0.75), showing significant differences compared to the 10 g/kg Trp group (*t*-test, $p = 0.012$). The FCR value for the 10 g/kg Trp treatment was 12.43 ± 1.60, showing a significant difference compared to the CD group (5.66 ± 0.62) (*t*-test, $p = 0.030$). FI and K did not show any differences (Table 6).

**Table 6.** Growth performance and feed utilization indexes of *A. tropicus* cannibal larvae fed with 10 g/kg Trp and CD (mean ± standard deviation, SD).

| | CD | 10 g/kg Trp |
|---|---|---|
| Initial weight (g) | 0.049 ± 0.004 | 0.048 ± 0.006 |
| Final weight (g) | 0.179 ± 0.03 * | 0.080 ± 0.02 |
| Initial total length (cm) | 2.16 ± 0.11 | 2.19 ± 0.08 |
| Final total length (cm) | 3.68 ± 0.34 * | 2.72 ± 0.18 |
| S (%) | 23.33 ± 5.77 | 75.0 ± 7.07 * |
| FI (g/d) | 0.72 ± 0.07 | 0.76 ± 0.08 |
| AWG (g/fish) | 0.13 ± 0.03 * | 0.03 ± 0.01 |
| SGR (%/d) | 14.62 ± 2.87 * | 5.81 ± 3.18 |
| FCR | 5.66 ± 0.62 | 12.43 ± 1.60 * |
| PER | 0.17 ± 0.01 * | 0.05 ± 0.03 |
| K | 0.35 ± 0.02 | 0.39 ± 0.01 |
| CV (%) | 55.64 ± 6.84 * | 26.74 ± 14.20 |
| SH | 6.07 ± 0.75 * | 1.90 ± 1.02 |

Significant differences among the diets are indicated by an asterisk mark ($p < 0.05$). FI: feed intake; AWG: absolute weight gain; SGR: specific growth rate; S: survival; FCR: feed conversion rate; PER: protein efficiency rate; K: condition factor; CV: coefficient of variation; SH: size heterogeneity.

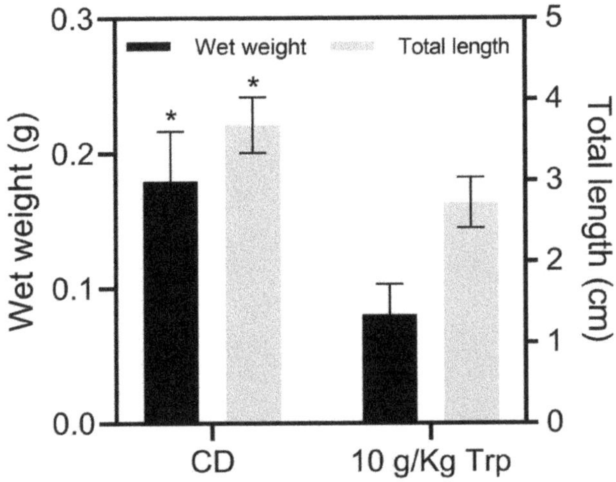


**Figure 5.** Growth in the weight (g) and total length (cm) of *A. tropicus* cannibal larvae fed with 10 g/kg Trp and CD. Values are mean ± SD. Significant differences among the diets are indicated by an asterisk mark ($p < 0.05$).

*3.4. Cannibalism*

The percentage of cannibalism was 20.0 ± 10.0% in the fish fed with 10 g/kg Trp, which was significantly different compared to the CD treatment (76.66 ± 5.77%) (*t*-test, $p = 0.0011$) (Figure 6). At the end of the bioassay, the modification of the distribution of the weights and sizes of the fish by treatment was observed, directly related to the results of cannibalism (Figure 7).

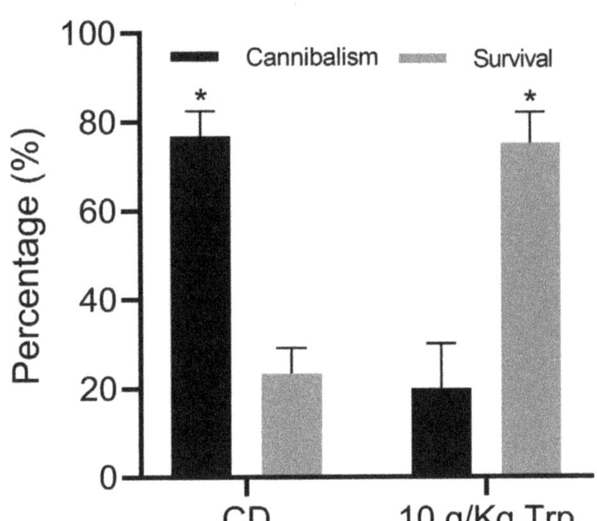


**Figure 6.** Survival and cannibalism of *A. tropicus* cannibal larvae fed with 10 g/kg Trp and CD. Values are mean ± SD. Significant differences among the diets are indicated by an asterisk mark ($p < 0.05$).

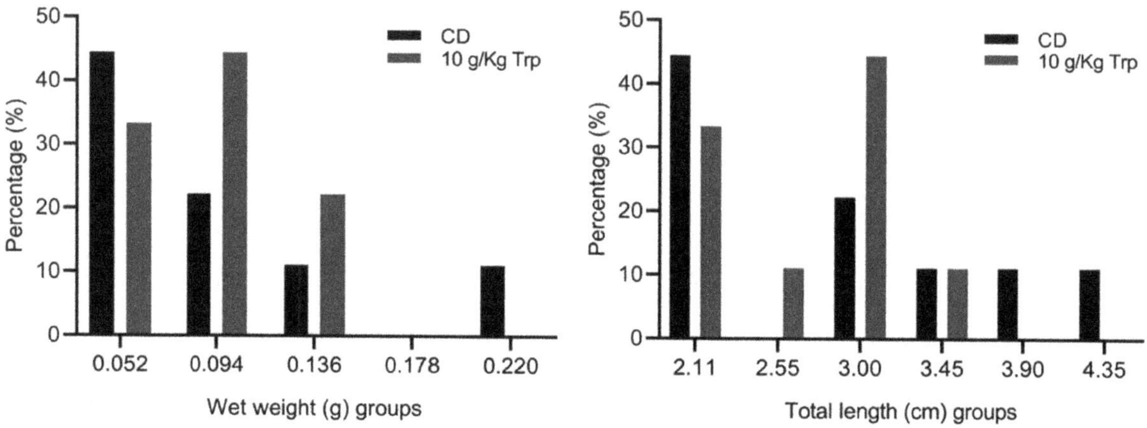


**Figure 7.** Wet weight (g) and total length (cm) class distribution by cannibalism effect in *A. tropicus* cannibal larvae fed 10 g/kg Trp and CD. Values are mean ± SD.

### *3.5. Digestive Enzyme Activity*

In the first stage, larvae from the CD treatment showed a greater activity in acid protease ($p < 0.05$) compared to those from the 10 and 20 g/kg Trp groups. Also, the CD treatment obtained a greater activity for alkaline protease, with a significant difference with all Trp treatments ($p < 0.05$). For trypsin, 20 g/kg Trp ($p < 0.05$) was statistically different compared to the CD treatment. The 10 and 20 g/kg Trp treatments showed the greatest activity for lipase compared to the other treatments ($p < 0.05$). Leucine aminopeptidase did not show any difference among treatments ($p > 0.05$).

When comparing the digestive enzyme activity between cannibals and non-cannibals, the acid protease activity was greater for cannibals and non-cannibals of the 10 g/kg Trp treatment ($p < 0.05$). CD cannibal and non-cannibal larvae presented a greater alkaline protease activity ($p < 0.05$). Trypsin enzymatic activity between cannibals and non-cannibals was higher in the cannibals of the CD treatment ($p < 0.05$). All treatments with Trp led to a greater leucine aminopeptidase activity than CD; however, no difference was recorded between cannibals and non-cannibals. The cannibal larvae from the 10 g/kg Trp treatment showed the greatest activity for lipase ($t$-test, $p < 0.05$).

In the second stage, the larvae treated with Trp recorded a greater activity in acidic and alkaline proteases and leucine aminopeptidase ($t$-test, $p < 0.05$). The larvae fed with CD recorded a higher trypsin activity ($t$-test, $p < 0.05$). No statistical differences were observed in lipase activity for CD and the 10 g/kg Trp treatment (Table 7).

**Table 7.** Digestive enzymatic activities (mean ± standard deviation, SD) of *A. tropicus* larvae fed with different concentrations of Trp and CD.

| | **Activities** | | **Trp (g/kg)** | | |
|---|---|---|---|---|---|
| | (u/mg Protein) | CD | 10 | 20 | 30 |
| First stage | Acid protease | 12.233 ±0.924 $^{a}$ | 2.505 ± 2.272 $^{b}$ | 3.546 ± 1.207 $^{b}$ | 7.645 ± 0.792 $^{a,b}$ |
| | Alkaline protease | 17.374 ± 1.550 $^{a}$ | 8.160 ± 2.257 $^{b}$ | 7.388 ± 0.852 $^{b}$ | 7.784 ± 0.859 $^{b}$ |
| | Trypsin | 1.334 ± 0.088 $^{a}$ | 0.682 ± 0.067 $^{a,b}$ | 0.640 ± 0.018 $^{b}$ | 0.645 ± 0.236 $^{a,b}$ |
| | Leucine aminopeptidase | 0.366 ± 0.014 | 0.290 ± 0.007 | 0.317 ± 0.057 | 0.337 ± 0.107 |
| | Lipase | 1.711 ± 0.654 $^{b}$ | 4.529 ± 0.318 $^{a}$ | 4.514 ± 0.099 $^{a}$ | 2.413 ± 0.079 $^{b}$ |

**Table 7.** *Cont.*

| | Activities | | Trp (g/kg) | | | |
|---|---|---|---|---|---|---|
| | | | CD | | 10 | |
| Second stage | Acid protease | | 20.492 ± 4.162 | | 35.869 ± 3.670 * (0.033) | |
| | Alkaline protease | | 7.043 ± 0.835 | | 14.395 ± 2.342 * (0.028) | |
| | Trypsin | | 0.755 ± 0.079 | | 1.032 ± 0.086 * (0.0151) | |
| | Leucine aminopeptidase | | 0.432 ± 0.017 | | 1.333 ± 0.203 * (0.0167) | |
| | Lipase | | 3.178 ± 0.640 | | 3.072 ± 0.400 | |
| | | | CD | 10 | 20 | 30 |
| Cannibals vs. Non-cannibals | Acid protease | C | 28.261 ± 0.028 [b,*] | 43.110 ± 0.045 [a,*] | 18.780 ± 0.102 [d] | 25.451 ± 0.017 [c,*] |
| | | N | 23.576 ± 0.062 [b] | 39.280 ± 0.049 [a] | 19.469 ± 0.031 [d,*] | 23.228 ± 0.020 [d] |
| | Alkaline protease | C | 15.887 ± 0.016 [a] | 10.241 ± 0.218 [c,*] | 14.263 ± 0.035 [b,*] | 9.963 ± 0.072 [c,*] |
| | | N | 18.021 ± 0.128 [a,*] | 9.722 ± 0.186 [b] | 8.393 ± 0.166 [c] | 9.762 ± 0.016 [b] |
| | Trypsin | C | 1.919 ± 0.318 [a,*] | 1.491 ± 0.143 [b,*] | 1.203 ± 0.082 [b,*] | 1.154 ± 0.144 [b,*] |
| | | N | 1.493 ± 0.232 [a] | 1.170 ± 0.078 [b] | 0.851 ± 0.087 [c] | 0.863 ± 0.069 [c] |
| | Leucine aminopeptidase | C | 0.284 ± 0.044 [c] | 0.527 ± 0.079 [a] | 0.684 ± 0.164 [a,b] | 0.382 ± 0.051 [b,c] |
| | | N | 0.184 ± 0.086 [c] | 0.494 ± 0.045 [a,b] | 0.514 ± 0.053 [a] | 0.379 ± 0.042 [b] |
| | Lipase | C | 0.900 ± 0.011 [b] | 3.288 ± 0.038 [a] | 0.481 ± 0.025 [c] | 0.588 ± 0.032 [c] |
| | | N | 3.324 ± 0.019 [b,*] | 3.889 ± 0.001 [a,*] | 1.759 ± 0.016 [c,*] | 1.502 ± 0.012 [d,*] |

Significant differences among the diets are indicated by different letters ($p < 0.05$). Significant differences between cannibals and non-cannibals are indicated by * ($p < 0.05$).

*3.6. Gene Expression*

When comparing cannibal and non-cannibal larvae, all cannibals in the 20 g/kg Trp treatment showed overexpression in all genes, followed by 30 g/kg Trp with overexpression only in *avpi1*, *crh*, *htr1a*, *sstr1*, *th*, and t *ph1* genes. In non-cannibal larvae, the 10 g/kg Trp treatment showed a greater expression, with statistical differences compared to the other treatments and with cannibal larvae from the same treatment (*avpi1*, *crh*, *hdc*, *htr1a*, *sstr1*, and *th*). *tph1* was overexpressed in all treatments and was significantly different compared to non-cannibals (Figure 8).

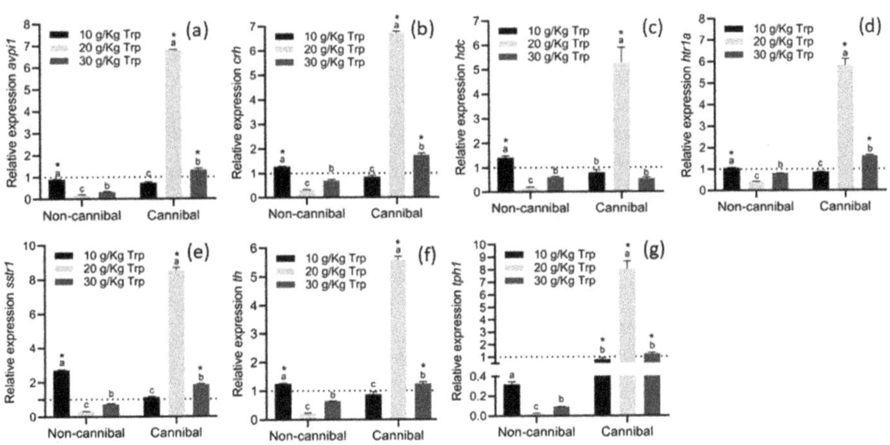


**Figure 8.** Relative gene expression of *avpi1* (**a**), *crh* (**b**), *hdc* (**c**), *htr1a* (**d**), *sstr1* (**e**), *th* (**f**), and *tph1* (**g**) in *A. tropicus* larvae fed with different Trp concentrations. Relative mRNA levels of the indicated genes were measured by RT-qPCR using β-actin as the reference gene. Data are presented as fold-changes in the mRNA levels, in comparison to the sample with CD (dotted line) (n = 3, mean ± SD). Significant differences with respect to the CD group (dotted line) are indicated by different letters ($p < 0.05$) Significant differences between the cannibal and non-cannibal larvae are indicated by * ($p < 0.05$).

In the second stage, larvae from the 10 g/kg group showed overexpression of *avpi1*, *crh*, and *htr1a* genes ($p < 0.05$) compared to those from the CD group. The same was observed for *gnrh* ($p > 0.05$). A subexpression was observed in the *tph1, th, sstr1*, and *hdc* genes ($p < 0.05$) of larvae fed with 10 g/kg Trp (Figure 9).

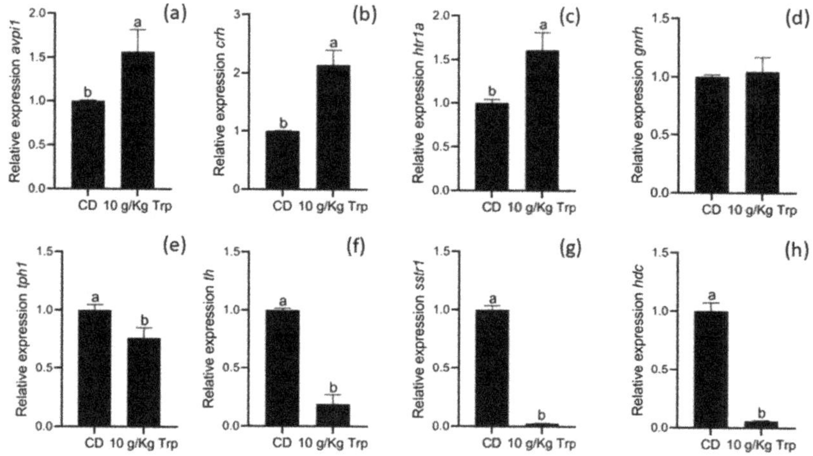

**Figure 9.** Relative gene expression of *avpi1* (**a**), *crh* (**b**), *htr1a* (**c**), *gnrh* (**d**), *tph1* (**e**), *th* (**f**) *sstr1* (**g**), and *hdc* (**h**) in *A. tropicus* cannibal larvae fed with 10 g/kg Trp. Relative mRNA levels of the indicated genes were measured by RT-qPCR using β-actin as the reference gene. Data are presented as fold-changes in the mRNA levels, in comparison to the sample with CD (n = 3, mean ± SD). Significant differences with respect to the CD group are indicated by different letters ($p < 0.05$).

### 3.7. Behavior

In total, 12 videos were obtained (185.8 min), and a total of 12 events were recorded, of which 9 were defensive behaviors (escapes), where there was no contact between fish. The remaining three events were direct attacks (with contact). The organisms with rocks as refuge were present in three events, followed by without refuge in six events; finally, only three events were recorded for the organisms with vegetation as refuge. Regarding the effect of diet, larvae fed with 10 g/kg Trp displayed 25% of aggressive behaviors and those fed with CD 75%. Regarding the number of attacks, cannibals fed with 10 g/kg Trp did not display aggressive behaviors, unlike those fed with CD. The use of artificial vegetation also led to zero aggressive behaviors in both treatments (Table 8).

**Table 8.** Attack types and cannibalism behavior of *A. tropicus* larvae fed with 10 g/kg Trp and CD in combination with different shelters.

| | | Lateral Attack | | | Frontal Attack | | |
|---|---|---|---|---|---|---|---|
| | **Escape** | **Head** | **Middle** | **Tail** | **Head** | **Middle** | **Total** |
| | | | Without shelter | | | | |
| CD | 2 | 2 | 0 | 0 | 0 | 0 | 4 |
| 10 g/kg Trp | 2 | 0 | 0 | 0 | 0 | 0 | 2 |
| Total | 4 | 2 | 0 | 0 | 0 | 0 | 6 |
| | | | Rocks | | | | |
| CD | 2 | 0 | 1 | 0 | 0 | 0 | 3 |
| 10 g/kg Trp | 0 | 0 | 0 | 0 | 0 | 0 | 0 |
| Total | 2 | 0 | 1 | 0 | 0 | 0 | 3 |

**Table 8.** *Cont.*

| | Escape | Lateral Attack Head | Lateral Attack Middle | Lateral Attack Tail | Frontal Attack Head | Frontal Attack Middle | Total |
|---|---|---|---|---|---|---|---|
| | | | Artificial Vegetation | | | | |
| CD | 2 | 0 | 0 | 0 | 0 | 0 | 2 |
| 10 g/kg Trp | 1 | 0 | 0 | 0 | 0 | 0 | 1 |
| Total | 3 | 0 | 0 | 0 | 0 | 0 | 3 |
| SUMA total | 9 | 2 | 1 | 0 | 0 | 0 | 12 |

## 4. Discussion

### 4.1. Growth

Trp is an essential amino acid (AA) in fish; due to this, it is necessary to know the requirement in important species (*Salmo gairdneri*, *Oncorhynchus mykiss*, and *Rhamdia quelen*). This value ranges from around 0.1% to 0.5% [33]. In this sense, the formulation of the CD diet used contained 1.06% of Trp according to the total amino acid analysis carried out. The relationship of Trp functions in the development of fish larvae has been well studied; it has been demonstrated that a low Trp administration is related to the presence of deformities (scoliosis, lordosis, and eye cataracts) [34]. Furthermore, its administration has been related to the modulation of aggressiveness, feed intake, immune system, stress, oxidative damage, and feed efficiency ratio [35–38]. Administering a high concentration of Trp can activate the kynurenine pathway, thus increasing kynurenic acid concentrations, which have been linked to negative physiological effects, such as increased stress and a decreased immune system [39]. In coho salmon (*Oncorhynchus kisutch*), exposure to 50 µL/mg of Trp compared to 5 µL/mg for 3 h significantly increases the expression of kynurenine aminotransferase 2 (*KIAT 2*) [40]. Therefore, the administration of its optimal level must be sought for each species.

The effect of Trp administration on growth has been reported for Asian seabass juveniles (*Lates calcarifer*), in which Trp reduced growth and feed intake, although it increased serotonin levels in the brain [41]. In grouper juveniles (*Epinephelus coioides*), administering 0.25%, 0.5%, and 1% of tryptophan in the diet resulted in a lower height and weight than those in the control [37]. This same effect was observed in pikeperch (*Sander lucioperca*) larvae: using 5, 10, and 20 g Trp per kg, the weight and height were lower than those in the control treatment [42]; and in Pabda (*Ompok bimaculatus*) fry, the fish growth with Trp (2, 4, and 6 ppm) was lower than that in the control [17]. A similar result but only with high levels of Trp was found in Indian major carp (*Cirrhinus mrigala*), and in Indian catfish (*Heteropneustes fossilis*), a high concentration of Trp in the diet decreased growth [43,44]. The *A. tropicus* larvae did not show a significant difference in growth in the first stage of this study, only a slight tendency of improved growth in weight and size compared to those fed with CD. However, in the second stage, the cannibal larvae fed with Trp (10 g/kg) had a lower growth and final size than the CD larvae. This difference in growth between the two stages of the study can be attributed to cannibalism, due to the "jumper" effect. This effect consists of a rapid gain in weight and size when a cannibal larva consumes another larva [45]. This can lead to an increase in the differences between sizes and weights. This statement was proved in our study, in which the values of CV and SH were higher in the CD group, proving that the high cannibalism of this treatment (76.66 ± 5.77%) results in a low survival (23.33 ± 5.77%) and generating a great heterogeneity in size and weight of the fish fed with CD, unlike the larvae fed with Trp (cannibalism: 20.0 ± 10.0%; survival: 75.0 ± 7.07%). Similar results have been reported in Asian seabass larvae (*Lates calcarifer*), where the coefficient of size variation (%) and size heterogeneity decreased with the increase in the level of Trp supplementation [46]. This heterogeneity in size may also be a trigger for cannibalism, for example, in the Atlantic cod (*Gadus morhua*), African catfish (*Heterobranchus longifili*), giant grouper (*Epinephelus lanceolatus*), and black rockfish (*Sebastes schlegelii*) [47–50].

### 4.2. Cannibalism

Although a decreasing trend in cannibalism due to the administration of Trp was seen in the first stage, it was in the second where it was significantly clearer. These results coincide with those for Asian seabass (*Lates calcarifer*) fry, which when using 0.5, 1.0, 1.5, and 2% of Trp displayed a decreased cannibalism and increased survival compared to the control, with the lowest percentage (0.5) being the best treatment [46]. In grouper (*Epinephelus coioides*) juveniles, the use of 0.25%, 0.5%, and 1% of tryptophan in the diet decreased cannibalism, increasing the serotonin concentration (5-HT) in the brain [37]. In pikeperch (*Sander lucioperca*) larvae fed with 5, 10, and 20 g Trp per kilogram, the levels of 5-HT in fish tissue increased and cannibalism levels decreased [42]. Also, in Pabda (*Ompok bimaculatus*) fry, Trp (2, 4, and 6 ppm) decreased cannibalism and increased survival compared to the control treatment [17]. In Atlantic cod (*Gadus morhua*), Trp (28 g/kg) supplemented to juveniles decreased aggressivity [51]. In rainbow trout (*Oncorhynchus mykiss*), the increase in the concentration of Trp in the plasma and brain decreased the aggressive behavior in dominant fish [52,53], which is related to the action of the 5-HT neurotransmitter. Serotonin (5-HT) is a neurotransmitter that has been related to behaviors such as aggression, reaction to stress, feeding, maturation, and sexual behavior [54].

The synthesis of 5-HT takes place in serotonergic neurons, where Trp serves as a precursor. The enzyme tryptophan hydrolase hydroxylates tryptophan into L-5-hydroxytryptophan, which is subsequently decarboxylated by the enzyme L-amino acid decarboxylase, generating 5-hydroxytryptamine (5-HT). Continuing the process, 5-HT is degraded by the enzyme monoamine oxidase, transforming it into 5-hydroxyindole acetaldehyde. At the end of the reaction, the enzyme aldehyde dehydrogenase produces 5-hydroxyindoleacetic acid (5-HIAA) [54–56]. The bioavailability of Trp in the brain of organisms is important as it allows 5-HT synthesis to take place. In this sense, Trp competes with other amino acids (AAs) (valine, isoleucine, leucine, tyrosine, phenylalanine, and methionine) to enter the brain of organisms, which makes the balance between Trp, the other AAs, and carbohydrates important (since carbohydrates promote the uptake of AAs, except for Trp) and thus generates adequate concentrations of AAs in the plasma, allowing the flow of Trp to the brain [55] and thereby the synthesis of 5-HT. The use of Trp as a mitigator of cannibalism in fish is based on the principle of increasing the bioavailability of this amino acid, which functions as an essential substrate for the synthesis of 5-HT and, in turn, reduces aggressiveness in fish, promoting the reduction in cannibalism. On the other hand, the differences between the total weight and size of a cannibal fish and its prey were 16.39 ± 10.864% and 15.23 ± 5.68%, respectively, similar percentages to those reported by Sepúlveda-Quiroz et al. [57].

### 4.3. Digestive Enzymes

The interaction between Trp and digestive enzymes has been studied in several works; however, this interaction related to cannibalism has not yet been addressed. For *A. tropicus* larvae, it has been reported that the functionality of its digestive system and the differentiation of its organs are completely developed at 9 DAH [58] and, in addition to their anatomy (mouth width and depth, lengths of upper and lower jaw, and mouth depth angle), allow them to capture and ingest their own conspecifics (intracohort cannibalism) [57]. With the administration of Trp, the enzymes acid and alkaline proteases, trypsin, and leucine aminopeptidase registered a lower activity than with that of CD, except for lipases. It has been shown in vitro that Trp can be an activator of amylase, lipase, and trypsin [59]. An increase in the enzymatic activity is reflected in a greater hydrolysis of macronutrients, releasing a greater number of micronutrients [60], which are used by the body, generating a greater absorption of these microelements, resulting in optimal growth in fish [61]. In our study, cannibal and non-cannibal larvae presented significant differences in enzymatic activity; trypsin activity was higher in all cannibals, contrasting with the activity of lipase enzymes, which were more expressed in non-cannibal larvae. The function of trypsin is to hydrolyze proteins by breaking peptide bonds [62]. In turn, lipases promote the digestion of lipids, participating in the denaturation of triacylglycerol to diacylglycerol, subsequently converting it to monoacylglycerol [63].

In the second stage, cannibal larvae fed with Trp (10 g/kg) showed a greater enzymatic activity (acid and alkaline proteases and leucine aminopeptidase) than the CD cannibal larvae. The activity of trypsin was greater in the cannibal larvae of the CD treatment. In juveniles of Jian carp (*Cyprinus carpio*), when fed with a concentration of 3.8 g/kg of Trp in the diet, the organisms increased their growth and obtained a greater activity of digestive enzymes (trypsin, lipase, and α-amylase) and brush border enzymes and an increase in the height of the intestinal folds [64]. On the other hand, Trp has been considered as a component that can improve digestive enzymatic activity through two components, melatonin (a Trp metabolite) and cholecystokinin (a regulatory hormone), both of which act on the secretion of pancreatic enzymes [65,66], such as trypsin, chymotrypsin, lipase, and amylase [67]. In the case of the administration of Trp in juvenile silver catfish (*Rhamdia quelen*), there was no difference in trypsin and chymotrypsin; however, by performing a polynomial regression, an increase in acid protease activity was identified with respect to the increase in Trp (1–3.1 g/kg) [68]. Changes in the physiology of two salmon species (*Salmo salar* and *Oncorhynchus kisutch*) have been observed following Trp inclusion, with modifications in digestive enzymatic activity, increased 5-HT concentrations, and decreased cortisol levels [69,70].

### *4.4. Gene Expression*

The gene expression results show a significant difference between the cannibal and non-cannibal larvae, where cannibals exposed to a high Trp treatment have a greater overexpression. Trp functions as a substrate in the synthesis of 5-HT, a neurotransmitter that regulates aggressive behavior, among other aspects [54,56]. In zebra fish (*Danio rerio*), seven neurological pathways were identified (hypothalamo-neurohypophysial system (HNS), serotonin (5-HT), somatostatin, dopamine, hypothalamo-pituitary-interrenal (HPI), hypothalamo-pituitary-gonadal (HPG), and histamine) in relation to the expressed genes for aggressiveness [28]. Among the genes that participate in these metabolic pathways are those used in this study (*sstr1*, *th*, *hdc*, *crh*, *htr1a*, *gnrh1*, *avpi1*, and *tph1*) and whose relation to aggressive behavior has already been reported in humans, mice, and fish [29].

In the second stage, the cannibal larvae of *A. tropicus* fed with Trp showed an overexpression of *avpi1* (HNS), *crh* (HPI), and *htr1a* (5-HT), as well as a subexpression of *tph1*(5-HT), *th* (dopamine), *sstr1* (somatostatin), and *hdc* (histamine) with respect to the cannibal larvae fed with CD. The expression of *avpil* (arginine vasopressin-like) is related to behaviors such as aggression and social interactions; in particular, overexpression has been detected in dominant male zebra fish (*Danio rerio*) [28,71]. On the other hand, the HPI pathway is responsible for different processes, such as stress response [39]; in this sense, the corticotropin-releasing hormone (CRH, or corticotropin-releasing factor (CRF)) works as an activator of this pathway [72]. In rainbow trout (*Oncorhynchus mykiss*), the use of CRF through injections decreased the number of attacks, increased locomotion and head movements, and increased the concentrations of serotonin, 5-HIAA, and dopamine [73,74]. When comparing dominant and subordinate males of *Astatotilapia burtoni*, the relative abundance of mRNA of 5-HT transporters, such as *htr1a* and *htr2a*, in the telencephalon was higher in subordinate than in dominant males, and was related to an increase in 5-HT production [75]. In this sense, the use of a specific agonist (8-OH-DPAT) in HTR1A receptors decreases aggression in fighting fish (*Betta splendens*) [76]. In male zebra fish (*Danio rerio*), a differentiation between dominants and subordinates was identified with respect to the genes involved in sexual behavior (*cyp19a1b*, *cyp17*, *hsd11b2*, *hsd17b3*, and *ar*) and aggressiveness (*avplr1b*, *tph1b*, *htr1a*, *sst1*, *sstr1*, *th*, and *slc6a3*), with the dominant males being the ones with the highest expression [77]. Although the social components have not yet been studied in *A. tropicus* larvae, the results obtained indicate that the HNS, 5-HT, and HPI pathways were modified using Trp, and thus a decrease in cannibalism was observed.

### 4.5. Behavior

The effect of Trp administration can be seen in the absence of aggressive behavior in the larvae of *A. tropicus* under the influence of refuge. Regardless of the presence or absence of any type of shelter (rocks and artificial vegetation), the larvae fed with Trp (10 g/kg) did not attack, unlike the larvae fed with CD. Of the three scenarios analyzed, the artificial vegetation did not lead to attacks among the larvae fed with CD, as reported by Sepúlveda-Quiroz et al. [57]. The use of enriched environments attempts to replicate the conditions of the natural environment in the culture ponds, improving stress levels by reducing aggression, cannibalism, energy expenditure, injuries, and diseases [78–81].

## 5. Conclusions

The use of 10 g/kg of Trp in the diet of *A. tropicus* larvae reduces cannibalism, improving survival, specifically proven in larvae with cannibalistic behavior. Both cannibal and non-cannibal larvae showed a difference in digestive enzyme activity and expression of aggressiveness genes. The inclusion of tryptophan generates the activation of the HNS, 5-HT, and HPI pathways, demonstrated by the overexpression of the *avpi1*, *crh*, and *htr1a* genes. Furthermore, cannibal behaviors did not occur with the use of Trp, regardless of the type of shelter, although artificial vegetation was better than other shelters. It is recommended that studies continue to focus on explaining other effects of Trp in *A. tropicus*.

**Author Contributions:** C.A.S.-Q., C.S.A.-V. and R.M.-G. designed and planned the bioassays; C.A.S.-Q G.M.P.-J. and G.G.A.-A. conducted the experiment and sample collection; C.A.S.-Q., O.M.-P., L.D.J.-M. and G.M.P.-J. carried out the biochemical analyses; C.A.S.-Q., L.D.J.-M. and C.S.A.-V. conducted the molecular analyses; C.A.S.-Q., C.A.A.-G., M.A.G.-E. and D.T.-R. performed the statistical analysis; C.A.S.-Q., C.S.A.-V., C.A.A.-G. and R.M.-G. wrote the paper. All authors have read and agreed to the published version of the manuscript.

**Funding:** This work was supported by the National Council of Science and Technology of Mexico "Estudio de la fisiología digestiva en larvas y juveniles de pejelagarto (*Atractosteus tropicus*) con base en técnicas histológicas, bioquímicas y moleculares" CB-2016-01-282765. And for the thesis fellowship for Doctoral graduate studies (CVU 858306).

**Institutional Review Board Statement:** The study was conducted in accordance with the Declaration of Helsinki, and the protocol was approved by Secretaría de Agricultura, Ganadería, Desarrollo Rural, Pesca y Alimentación (NOM-062-ZOO-1999, 2001) [20].

**Data Availability Statement:** The data presented in this study are available upon request from the corresponding authors.

**Acknowledgments:** The authors thanks the technicians and lab members from DACBiol, DAMJM, UABC and CIBNOR for technical assistance.

**Conflicts of Interest:** The authors have no conflicts of interest to declare that are relevant to the content of this article.

## References

1. Pereira, L.S.; Agostinho, A.A.; Winemiller, K.O. Revisiting cannibalism in fishes. *Rev. Fish Biol. Fish.* **2017**, *27*, 499–513. [CrossRef]
2. Aguilera, C.; Mendoza, R.; Iracheta, I.; Márquez, G. Digestive enzymatic activity on Tropical gar (*Atractosteus tropicus*) larvae fed different diets. *Fish Physiol. Biochem.* **2012**, *38*, 679–691. [CrossRef] [PubMed]
3. Márquez, C.G. Biología y tecnología para el cultivo del pejelagarto Atractosteus tropicus en el sureste de México. In *Redes nacionales de Investigación en Acuicultura. Memorias de la IV Reunión. Secretaría de Medio Ambiente Recursos Naturales y Pesca (SEMARNAP)*; Alvarez, P., Guzman, M., Contreras, S., Silva, A., Eds.; Instituto Nacional de la Pesca: México City, Mexico, 2000; pp. 265–267.
4. Márquez-Couturier, G.; Vázquez-Navarrete, C.J. Estado del arte de la biología y cultivo de pejelagarto (*Atractosteus tropicus*). *Agroproductividad* **2015**, *8*, 44–51.
5. Nelson, J.S. *Fishes of the World*, 4th ed.; John Wiley & Sons: Hoboken, NJ, USA, 2006; 601p, ISBN 978-0471250319.
6. Frías-Quintana, C.; Álvarez-González, C.; Tovar-Ramírez, D.; Martínez-García, R.; Camarillo-Coop, S.; Peña, E.; Galaviz, M. Use of potato starch in diets of tropical gar (*Atractosteus tropicus*, Gill 1863) larvae. *Fishes* **2017**, *2*, 3. [CrossRef]

7. Palma-Cancino, D.J.; Martínez-García, R.; Álvarez-González, C.A.; Camarillo-Coop, S.; Peña-Marín, E.S. Evaluation of feeding strategies in tropical gar (*Atractosteus tropicus* Gill) larvae: Growth, survival and cannibalism. *Ecosist. Recur. Agropecu.* **2019**, *6*, 273–281. [CrossRef]
8. Jiménez-Martínez, L.D.; Tovar-Ramírez, D.; Álvarez-González, C.A.; Peña-Marín, E.; Camarillo-Coop, S.; Martínez-García, R.; Palacios-Mechetnov, E.; Martínez-Yáñez, R.; Concha-Frias, B. Assessment of dietary lipid sources in tropical gar, *Atractosteus tropicus* larvae: Growth parameters and intermediary lipogenic gene expression. *Aquac. Res.* **2020**, *51*, 2629–2640. [CrossRef]
9. Leathwood, P.D. Tryptophan availability and serotonin synthesis. *Proc. Nutr. Soc.* **1987**, *46*, 143–156. [CrossRef]
10. Smith, T.K.; Seddon, I.R. Synergism demonstrated between *Fusarium mycotoxins. Feedstuffs* **1998**, *70*, 12–17.
11. Akiyama, T.; Shiraaishi, M.; Yamamoto, T.; Unuma, T. Effect of dietary tryptophan on maturation of ayu *Plecoglossus altivelis. Fish. Sci.* **1996**, *62*, 776–782. [CrossRef]
12. Comai, S.; Bertazzo, A.; Brughera, M.; Crotti, S. Tryptophan in health and disease. *Adv. Clin. Chem.* **2020**, *95*, 165–218. [CrossRef]
13. Maler, L.; Ellis, W.G. Inter-male aggressive signals in weakly electric fish are modulated by monoamines. *Behav. Brain Res.* **1987**, *25*, 75–81. [CrossRef] [PubMed]
14. Munro, A.D. Effects of melatonin, serotonin, and naloxone on aggression in isolated cichlid fish (*Aequiidens pulcher*). *J. Pineal Res.* **1986**, *3*, 257–262. [CrossRef] [PubMed]
15. Adams, C.F.; Liley, N.R.; Gorzalka, B.B. PCPA increases aggression in male firemouth Cichlids. *Pharmacology* **1996**, *55*, 328–330. [CrossRef] [PubMed]
16. Biswas, P.; Rawat, P.; Patel, A.B.; Jena, A. Dietary supplementation of L-tryptophan: Effect on growth and survival of Pabda, *Ompok bimaculatus* (Bloch) fry. *J. Appl. Aquac.* **2018**, *31*, 322–336. [CrossRef]
17. Biswas, P.; Rawat, P.; Jena, A.K.; Patel, A.B.; Pandey, P.K. Effect of L-tryptophan on growth and survival of Pabda Fry, *Ompok bimaculatus* (Bloch, 1794). *Fish. Technol. Soc. Fish. Technol.* **2019**, *56*, 29–33.
18. Álvarez-González, C.A.; Civera-Cerecedo, R.; Ortiz-Galindo, J.L.; Dumas, S.; Moreno-Legorreta, M.; Grayeb-Del, A.T. Effect of dietary protein level on growth and body composition of juvenile spotted sand bass, *Paralabrax maculatofasciatus*, fed practical diets. *Aquaculture* **2001**, *194*, 151–159. [CrossRef]
19. Association of Official Analytical Chemists (AOAC). *Official Methods of Analysis*, 17th ed.; Association of Official Analytical Chemists: Arlington, VA, USA, 2000.
20. NOM-062-ZOO-1999, 2001. Norma Oficial Mexicana: Especificaciones técnicas para la producción, cuidado y uso de los animales de laboratorio. Available online: https://www.gob.mx/senasica/documentos/nom-062-zoo-1999 (accessed on 7 June 2022).
21. Shuangyao, W.; Jiang, Z.; Mingguang, M.; Shoukang, M.; Yang, S.; Youzhen, S. Effects of seawater pH on survival, growth, energy budget and oxidative stress parameters of juvenile turbot scophthalmus maximus. *Iran. J. Fish. Sci.* **2018**, *17*, 675–689. [CrossRef]
22. Bradford, M.M. A rapid and sensitive method for the quantization of microgram quantities of protein utilizing the principle of protein-dye binding. *Anal. Biochem.* **1976**, *72*, 248–254. [CrossRef]
23. Sarath, G.; de La Motte, R.S.; Wagner, F.W. Protease assay methods. In *Proteolytic Enzymes: A Practical Approach*; Beynon, R.J., Bonde, J.S., Eds.; Oxford University Press: Oxford, UK, 1989; pp. 25–54.
24. Erlanger, B.F.; Kokowsky, N.; Cohen, W. The preparation and properties of two new chromogenic substrates of trypsin. *Arch. Biochem. Biophys.* **1961**, *95*, 271–278. [CrossRef]
25. Maroux, S.; Louvard, D.; Barath, J. The aminopeptidase from hog intestinal brush border. *Biochim. Biophys Acta Enzymol.* **1973**, *321*, 282–295. [CrossRef]
26. Versaw, W.; Cuppett, S.L.; Winters, D.D.; Williams, L.E. An improved colorimetric assay for bacterial lipase in nonfat dry milk. *J. Food Sci.* **1989**, *54*, 232–254. [CrossRef]
27. Martínez-Burguete, T.; Peña-Marin, E.S.; García-Gasca, A.; Alvarez-González, C.A.; Llera-Herrera, R. Nutrigenomic marker discovery by de novo transcriptomic sequencing during early development of the tropical gar (*Atractosteus tropicus*). *Aquac. Res.* **2021**, *52*, 3829–3842. [CrossRef]
28. Filby, A.L.; Paull, G.C.; Hickmore, T.F.A.; Tyler, C.R. Unravelling the neurophysiological basis of aggression in a fish model. *BMC Genom.* **2010**, *11*, 498. [CrossRef]
29. de Abreu, M.S.; Giacomini, A.C.; Genario, R.; dos Santos, B.E.; da Rosa, L.G.; Demin, K.A.; Wappler-Guzzetta, E.A.; Kalueff, A.V. Neuropharmacology, pharmacogenetics and pharmacogenomics of aggression: The zebrafish model. *Pharmacol. Res.* **2019**, *141*, 602–608. [CrossRef] [PubMed]
30. Jiménez-Martínez, L.D.; Morales-Garcia, V.; Frias-Quintana, C.A.; Castillo-Collado, A.C.; Asencio-Alcudia, G.G.; Alvarez-Villagomez, C.S.; Peña-Marín, E.S.; Concha-Frias, B.; Alvarez-Gonzalez, C.A. Quality evaluation of reference gene expression on different tissues in adults of tropical gar *Atractosteus tropicus*. *Pak. J. Zool.* **2021**, *54*, 363–372. [CrossRef]
31. Livak, K.J.; Schmittgen, T.D. Analysis of relative gene expression data using realtime quantitative PCR and the $2^{-\Delta\Delta C(T)}$ method. *Methods* **2001**, *25*, 402–408. [CrossRef] [PubMed]
32. Friard, O.; Gamba, M. BORIS: A free, versatile open-source event-logging software for video/audio coding and live observations. *Methods Ecol. Evol.* **2016**, *7*, 1325–1330. [CrossRef]
33. Mai, K.; Xue, M.; He, G.; Xie, S.; Kaushik, S.J. Protein and amino acids. In *Fish Nutrition*; Academic Press: Cambridge, MA, USA, 2022; pp. 181–302.
34. Walton, M.J.; Coloso, R.M.; Cowey, C.B.; Adron, J.W.; Knox, D. The effects of dietary tryptophan levels on growth and metabolism of rainbow trout (*Salmo gairdneri*). *Br. J. Nutr.* **1984**, *51*, 279. [CrossRef]

35. Coloso, R.M.; Murillo-Gurrea, D.P.; Borlongan, I.G.; Catacutan, M.R. Tryptophan requirement of juvenile Asian sea bass *Lates calcarifer*. *J. Appl. Ichthyol.* **2004**, *20*, 43–47. [CrossRef]
36. Hoseini, S.M.; Mirghaed, A.T.; Mazandarani, M.; Zoheiri, F. Serum cortisol, glucose, thyroid hormones' and non-specific immune responses of Persian sturgeon, *Acipenser persicus* to exogenous tryptophan and acute stress. *Aquaculture* **2016**, *462*, 17–23. [CrossRef]
37. Hseu, J.; Lu, F.; Su, H.; Wang, L.; Tsai, C.; Hwang, P. Effect of exogenous tryptophan on cannibalism, survival and growth in juvenile grouper, *Epinephelus coioides*. *Aquaculture* **2003**, *218*, 251–263. [CrossRef]
38. Jiang, W.D.; Wen, H.L.; Liu, Y.; Jiang, J.; Kuang, S.Y.; Wu, P.; Zhao, J.; Tang, L.; Tang, W.N.; Zhang, Y.A.; et al. The tight junction protein transcript abundance changes and oxidative damage by tryptophan deficiency or excess are related to the modulation of the signalling molecules, NF-κB p65, TOR, caspase-(3,8,9) and Nrf2 mRNA levels, in the gill of young grass carp (*Ctenopharyngodon idellus*). *Fish Shellfish Immunol.* **2015**, *46*, 168–180. [CrossRef] [PubMed]
39. Wendelaar-Bonga, S.E. The stress response in fish. *Physiol. Rev.* **1997**, *77*, 591–625. [CrossRef] [PubMed]
40. Vargas-Chacoff, L.; Nualart, D.; Vargas-Lagos, C.; Dann, F.; Muñoz, J.L.P.; Pontigo, J.P. Tryptophan and cortisol modulate the kynurenine and serotonin transcriptional pathway in the kidney of *Oncorhynchus kisutch*. *Animals* **2023**, *13*, 3562. [CrossRef] [PubMed]
41. Chi, C.C.; Shiu, Y.L.; Lin, H.L.; Liu, C.H. Immune response and disease resistance of barramundi, *Lates calcarifer* (Bloch), offered diets supplemented with replete levels of tryptophan or whey. *J. World Aquac. Soc.* **2017**, *49*, 127–140. [CrossRef]
42. Król, J.; Zakęś, Z. Effect of dietary l-tryptophan on cannibalism, survival and growth in pikeperch *Sander lucioperca* (L.) post-larvae. *Aquac. Int.* **2015**, *24*, 441–451. [CrossRef]
43. Ahmed, I. Dietary amino acid l-tryptophan requirement of fingerling Indian catfish, *Heteropneustes fossilis* (Bloch), estimated by growth and haemato-biochemical parameters. *Fish Physiol. Biochem.* **2012**, *38*, 1195–1209. [CrossRef] [PubMed]
44. Ahmed, I.; Khan, M.A. Dietary tryptophan requirement of fingerling Indian major carp, *Cirrhinus mrigala* (Hamilton). *Aquac. Res.* **2005**, *36*, 687–695. [CrossRef]
45. Baras, E.; Silva Del Aguila, D.V.; Montalvan Naranjos, G.V.; Dugué, R.; Koo, F.C.; Duponchelle, F.; Renno, J.F.; Garcia-Dávila, C.; Nuñez, J. How many meals a day to minimize cannibalism when rearing larvae of the Amazonian catfish *Pseudoplatystoma punctifer*? The cannibal's point of view. *Aquat. Living Resour.* **2011**, *24*, 379–390. [CrossRef]
46. Kumar, P.; Kailasam, M.; Sethi, S.N.; Sukumaran, K.; Biswas, G.; Subburaj, R.; Thiagarajan, G.; Ghoshal, T.; Vijayan, K. Effect of dietary L-tryptophan on cannibalism, growth and survival of Asian seabass, *Lates calcarifer* (Bloch, 1790) fry. *Indian J. Fish.* **2017**, *64*, 28–32. [CrossRef]
47. Baras, E.; Dugué, R.; Legendre, M. Do cannibalistic fish forage optimally? An experimental study of prey size preference, bioenergetics of cannibalism and their ontogenetic variations in the African catfish *Heterobranchus longifilis*. *Aquat. Living Resour.* **2014**, *27*, 51–62. [CrossRef]
48. Folkvord, A. Ontogeny of cannibalism in larval and juvenile fishes with special emphasis on Atlantic cod. In *Early Life History and Recruitment in Fish Populations*; Chambers, R.C., Edward, A., Trippel, E.A., Eds.; Chapman & Hall: London, UK, 1992. [CrossRef]
49. Hseu, J.R.; Hwang, P.P.; Ting, Y.Y. Morphometric model and laboratory analysis of intracohort cannibalism in giant grouper *Epinephelus lanceolatus* fry. *Fish. Sci.* **2004**, *70*, 482–486. [CrossRef]
50. Xi, D.; Zhang, X.; Lü, H.; Zhang, Z. Cannibalism in juvenile black rockfish, *Sebastes schlegelii* (Hilgendorf, 1880), reared under controlled conditions. *Aquaculture* **2017**, *479*, 682–689. [CrossRef]
51. Höglund, E.; Bakke, M.J.; ØVerli, Y.; Winberg, S.; Nilsson, G.E. Suppression of aggressive behaviour in juvenile Atlantic cod (*Gadus morhua*) by l-tryptophan supplementation. *Aquaculture* **2005**, *249*, 525–531. [CrossRef]
52. Lepage, O.; Vílchez, I.M.; Pottinger, T.G.; Winberg, S. Time-course of the effect of dietaryl-tryptophan on plasma cortisol levels in rainbow trout *Oncorhynchus mykiss*. *J. Exp. Biol.* **2003**, *206*, 3589–3599. [CrossRef]
53. Lepage, O.; Larson, E.T.; Mayer, I.; Winberg, S. Serotonin, but not melatonin, plays a role in shaping dominant–subordinate relationships and aggression in rainbow trout. *Horm. Behav.* **2005**, *48*, 233–242. [CrossRef] [PubMed]
54. Sahu, S.; Ngasotter, S.; Mog, M.; Tesia, S.; Sharma, S.; Dayakar, B.; Waikhom, D. A review on physiological, behavioral and metabolic role of dietary tryptophan in fish. *Int. J. Chem. Stud.* **2020**, *8*, 2411–2417. [CrossRef]
55. Höglund, E.; ØVerli, Y.; Winberg, S. Tryptophan metabolic pathways and brain serotonergic activity: A comparative review. *Front. Endocrinol.* **2019**, *10*, 158. [CrossRef]
56. Winberg, S.; Nilsson, G.E. Roles of brain monoamine neurotransmitters in agonistic behaviour and stress reactions, with particular reference to fish. *Comp. Biochem. Physiol. Part C Toxicol. Pharmacol.* **1993**, *106*, 597–614. [CrossRef]
57. Sepúlveda-Quiroz, C.; Alvarez-Villagomez, C.; Mendoza-Porras, O.; Peña-Marín, E.; Maytorena-Verdugo, C.; Pérez-Jiménez, G.; Jesus-Contreras, R.; Álvarez-González, C.; Martínez-García, R. Attack behavior leading cannibalism in tropical gar (*Atractosteus tropicus*) larvae under different tank colors and shelter type. *Aquaculture* **2023**, *563*, 738991. [CrossRef]
58. Frías-Quintana, C.A.; Márquez-Couturier, G.; Álvarez-González, C.A.; Tovar-Ramírez, D.; Nolasco-Soria, H.; Galaviz-Espinosa, M.A.; Gisbert, E. Development of digestive tract and enzyme activities during the early ontogeny of the tropical gar *Atractosteus tropicus*. *Fish Physiol. Biochem.* **2015**, *41*, 1075–1091. [CrossRef] [PubMed]
59. Svatos, A. The effect of amino acids, sugars and fats on activation of pancreatic enzymes. *Cas. Lek. Cesk.* **1994**, *133*, 140–143. [PubMed]
60. Bone, Q.; Moores, R.H. *Biology of Fishes*, 3rd ed.; Taylor & Francis: London, UK, 2008. [CrossRef]

61. Hakim, Y.; Uni, Z.; Hulata, G.; Harpaz, S. Relationship between intestinal brush border enzymatic activity and growth rate in tilapias fed diets containing 30% or 48% protein. *Aquaculture* **2006**, *257*, 420–428. [CrossRef]
62. Moyano, F.J.; Díaz, M.; Alarcón, F.J.; Sarasquete, M.C. Characterization of digestive enzyme activity during larval development of gilthead seabream (*Sparus aurata*). *Fish Physiol. Biochem.* **1996**, *15*, 121–130. [CrossRef] [PubMed]
63. Cahu, C.; Zambonino Infante, J. Substitution of live food by formulated diets in marine fish larvae. *Aquaculture* **2001**, *200*, 161–180. [CrossRef]
64. Tang, L.; Feng, L.; Sun, C.Y.; Chen, G.F.; Jiang, W.D.; Hu, K.; Liu, Y.; Jiang, J.; Li, S.H.; Kuang, S.Y.; et al. Effect of tryptophan on growth, intestinal enzyme activities and TOR gene expression in juvenile Jian carp (*Cyprinus carpio* var. Jian): Studies in vivo and in vitro. *Aquaculture* **2013**, *412–413*, 23–33. [CrossRef]
65. Aldman, G.; Grove, D.; Holmgren, S. Duodenal acidification and intra-arterial injection of CCK8 increase gallbladder motility in the rainbow trout, *Oncorhynchus mykiss*. *Gen. Comp. Endocrinol.* **1992**, *86*, 20–25. [CrossRef]
66. Jaworek, J. Ghrelin and metonin in the regulation of pancreatic exocrine secretion and maintaining of integrity. *J. Physiol. Pharmacol.* **2006**, *57*, 83–96.
67. Zambonino-Infante, J.; Cahu, C. Ontogeny of the gastrointestinal tract of marine fish larvae. *Comp. Biochem. Physiol. Part C Toxicol. Pharmacol.* **2001**, *130*, 477–487. [CrossRef]
68. Pianesso, D.; Radünz Neto, J.; da Silva, L.; Goulart, F.; Adorian, T.; Mombach, P.; Loureiro, B.; Dalcin, M.; Rotili, D.; Lazzari, R. Determination of tryptophan requirements for juvenile silver catfish (*Rhamdia quelen*) and its effects on growth performance, plasma and hepatic metabolites and digestive enzymes activity. *Anim. Feed Sci. Technol.* **2015**, *210*, 172–183. [CrossRef]
69. Mardones, O.; Oyarzún-Salazar, R.; Labbé, B.; Míguez, J.M.; Vargas-Chacoff, L.; Muñoz, J. Intestinal variation of serotonin, melatonin, and digestive enzymes activities along food passage time through GIT in *Salmo salar* fed with supplemented diets with tryptophan and melatonin. *Comp. Biochem. Physiol. Part A Mol. Integr. Physiol.* **2022**, *266*, 111159. [CrossRef] [PubMed]
70. Mardones, O.; Devia, E.; Labbé, B.; Oyarzún-Salazar, R.; Vargas-Chacoff, L.; Muñoz, J. Effect of L -tryptophan and melatonin supplementation on the serotonin gastrointestinal content and digestive enzymatic activity for *Salmo salar* and *Oncorhynchus kisutch*. *Aquaculture* **2018**, *482*, 203–210. [CrossRef]
71. Caldwell, H.K.; Lee, H.J.; Macbeth, A.H.; Young, W.S. Vasopressin: Behavioral roles of an "original" neuropeptide. *Prog. Neurobiol.* **2008**, *84*, 1–24. [CrossRef] [PubMed]
72. Conrad, J.L.; Weinersmith, K.L.; Brodin, T.; Saltz, J.B.; Sih, A. Behavioural syndromes in fishes: A review with implications for ecology and fisheries management. *J. Fish Biol.* **2011**, *78*, 395–435. [CrossRef] [PubMed]
73. Carpenter, R.E.; Watt, M.J.; Forster, G.L.; Øverli, Y.; Bockholt, C.; Renner, K.J.; Summers, C.H. Corticotropin releasing factor induces anxiogenic locomotion in trout and alters serotonergic and dopaminergic activity. *Horm. Behav.* **2007**, *52*, 600–611. [CrossRef] [PubMed]
74. Carpenter, R.; Korzan, W.; Bockholt, C.; Watt, M.; Forster, G.; Renner, K.; Summers, C. Corticotropin releasing factor influences aggression and monoamines: Modulation of attacks and retreats. *Neuroscience* **2009**, *158*, 412–425. [CrossRef] [PubMed]
75. Loveland, J.L.; Uy, N.; Maruska, K.P.; Carpenter, R.E.; Fernald, R.D. Social status differences regulate the serotonergic system of a cichlid fish, *Astatotilapia burtoni*. *J. Exp. Biol.* **2014**, *217*, 2680–2690. [CrossRef]
76. Clotfelter, E.D.; O'Hare, E.P.; McNitt, M.M.; Carpenter, R.E.; Summers, C.H. Serotonin decreases aggression via 5-HT1A receptors in the fighting fish Betta splendens. *Pharmacol. Biochem. Behav.* **2007**, *87*, 222–231. [CrossRef]
77. Filby, A.L.; Paull, G.C.; Searle, F.; Ortiz-Zarragoitia, M.; Tyler, C.R. Environmental estrogen-induced alterations of male aggression and dominance hierarchies in fish: A mechanistic analysis. *Environ. Sci. Technol.* **2012**, *46*, 3472–3479. [CrossRef]
78. Hecht, T.; Appelbaum, S. Observations on intraspecific aggression and coeval sibling cannibalism by larval and juvenile *Claias gariepinus* (Clariidae: Pisces) under controlled conditions. *J. Zool.* **1988**, *214*, 21–44. [CrossRef]
79. Näslund, J.; Johnsson, J.I. Environmental enrichment for fish in captive environments: Effects of physical structures and substrates. *Fish Fish.* **2014**, *17*, 1–30. [CrossRef]
80. Qin, J.G.; Mittiga, L.; Ottolenghi, F. Cannibalism reduction in juvenile Barramundi *Lates calcarifer* by providing refuges and low light. *J. World Aquac. Soc.* **2004**, *35*, 113–118. [CrossRef]
81. Zhang, Z.; Xu, X.; Wang, Y.; Zhang, X. Effects of environmental enrichment on growth performance, aggressive behavior and stress-induced changes in cortisol release and neurogenesis of black rockfish *Sebastes schlegelii*. *Aquaculture* **2020**, *528*, 735483. [CrossRef]

Article

# Effects of *Myo*-Inositol on the Growth Performance, Digestive Enzyme Activity, and Antioxidation of Juvenile *Hucho taimen*

**Meiyan Zhang [1,2,3], Xing Yang [1,3], Changan Wang [2], Baodi Shang [1], Feng Zhao [1], Hong Xu [4,5] and Qiyou Xu [4,5,*]**

1 Guizhou Fisheries Research Institute, Guiyang 550025, China; zhangmeiyan1099@126.com (M.Z.)
2 Heilongjiang River Fisheries Research Institute, Chinese Academy of Fishery Sciences, Harbin 150070, China
3 School of Fisheries, Nanjing Agricultural University, Wuxi 214081, China
4 School of Life Science, Huzhou University, Huzhou 313000, China
5 Zhejiang Provincial Key Laboratory of Aquatic Resources Conservation and Development, Huzhou University, 759 Erhuan Road (E), Huzhou 313000, China
* Correspondence: 02655@zjhu.edu.cn

**Abstract:** *Hucho taimen* is a cold-water fish with high economic value. *Myo*-inositol (MI) can accelerate lipid metabolism and promote growth in fish species. The present study aimed to assess the effect of MI on the growth performance, digestive enzyme activity, and antioxidation of juvenile *H. taimen*. Accordingly, an 8-week feeding trial was conducted. The results demonstrated that increasing MI concentration promoted growth performance in *H. taimen*. Among the MI concentrations tested, a dose of 328 mg MI/kg corresponded with the lowest feed conversion ratio (FCR) and the highest growth rate. Compared with fish fed a diet of 128 mg MI/kg, the lipase activity in the pyloric caeca significantly increased in fish fed 528 mg MI/kg, while superoxide dismutase (SOD) activity was significantly higher in fish fed 728 mg MI/kg. Consistently, the 128 mg MI/kg diet presented the highest malonaldehyde (MDA) levels. In conclusion, our study revealed that enhanced growth performance, digestive enzyme activity, and antioxidant capacity increased as MI concentration increased. The optimum level of dietary MI in *H. taimen* was 270–321 mg/kg, based on the FCR and specific growth rate (SGR) on the broken-line regression.

**Keywords:** *myo*-inositol; growth performance; digestive enzyme; antioxidant; *Hucho taimen*

**Key Contribution:** In this research, analysis of the effects of MI on juvenile *H. taimen* was carried out. The results showed that growth performance, digestive enzyme activity, and antioxidant capacity were enhanced with the increasing of MI concentration to a specified level. The optimum level of dietary MI in *H. taimen* was proven to be 270–321 mg/kg; this is the level associated with the lowest FCR and highest SGR on the broken-line regression.

**Citation:** Zhang, M.; Yang, X.; Wang, C.; Shang, B.; Zhao, F.; Xu, H.; Xu, Q. Effects of *Myo*-Inositol on the Growth Performance, Digestive Enzyme Activity, and Antioxidation of Juvenile *Hucho taimen*. *Fishes* **2023**, *8*, 567. https://doi.org/10.3390/fishes8120567

Academic Editors: Francisco J. Moyano and Marina Paolucci

Received: 28 September 2023
Revised: 13 November 2023
Accepted: 18 November 2023
Published: 22 November 2023



## 1. Introduction

*Hucho taimen* belongs to *Salmoniformes*, *Salmonidae*, *Salmon* subfamily. It is a cold-water fish and its growth rate (WG) is slower than most other fish; the optimal temperature for it is 13–16 °C [1]. It is large: its body weight can be up to 50 kg and its body length can reach 1 meter or more. It is a precious fish with delicious, nutritious meat, and it has a high economic value. In recent years, due to overfishing and serious damage to the ecological environment, wild *H. taimen* resources have plummeted. It is now included in the "Chinese red list of endangered animal species". At present, it has been successfully domesticated and is promoted for breeding in more than a dozen provinces in China.

Inositol, also known as cyclohexanol, has nine cis-trans isomers. The most common is *myo*-inositol (MI), which is a type of vitamin B [2] that can promote lipid metabolism, reduce blood lipids, and protect liver health. The main function of MI in aquatic animals is to improve feed utilization, accelerate growth, and promote lipid metabolism in the liver and other tissues. At present, feed company incorporate MI as a nutritional additive in aquatic animal feed.

Symptoms of MI deficiency have been found in young grass carp (*Ctenopharyngodon idella*) [3], juvenile Chinese mitten crab (*Eriocheir sinensis*) [4], and juvenile Pacific white shrimp (*Litopenaeus vannamei*) [5]. Symptoms manifested as anorexia, anemia, poor quality in growth and development, skin erosion, darkened skin color, slow gastric emptying, decreased cholinesterase, and the decreased activity of some transaminases. When Jian carp (*Cyprinus carpio* var. Jian) were deficient in MI, they suffered from fin erosion, skin bleeding, tail handle exposure (in severe cases), intestinal epithelial edema, and even shedding and necrosis [6,7].

Adding an appropriate level of MI to feed can improve growth performance, as has been demonstrated in Wuchang bream (*Megalobrama amblycephala*) [8], juvenile sunfish (*Morone chrysops* × *Morone saxatilis*) [9], and juvenile *Ctenopharyngodon idella* [10]. Khosravi found that an appropriate level of MI in feed could significantly promote the growth performance of striped beakfish (*Oplegnathus fasciatus*); the appropriate supplemental level of MI was found to be 94.3 mg/kg [11]. The growth performance and feed efficiency of olive flounder (*Paralichthys olivaceus*) were significantly improved by adding 800 mg MI/kg to their diet [12]. In *Ctenopharyngodon idella*, adding 100–150 mg MI/kg to its whole plant protein source diet was shown to significantly improve its growth performance and feed efficiency [13]. Within the large-scale and high-density culture of Nile tilapia (*Oreochromis niloticus*), the addition of 0.03% MI to feed was proven to be helpful to the growth and the mobilization of lipid, though the amount of MI added should be controlled strictly [14].

In the research of crustaceans, such as the giant tiger prawn (*Penaeus monodon*), Chinese white shrimp (*Penaeus chinensis*), and kuruma shrimp (*Penaeus japonicus*) [15], it was found that a lack of MI in feed led to a significant decrease in survival rate. Adding the appropriate levels of MI to feed could significantly improve the growth performance of *Penaeus monodon*, *Penaeus chinensis* and *Penaeus japonicus*. The suitable supplemental level of MI in the feed of *Penaeus monodon* was 3275–3514 mg/kg [16]; the level for *Penaeus chinensis* was 4000 mg MI/kg [17]. For *Eriocheir sinensis*, that level was 1613–1707 mg MI/kg [18]. In the study of *Litopenaeus vannamei*, 600 mg MI/kg was found to be sufficient to meet its needs in terms of body growth and development; the appropriate level of MI significantly improved its resistance to salinity stress [19]. These studies have shown that the level of MI required for aquatic animals to thrive may be related to different feed formulations, growth stages, breeding varieties, feeding habits, and breeding environments and models; however, the regulatory mechanism of MI in growth performance is still not completely clear.

Juvenile *H. taimen* can accurately assess their own nutritional needs, including determining their MI requirements. In this paper, the MI requirement of juvenile *H. taimen* was comprehensively evaluated based on growth performance and antioxidant indexes, in order to provide theoretical guidance for the production of *H. taimen* feed under this lipid level.

## 2. Materials and Methods

### 2.1. Experimental Design and Feeding Management

The MI used in this experiment, which was purchased from Sigma, USA, was in powder form and its purity was ≥99%. Six treatment groups were set in the experiment; the MI content in the basic diet was 128 mg/kg without additional MI. The MI supplementation in each group was 0 (G1), 100 (G2), 200 (G3), 400 (G4), 600 (G5), 800 (G6) mg/kg, respectively, and the inositol content values of each group was calculated as 128 (G1), 228 (G2), 328 (G3), 528 (G4), 728 (G5) and 928 (G6) mg/kg, respectively. Feed composition and nutritional level are shown in Table 1. After crushing, the raw materials were mixed evenly according to proportion, granulated (1.0 mm), dried naturally, and stored in a refrigerator at −20 °C.

**Table 1.** Ingredients and chemical composition of the basal diets (air-dry basis, %).

| Ingredients | Content |
|---|---|
| Casein | 34.00 |
| Fish meal | 23.00 |
| Gelatin | 10.00 |
| Dextrin | 15.30 |
| Fish oil | 11.00 |
| Phospholipids | 2.00 |
| $Ca(H_2PO_4)_2$ | 1.00 |
| CMC | 2.60 |
| Vitamin premix [1] | 0.30 |
| Mineral premix [2] | 0.20 |
| Additive [3] | 0.60 |
| Total | 100.00 |
| Nutrient levels | |
| Crude protein | 52.3 |
| Crude lipid | 12.4 |
| MI (mg/kg) | 128 |

[1] The vitamin premix provided the following diet: VK 5 mg·kg$^{-1}$, VA 15,000 IU·kg$^{-1}$, $VD_3$ 3000 IU·kg$^{-1}$, $VB_1$ 15 mg·kg$^{-1}$, $VB_2$ 30 mg·kg$^{-1}$, $VB_6$ 15 mg·kg$^{-1}$, $VB_{12}$ 0.5 mg·kg$^{-1}$, VC 1000 mg·kg$^{-1}$, VE 60 mg·kg$^{-1}$, Nicotinic acid 175 mg·kg$^{-1}$, Folic acid 5 mg·kg$^{-1}$, Biotin 2.5 mg·kg$^{-1}$, and Pantothenic acid 50 mg·kg$^{-1}$. [2] The mineral premix provided the following diet: Fe (as ferrous sulfate) 25 mg·kg$^{-1}$, Cu (as copper sulfate) 3 mg·kg$^{-1}$, Mn (as manganese sulfate) 15 mg·kg$^{-1}$, I (as potassium iodide) 0.6 mg·kg$^{-1}$, and Zn (as zinc sulfate) 60 mg·kg$^{-1}$. [3] The additive provided the following diet per kg: Choline chloride 2 g, Antimildew 0.5 g, Magnesia 2 g, Antioxidant 0.5 g, and DMPT 1 g.

A total of 540 *H. taimen* (2.83 ± 0.44 g) were randomly assigned to six groups with three replicates per group and 30 fish per replicate. Body weight and body length were recorded after 7 days of pre-feeding with commercial feed and 24 h of starvation. *H. taimen* was fed four times a day (07:00, 10:30, 14:00, and 17:00). The daily ration was determined based on the weight of each treatment of fish, which were weighed every 2 weeks and fed at 3–8% of the total weight of each treatment. The feeding trial lasted for 8 weeks [20–24]. The trial employed a flow-through culture system; no aerators were used. The current speed of the flowing water was $[(0.3 \times 10^{-3})–(0.4 \times 10^{-3})\ m^3/h]$, the water temperature was 9.5–15.2 °C, the dissolved oxygen was kept at more than 8.0 mg/L$^{-1}$, and the pH of the water was 7.5–7.8.

*H. taimen* was weighed at the beginning (initial body weight) and at the end (final body weight) of the 8-week feeding trial. The balance (BP221S) used for weighing the fish was made by Sartorius; its maximum range is 210 g and its precision is 0.1 mg. The weight gain, specific growth rate, feed conversion ratio, condition factor, viscerosomatic index, and hepatosmatic index of each treatment group were recorded and analyzed. These parameters were calculated as:

Weight gain (WG, %) = $(W_f - W_i)/W_i \times 100$;
Specific growth rate (SGR, %/day) = $(\ln W_f - \ln W_i)/\text{days}) \times 100$;
Feed conversion ratio (FCR) = Feed consumed (g)/$(W_f - W_i)$;
Condition factor (CF, g·cm$^{-3}$) = $W_f \times 100/L_f^3$;
Viscerosomatic index (VSI, %) = $(W_v/W_f) \times 100$;
Hepatosmatic index (HSI, %) = $(W_h/W_f) \times 100$.

where $W_f$ and $W_i$ are the initial and final body weights, $L_f$ is the initial body length, $W_v$ is the total weight of fish viscera, and $W_h$ is the weight of the fish liver.

*2.2. Sample Collection*

At the end of the feeding trial, starvation was enforced for 24 h before sampling. Three *H. taimen* per tank were collected randomly and frozen (−40 °C) for whole-body composition analyses. In addition, three *H. taimen* were selected per tank and placed on an ice tray for dissection. The liver, pyloric caeca, and intestinal tract were all sampled; the crude enzyme solution of the tissue was extracted in each case and stored at −80 °C for further analysis.

### 2.3. Whole-Body Composition Analyses

Moisture, crude protein, crude lipid, and ash were measured [25]. The moisture content was measured using the 105 °C drying method. The crude protein was determined by the Kjeldahl method. The crude lipid was analyzed by the Soxhlet extraction method. Last, ash content was determined by using a high-temperature combustion method applied to the fish body.

### 2.4. Biochemical Analysis

The liver, pyloric caeca, and intestinal tract were sampled. After homogenization, the samples were centrifuged for 10 min at 3500 r/min to obtain the supernatant at 4 °C, then the supernatant was collected. Amylase activity was determined using the iodine-starch colorimetric method. After the color reaction, absorbance was measured at 660 nm. Protease activity was tested by the Folin-phenol method. MDA was determined by the TBA method, and there was a maximum absorption peak at 532 nm. One unit of SOD is the amount of SOD that corresponds to an inhibition rate of 50% in 1 mL of tissue protein in a 1 ml reaction solution. ATP per milligram of tissue protein per hour produces 1 umol of inorganic phosphorus as a unit of $Na^+$,$K^+$-ATPase activity. Amylase, protease, lipase, alkaline phosphatase (AKP), $Na^+$,$K^+$-ATPase, SOD, and MDA were measured by kits that were produced by Nanjing Jiancheng Bioengineering Institute.

### 2.5. Statistical Analysis

The experimental data were analyzed by one-way ANOVA in SPSS 17.0; $p < 0.05$ was considered to be statistically significant. Experimental data were expressed as mean ± standard deviation ($\overline{X}$ ± SD) across at least three independent experiments. The optimal MI content for *H. taimen* was estimated using the broken-line model. The WG and FCR of *H. taimen*, as well as MI content, were estimated.

## 3. Results

### 3.1. Growth Performance

Compared to the control group, other groups significantly increased WG ($p < 0.05$). The FCR of the G3 group was significantly lower than that of the others ($p < 0.05$). There was no significant difference in HSI among all groups ($p > 0.05$). Compared to the control group, the G3 group significantly decreased the CF ($p < 0.05$) and significantly increased the SGR ($p < 0.05$); additionally, the G4 group significantly increased the VSI ($p < 0.05$) (Table 2).

**Table 2.** Effects of different levels of MI on the growth performance of *H. taimen*.

| Treatments | Initial Weight (g) | Final Weigh (g) | WG (%) | SGR (%) | FCR | CF ($g \cdot cm^{-3}$) | HSI (%) | VSI (%) |
|---|---|---|---|---|---|---|---|---|
| G1 | 2.45 ± 0.27 | 4.17 ± 0.76 | 70.20 ± 1.91 $^{a}$ | 0.94 ± 0.11 $^{a}$ | 1.98 ± 0.20 $^{bc}$ | 0.60 ± 0.10 $^{b}$ | 1.03 ± 0.18 | 7.11 ± 0.33 $^{a}$ |
| G2 | 2.55 ± 0.14 | 4.53 ± 1.02 | 77.65 ± 1.00 $^{b}$ | 1.03 ± 0.14 $^{ab}$ | 1.94 ± 0.14 $^{ab}$ | 0.62 ± 0.03 $^{bc}$ | 1.12 ± 0.28 | 7.34 ± 0.98 $^{ab}$ |
| G3 | 2.75 ± 0.20 | 5.02 ± 0.68 | 82.55 ± 2.99 $^{bc}$ | 1.11± 0.21 $^{b}$ | 1.89 ± 0.09 $^{a}$ | 0.58 ± 0.08 $^{a}$ | 1.61 ± 0.08 | 7.88± 0.70 $^{ab}$ |
| G4 | 2.55 ± 0.19 | 4.74 ± 0.41 | 86.00 ± 2.16 $^{c}$ | 1.07 ± 0.12 $^{ab}$ | 2.00 ± 0.07 $^{bc}$ | 0.62 ± 0.08 $^{bc}$ | 1.43 ± 0.15 | 9.88 ± 1.00 $^{b}$ |
| G5 | 2.54± 0.21 | 4.58± 1.27 | 80.31 ± 1.40 $^{b}$ | 1.05 ± 0.18 $^{ab}$ | 2.06 ± 0.15 $^{c}$ | 0.61 ± 0.07 $^{b}$ | 1.36 ± 0.04 | 7.59 ± 0.32 $^{ab}$ |
| G6 | 2.91± 0.38 | 5.14± 1.25 | 76.63 ± 3.51 $^{b}$ | 1.02 ± 0.09 $^{ab}$ | 2.06 ± 0.04 $^{c}$ | 0.63 ± 0.05 $^{bc}$ | 1.53 ± 0.13 | 7.43 ± 1.11 $^{ab}$ |

Note: within a column, values ($\overline{X}$ ± SD) with different superscripts are significantly different ($p < 0.05$). G1, MI 128 mg/kg of feed; G2, MI 228 mg/kg of feed; G3, MI 328 mg/kg of feed; G4, MI 528 mg/kg of feed; G5, MI 728 mg/kg of feed; G6, MI 928 mg/kg of feed.

Based on the broken-line model of FCR, SGR, and MI levels, it was concluded that the optimal level for *H. taimen* was 270–321 mg MI/kg (Figures 1 and 2).

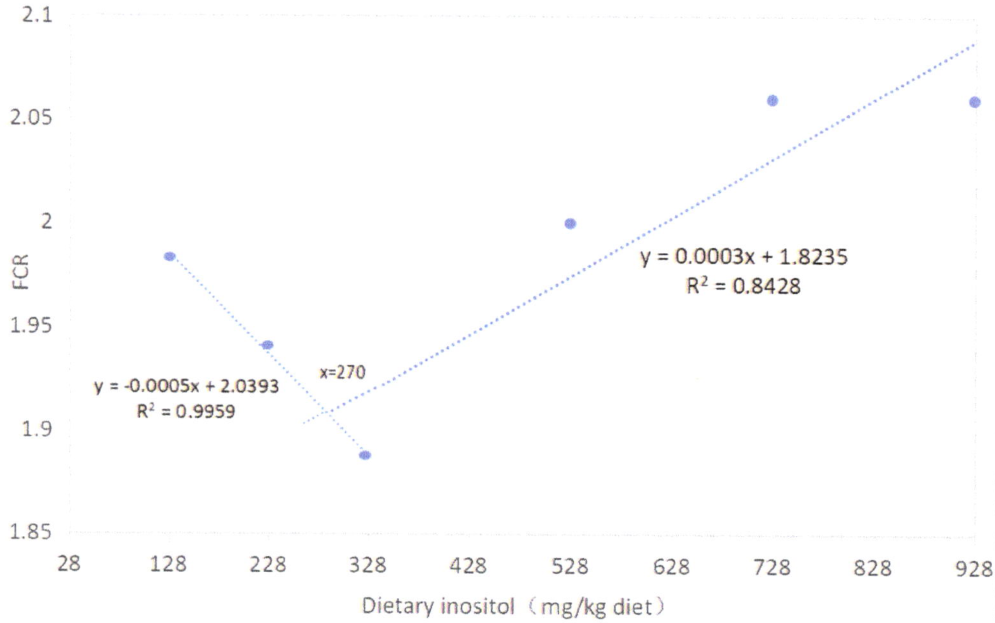


**Figure 1.** Effects of MI on the FCR of *H. taimen*.

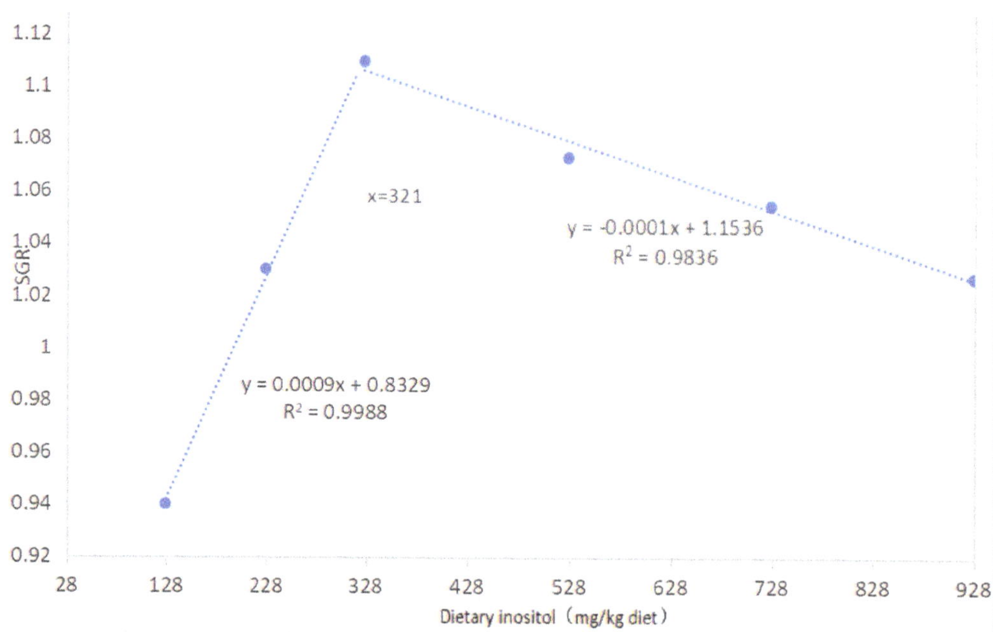


**Figure 2.** Effects of MI on the SGR of *H. taimen*.

### 3.2. Whole-Body Composition

No significant difference was found in terms of moisture and crude protein content among all groups ($p > 0.05$). The whole-body crude lipid contents of treatment groups were all higher than that of the control group, but only that of the G3 group was significantly increased ($p < 0.05$). Compared to the control group, whole-body ash content was significantly increased in the G3 and G5 groups ($p < 0.05$) (Table 3).

**Table 3.** Effects of different levels of MI on the whole-body composition of *H. taimen*.

| Treatments | Moisture (%) | Crude Protein (%) | Crude Lipid (%) | Ash (%) |
|---|---|---|---|---|
| G1 | 80.19 ± 0.84 | 14.97 ± 1.04 | 2.10 ± 0.02 [a] | 2.21 ± 0.08 [a] |
| G2 | 80.74 ± 0.84 | 13.53 ± 1.10 | 2.16 ± 0.18 [ab] | 2.29 ± 0.16 [a] |
| G3 | 80.32 ± 0.40 | 13.70 ± 1.42 | 2.41 ± 0.19 [b] | 2.55 ± 0.51 [b] |
| G4 | 79.90 ± 0.12 | 14.38 ± 0.55 | 2.26 ± 0.02 [ab] | 2.33 ± 0.28 [a] |
| G5 | 80.30 ± 1.98 | 13.12 ± 0.75 | 2.28 ± 0.01 [ab] | 2.80 ± 0.24 [c] |
| G6 | 80.49 ± 0.25 | 14.13 ± 0.11 | 2.32 ± 0.03 [ab] | 2.22 ± 0.36 [a] |

Note: within a column, values ($\overline{X}$ ± SD) with different superscripts are significantly different ($p < 0.05$). G1, MI 128 mg/kg of feed; G2, MI 228 mg/kg of feed; G3, MI 328 mg/kg of feed; G4, MI 528 mg/kg of feed; G5, MI 728 mg/kg of feed; G6, MI 928 mg/kg of feed.

### 3.3. Digestive Enzyme Activity

As shown in Table 4, there were no significant differences in the amylase activity of the liver and intestine among all groups ($p > 0.05$). The MI content in the diet had a significant effect on the amylase activity of the pyloric caeca ($p < 0.05$); the amylase activity of the G3 group was significantly increased ($p < 0.05$). There were no significant differences in the protease activity of the liver and pyloric caeca among all groups ($p > 0.05$). Compared to the control group, the protease activity of the intestine was significantly increased in the G4 and G5 groups ($p < 0.05$). The lipase activity of the G4 group was higher than that of the control group ($p < 0.05$); however, there was no significant difference in the lipase activity of the intestine among all groups ($p > 0.05$) (Table 4).

**Table 4.** Effects of different levels of MI on the digestive enzyme activity of *H. taimen*.

| Tissue | Enzyme | G1 | G2 | G3 | G4 | G5 | G6 |
|---|---|---|---|---|---|---|---|
| Liver | Amylase | 30.37 ± 1.76 | 30.67 ± 4.21 | 32.29 ± 0.68 | 34.85 ± 2.89 | 33.38 ± 3.06 | 30.82 ± 1.77 |
| | Protease | 24.60 ± 2.04 | 30.30 ± 5.54 | 31.55 ± 3.70 | 28.95 ± 1.26 | 27.96 ± 5.24 | 30.31 ± 4.61 |
| | Lipase | 8.54 ± 1.12 [b] | 6.44 ± 0.19 [a] | 7.61 ± 0.41 [ab] | 8.14 ± 0.82 [b] | 7.40 ± 0.38 [ab] | 8.02 ± 1.52 [b] |
| Pyloric caeca | Amylase | 195.98 ± 12.01 [a] | 205.04 ± 8.67 [ab] | 236.07 ± 21.91 [c] | 222.71 ± 5.81 [bc] | 224.36 ± 7.13 [bc] | 218.60 ± 15.49 [abc] |
| | Protease | 98.81 ± 9.59 | 105.28 ± 10.13 | 118.55 ± 14.21 | 120.66 ± 19.28 | 108.24 ± 8.38 | 113.07 ± 4.35 |
| | Lipase | 41.97 ± 0.42 [a] | 42.28 ± 1.92 [a] | 42.46 ± 1.32 [a] | 47.46 ± 2.91 [b] | 44.80 ± 4.75 [ab] | 41.94 ± 1.29 [a] |
| Intestine | Amylase | 161.64 ± 4.44 | 166.58 ± 11.02 | 176.06 ± 14.47 | 164.21 ± 13.04 | 160.95 ± 6.76 | 160.07 ± 4.75 |
| | Protease | 43.66 ± 5.39 [a] | 44.98 ± 5.02 [a] | 50.87 ± 8.44 [ab] | 56.66 ± 2.09 [b] | 56.21 ± 1.21 [b] | 49.83 ± 3.24 [ab] |
| | Lipase | 34.59 ± 6.51 | 34.28 ± 3.50 | 33.38 ± 1.66 | 30.26 ± 2.70 | 29.61 ± 0.94 | 33.18 ± 3.03 |

Note: within a column, values ($\overline{X}$ ± SD) with different superscripts are significantly different ($p < 0.05$). G1, MI 128 mg/kg of feed; G2, MI 228 mg/kg of feed; G3, MI 328 mg/kg of feed; G4, MI 528 mg/kg of feed; G5, MI 728 mg/kg of feed; G6, MI 928 mg/kg of feed.

### 3.4. Biochemical Index

The AKP and $Na^+$,$K^+$-ATPase activities of the pyloric caeca did not show any significant differences among all groups ($p > 0.05$). The $Na^+$,$K^+$-ATPase activities of the intestine in the treatment groups were all higher than that of the control group, but only the G4 group was significantly higher ($p < 0.05$). Compared to the control group, the AKP activities of the intestine were significantly increased in the G3 group, G4 group, and G5 group ($p < 0.05$) (Table 5).

### 3.5. Antioxidant Indices

Among the treatment groups, SOD activity was significantly increased in the G5 group ($p < 0.05$). Compared to the control group, MDA content was decreased in all treatment groups ($p < 0.05$) (Table 6).

**Table 5.** Effects of different levels of MI on the biochemical index of *H. taimen*.

| Tissue | Biochemical Index | G1 | G2 | G3 | G4 | G5 | G6 |
|---|---|---|---|---|---|---|---|
| Pyloric caeca | AKP | 55.89 ± 4.78 | 56.60 ± 6.77 | 69.45 ± 9.50 | 65.55 ± 8.57 | 60.48 ± 2.91 | 61.63 ± 6.93 |
| | $Na^+,K^+$-ATPase (μmolpi/gprot/h) | 1.32 ± 0.23 | 1.23 ± 0.05 | 1.25 ± 0.16 | 1.39 ± 0.30 | 1.21 ± 0.12 | 1.18 ± 0.27 |
| Intestine | AKP | 42.16 ± 3.11 [a] | 49.58 ± 4.68 [ab] | 74.81 ± 11.62 [c] | 59.31 ± 8.73 [b] | 59.95 ± 8.06 [b] | 47.23 ± 8.55 [ab] |
| | $Na^+,K^+$-ATPase (μmolpi/gprot/h) | 0.54 ± 0.09 [a] | 0.79 ± 0.05 [a] | 0.74 ± 0.05 [a] | 1.28 ± 0.27 [b] | 0.84 ± 0.05 [a] | 0.74 ± 0.12 [a] |

Note: within a column, values ($\overline{X}$ ± SD) with different superscripts are significantly different ($p < 0.05$). G1, MI 128 mg/kg of feed; G2, MI 228 mg/kg of feed; G3, MI 328 mg/kg of feed; G4, MI 528 mg/kg of feed; G5, MI 728 mg/kg of feed; G6, MI 928 mg/kg of feed. AKP, alkaline phosphatase; $Na^+,K^+$-ATPase, $Na^+,K^+$-stimulated ATPase.

**Table 6.** Effects of different levels of MI on the antioxidant indices of *H. taimen*.

| Treatments | SOD ($NU \cdot mgprot^{-1}$) | MDA($nmol \cdot mgprot^{-1}$) |
|---|---|---|
| G1 | 68.97 ± 8.31 [a] | 9.67 ± 0.42 [b] |
| G2 | 81.57 ± 12.81 [ab] | 6.20 ± 0.77 [a] |
| G3 | 82.26 ± 8.77 [ab] | 6.17 ± 1.57 [a] |
| G4 | 82.89 ± 7.70 [ab] | 7.32 ± 1.78 [a] |
| G5 | 89.59 ± 13.62 [b] | 7.47 ± 0.84 [a] |
| G6 | 76.32 ± 6.66 [ab] | 7.37 ± 0.81 [a] |

Note: within a column, values ($\overline{X}$ ± SD) with different superscripts are significantly different ($p < 0.05$). G1, MI 128 mg/kg of feed; G2, MI 228 mg/kg of feed; G3, MI 328mg/kg of feed; G4, MI 528 mg/kg of feed; G5, MI 728 mg/kg of feed; G6, MI 928 mg/kg of feed. SOD, superoxide dismutase; MDA, malonaldehyde.

## 4. Discussion

This study demonstrated that WG was observably increased by increasing MI levels in *H. taimen*, a result similar to that seen in *Penaeus monodon* [16], *Ctenopharyngodon idella* [3], and juvenile *Eriocheir sinensis* [4]. Our study showed that the CF and FCR decreased with increasing MI levels. A possible reason for this is that MI deficiency affects the synthesis and secretion of low-density lipoprotein. In other words, lipid cannot be transported out of the liver in time, resulting in abnormal accumulation in the liver, which causes lipid deposition in fish bodies which, in turn, results in the increasing of CF [26]. Attaining an appropriate level of MI content could make better use of the lipid in the feed, thereby promoting fish health and improving feed utilization efficiency. Levine found that adding MI could improve the swimming ability of crucian carp (*Carassius auratus*) [27]. The reason for our results might be that the addition of an appropriate amount of MI to feed improved the utilization rate of lipid in that feed, which promotes growth in *H. taimen*. A study of Mozambique tilapia (*Oreochromis mossambicus*) showed that MI could regulate lipid and pyruvic acid metabolism, increasing phospholipid synthesis-related mRNA levels and improving antioxidant properties under long-term salinity stress [18]. In addition, the feeding habits of fish also affect the demand for MI content. Carnivorous fish need high-level lipid in their feed as they need more MI to meet the needs of lipid metabolism. For example, the optimal MI demand of blackhead sea bream (*Acanthopagrus schlegelii*) was shown to be 2000 mg/kg [28], while the demand of *Ctenopharyngodon idella* was only 166–214 mg MI/kg [10]. In the liver and kidneys of some aquatic animals, MI can be synthesized with glucose as a substrate through l-inositol-1-phosphate synthase and l-inositol-1-phosphatase. MI synthesis ability is determined by MI synthase activity [29]; moreover, aquatic animals can synthesize a certain amount of MI through intestinal microorganisms, which also affects their MI demands.

This study demonstrated that MI level did not affect whole-body crude protein content in *H. taimen*, a result similar to that seen in Atlantic salmon (*Salmo salar*) [30], juvenile *Litopenaeus vannamei* [5], and Pacific abalone (*Haliotis discus hannai* Ino) [31], indicating that MI has little effect on digestion and the absorption of protein in *H. taimen* and that the addition of MI has no significant effect on the protein deposition rate. The contents of ash and crude lipid decreased significantly with the increasing of MI content to a certain extent, which is consistent with the results of a study of juvenile *Cyprinus carpio* var. Jian [7]. A possible reason for the decreasing of whole-body crude lipid is that an excessive amount of MI promotes lipid decomposition in the liver. The combination of fatty acids and various enzymes on the liver cell membrane improves the lipid exchange capacity of the cell membrane, which is transported to various tissues for utilization, and the lipid content in the body can be effectively reduced; however, no significant differences in the whole-body composition of golden pompano (*Trachinotus ovatus*) was found between control and treatment groups [32]. The reason for these two differing sets of results could be that the capacity of MI to regulate lipid metabolism in different aquatic animals varies; therefore, the mechanism for the effect of MI on the body composition of aquatic animals requires further study.

The digestive capacity of fish is directly related to the development of digestive organs and digestive enzyme activity [33]. This research demonstrated that digestive enzyme activity increased with the increasing of MI content, which is consistent with previous findings in various species, including juvenile turbot (*Scohpthalmus maximus* L.) [34], juvenile *Cyprinus carpio* var. [35], *Litopenaeus vannamei* [36], and Japanese sturgeon (*Acipenser schrenckii*) [37]. A possible reason for this finding is that, when MI content is insufficient, the liver and pancreas of fish cannot develop normally; therefore, the digestive enzyme secreted by the pancreas is reduced. On the other hand, in our study, the amount of bile acid secreted by the liver was reduced, which affected the lipase activity of the fish. In addition, MI exists as part of the cell membrane in the form of phosphatidyl-MI. When MI is deficient, the intestinal structure of aquatic animals is affected which, in turn, would also affect digestive enzyme activity. In our study, after the MI was absorbed by *H. taimen*, it was stored in serum and the liver, and it could improve the activity of cholinesterase, which proved to be conducive to the improvement of amylase activity and WG in individual fish [38]; moreover, by increasing the intake of each fish, MI indirectly stimulated the secretion of digestive enzymes.

The AKP in the intestine participates in the absorption of nutrients such as lipid and glucose, etc. $Na^+,K^+$-ATPase activity can reflect intestinal absorption capacity [39]. This study has shown that, as MI content is increased, $Na^+,K^+$-ATPase and AKP activities increase, promoting the absorption of nutrients. This finding aligns with the results of similar testing on juvenile *Cyprinus carpio* var. [6]. A possible reason for the finding is that $Na^+,K^+$-ATPase is the main component of the $Na^+,K^+$ pump, which constitutes part of the cell membrane; MI is also a component of the membrane. Substances targeting MI on the membrane would, therefore, indirectly affect the activities of $Na^+,K^+$-ATPase [40]. The amount and existing state of MI may also affect the activities of $Na^+,K^+$-ATPase. Appropriately increasing MI content can help to increase the activities of $Na^+,K^+$-ATPase and AKP.

Antioxidant levels can reflect the health status of animals. SOD is an important antioxidant enzyme, which has the effect of mitigating anti-oxidative damage and maintaining cell structure [41]. When there are fewer antioxidant enzymes, the metabolic balance of free radicals will be destroyed, and free oxygen radicals will react with unsaturated fatty acids, resulting in lipid peroxidation and damage to cells and macromoleculars in cells, all of which cause damage to the body. Lipid peroxidation can produce a variety of metabolites, of which MDA is one of the most important. The content of MDA in the liver can reflect the progress of lipid peroxidation; the degree of oxidative damage to cells can also be shown indirectly [42]. In this experiment, when MI content was insufficient, the MDA content increased and the SOD content decreased, which is consistent with the results of a study on juvenile *Cyprinus carpio* var. [7]. A possible reason for this finding is that insufficient MI content affects the development of the liver, leading to a disorder of lipid metabolism. Another possible reason is that the synthesis and secretion of low-density lipoprotein were affected; because the lipid in the liver could not be transported in time, the lipid did not accumulate normally. This process has also been seen in research on *Carassius auratus* [43] and juvenile *Ctenopharyngodon idella* [44]. Increasing MI content to an appropriate level could reduce the production of oxygen free radicals in the body; furthermore, this could promote the elimination of oxygen free radicals which could, in turn, effectively reduce the damage of oxygen free radicals to the body and improve antioxidant capacity, thus enhancing growth performance in *H. taimen*.

## 5. Conclusions

When MI content was deficient, the growth performance, digestive enzyme activity, and antioxidant capacity of *H. taimen* were decreased. The appropriate MI content in feed can improve the WG, digestive enzyme activity, and antioxidant capacity of *H. taimen*. When the protein content in the feed was 52.3% and the lipid content was 12.4%, the optimal requirement of MI was determined to be 270–321 mg/kg, according to broken-line regression analysis based on the FCR and SGR of *H. taimen*.

**Author Contributions:** All authors contributed to the study. M.Z.: experiment execution, formal analysis, investigation, and writing—original draft; X.Y., C.W., B.S., F.Z. and H.X.: methodology, resources, writing–review, and supervision; Q.X.: methodology, supervision, writing—review, project administration, editing, and funding acquisition. All authors commented on previous versions of the manuscript. All authors have read and agreed to the published version of the manuscript.

**Funding:** This research was funded by the Subsidy project from NSFC of Guizhou Academy of Agricultural Sciences (2021-11); characteristic aquatic industry technology system of Guizhou province (GZCYTX2022-01101).

**Institutional Review Board Statement:** This study was conducted according to the guidelines of the Declaration of Helsinki, and approved by the Institutional Review Board of Heilongjiang River Fisheries Research Institute, Chinese Academy of Fishery Sciences (code: 20200615, date: 15 June 2020).

**Informed Consent Statement:** Not applicable.

**Data Availability Statement:** All data presented in the work are available from the corresponding authors, upon responsible request.

**Acknowledgments:** The authors are grateful to Yingying Wei, Jinfeng Huang, Liang Luo, and Jinnan Li for their assistance and expertise throughout the study.

**Conflicts of Interest:** The authors declare no conflict of interest.

## References

1. Zhang, H.; Liu, Y.; Zhao, X.; Wang, C.; Lu, S.; Liu, H.; Xu, Q. Effects of Aloe Powder on growth performance and serum indices of Taimen *Hucho taimen*. *Chin. J. Fish.* **2020**, *33*, 1–6.
2. Wu, Q.; Wang, S.; Zhen, Y.; Wang, T.; Zhang, X.; Sun, Z. Research progress of preparation process and application of inositol. *Heilongjiang Anim. Sci. Vet. Med.* **2023**, *20*, 41–46.
3. Li, S.; Jiang, W.; Feng, L.; Liu, Y.; Wu, P.; Jiang, J.; Kuang, S.; Tang, L.; Tang, W.; Zhang, Y.; et al. Dietary *myo*-inositol deficiency decreased the growth performances and impaired intestinal physical barrier function partly relating to nrf2, jnk, e2f4 and mlck signaling in young grass carp (*Ctenopharyngodon idella*). *Fish Shellfish Immunol.* **2017**, *67*, 475–492. [CrossRef] [PubMed]
4. Bu, X.; Lin, Z.; Liu, S.; Wang, C.; Wang, N.; Lei, Y.; Zhu, J.; Wang, X.; Qin, J.; Chen, L. Effects of *myo* -inositol on growth performance, body composition, antioxidant status, non-specific immunity and lipid metabolism of juvenile Chinese mitten crab (*Eriocheir sinensis*). *Aquac. Nutr.* **2020**, *26*, 1623–1635. [CrossRef]
5. Li, R.; Kou, S.; Yang, Q.; Tan, B.; Chi, S.; Yi, Y.; Zhu, D. Effects of *Myo*-Inositol on Growth, Non-Specific Immunity and Intestinal Microflora Composition of Juvenile *Litopenaeus Vannamei*. *Chin. J. Anim. Nutr.* **2023**, *35*, 1182–1194.
6. Feng, L.; Li, J.; Zhou, X.; Liu, Y. Effects of inoistol deficiency on digestive and immune function of juvenile jian carp(*Cyprinus carpio* var. Jian). *Chin. J. Anim. Nutr.* **2009**, *21*, 878–883.
7. Jiang, W.; Liu, Y.; Feng, L.; Zhou, X. Histopathological observation of juvenile *Cyprinus carpio* var. Jian with *myo*-inositol deficiency and determinations of antioxidative ability and intestinal microbial population. *Vet. Sci. China* **2009**, *31*, 64–69.
8. Cui, H.; Liu, B.; Ge, X.; Xie, J. Effects of dietary inositol on growth performance, physiological and biochemical indexes of serum and fat content in liver and muscle of juvenile Wuchang bream (*Megalobrama amblycephala*). *J. Shanghai Ocean Univ.* **2013**, *22*, 868–875.

9. Deng, D.; Hemre, G.; Wilson, R.P. Juvenile sunshine bass (*Morone chrysops* × *Morone saxatilis*) do not require dietary *myo*-inositol. *Aquaculture* **2002**, *213*, 387–393. [CrossRef]
10. Wen, H.; Zhao, Z.; Jiang, M.; Liu, A.; Wu, F.; Liu, W. Dietary *myo*-inositol requirement for grass carp, *Ctenopharyngodon Idella Fingerling*. *J. Fish. Sci. China* **2007**, *14*, 794–800.
11. Khosravi, S.; Lim, S.; Rahimnejad, S.; Kim, S.; Lee, B.; Kim, K.; Han, H.; Lee, K. Dietary *myo*-inositol requirement of parrot fish. *Oplegnathus fasciatus*. *Aquaculture* **2015**, *436*, 1–7. [CrossRef]
12. Lee, B.; Lee, K.; Lim, S.; Lee, S. Dietary *myo*-inositol requirement for olive flounder, *Paralichthys olivaceus* (*Temminch et Schlegel*). *Aquac. Res.* **2009**, *40*, 83–90. [CrossRef]
13. Lin, K.; Feng, J.; Yang, H.; Chen, Y.; Chen, X.; Luo, H.; Huang, W.; Wang, H.; Luo, L. Effects of inositol supplementation to practical dietary on growth performance, lipid metabolism and antioxidant activity of *Cteno pharyngodon Idella*. *J. Fish. China* **2018**, *42*, 1428–1437.
14. Wu, H.; Tang, Z.; Yang, H.; Luo, Y.; Huang, K.; Gan, X. Effects of dietary inositol on growth, liver and muscle fat content and biochemical indices of blood serum in tilapia (*Oreoc hromis niloticus* × *Oreochromis aureus*). *J. South. Agric.* **2011**, *42*, 1415–1419.
15. Kanazawa, A.; Teshima, S.; Tanaka, N. Nutritional requirements of prawn V: Requirements for choline and inositol. *Mem. Fac. Fish. Kagoshima Univ.* **1976**, *25*, 47–51.
16. Shiau, S.; Su, S. Dietary inositol requirement for juvenile grass shrimp, *Penaeus Mondon*. *Aquac.* **2004**, *241*, 1–8. [CrossRef]
17. Liu, T.; Li, A.; Zhang, J. Studies on vitamin nutrition for the shrimp penaeus chinensis-X: Studies on the choline chloride and inositol requirements in the shrimp *Penaeus chinensis*. *J. Ocean Univ. Qingdao* **1993**, *23*, 71–78.
18. Bu, X.; Zhu, J.; Liu, S.; Wang, C.; Xiao, S.; Lu, M.; Li, E.; Wang, X.; Qin, J.G.; Chen, L. Growth, osmotic response and transcriptome response of the euryhaline teleost, *Oreochromis mossambicus* fed different *myo*-inositol levels under long-term salinity stress. *Aquaculture* **2021**, *534*, 736294. [CrossRef]
19. Chen, S.; Guo, Y.; Espe, M.; Yang, F.; Fang, W.; Wan, M.; Niu, J.; Liu, Y.; Tian, L. Growth performance, haematological parameters, antioxidant status and salinity stress tolerance of juvenile Pacific white shrimp (*Litopenaeus vannamei*) fed different levels of dietary *myo*-inositol. *Aquac. Nutr.* **2018**, *24*, 1527–1539. [CrossRef]
20. Zhao, Z.; Zhao, F.; Cairang, Z.; Zhou, Z.; Du, Q.; Wang, J.; Zhao, F.; Wang, Q.; Li, Z.; Zhang, X. Role of Dietary Tea Polyphenols on Growth Performance and Gut Health Benefits in Juvenile Hybrid Sturgeon (*Acipenser baerii* ♀× *A. schrenckii* ♂). *Fish Shellfish Immunol.* **2023**, *139*, 108911–108921. [CrossRef] [PubMed]
21. Zhang, Q.; Yang, Q.; Li, M.G.; Li, F.; Qin, M.; Xie, Y.; Xu, J.; Liu, Y.; Tong, T. The Effects of Dietary Fermented Soybean Meal Supplementation on the Growth, Antioxidation, Immunity, and mTOR Signaling Pathway of Juvenile Coho Salmon (*Oncorhynchus kisutch*). *Fishes* **2023**, *8*, 448. [CrossRef]
22. Yuan, H.; Hu, N.; Zheng, Y.; Hou, C.; Tan, B.; Shi, L.; Zhang, S. A Comparison of Three Protein Sources Used in Medium-Sized *Litopenaeus vannamei*: Effects on Growth, Immunity, Intestinal Digestive Enzyme Activity, and Microbiota Structure. *Fishes* **2023**, *8*, 449. [CrossRef]
23. Chu, J.-H.; Weng, T.-S.; Huang, T.-W. The Effects of Replacing Fish Meal Protein with Black Soldier Fly Meal and Sodium Butyrate Supplementation on the Growth Performance, Lipid Peroxidation, and Intestinal Villi Status of Jade Perch, *Scortum Barcoo Fingerlings*. *Fishes* **2023**, *8*, 437. [CrossRef]
24. Yang, Z.; Liu, P.; Kong, Q.; Deng, Y.; Zhang, W.; Xu, G.; Tang, H. Effects of Co-Fermented Feed Using Lactobacillus acidophilus, *Limosilactobacillus reuteri* and *Lactiplantibacillus plantarum* on Growth, Antioxidant Capacity, Fatty Acids and Gut Microbiota of Largemouth Bass (*Micropterus salmoides*). *Fishes* **2023**, *8*, 433–450. [CrossRef]
25. Aoac, I. *Official Methods of Analysis of AOAC International*, 18th ed.; AOAC International: Maryland, MD, USA, 2010.
26. Yin, G.; Zhen, Y.; Wang, T.; Zhang, X.; Sun, Z. The Mechanism of Inositol and Its Application in Animal Production. *J. Econ. Anim.* **2023**, *11*, 1–7.
27. Levine, A.; Billington, C. Why do we eat? A neural systems approach. *Annu. Rev. Nutr.* **1997**, *17*, 597–619. [CrossRef]
28. Chen, S.; Ji, W.; Lv, Y.; Chang, Q.; Pan, S. Nutrient function of inositol on the growth of black sea bream, juvenile, *Sparus Marcocephalus*. *J. Mar. Sci.* **1995**, *5*, 13–15.
29. Burtle, G.; Lovell, R. Lack of response of channel catfish (*Lctalurus punctatus*) to dietary *myo*-inositol. *Can. J. Fish. Aquat. Sci.* **2011**, *46*, 218–222. [CrossRef]
30. Waagbø, S.; Lie, R. Effects of inositol supplementation on growth, chemical composition and blood chemistry in Atlantic salmon (*Salmo salar* L.)fry. *Aquac. Nutr.* **1998**, *4*, 53–59. [CrossRef]
31. Mai, K.; Wu, G.; Zhu, W. Abalone, Haloiotis discus hannai Ino, can synthesize *myo*-inositol de novo to meet physiological needs. *Nuturition* **2001**, *131*, 2898–2903.
32. Chen, X.; Wang, J.; Zhao, W. Effects of dietary inositol supplementation on growth, digestive performance, antioxidant capacity, and body composition of golden pompano(*Trachinotus ovatus*). *Front. Physiol.* **2022**, *13*, 850470. [CrossRef] [PubMed]
33. Peres, H.; Lim, C.; Klesius, P. Growth, chemical composition and resistance to Streptococcus iniae challenge of juvenile nile tilapia (*Oreochromis niloticus*) fed graded levels of dietary inositol. *Aquaculture* **2004**, *235*, 423–430. [CrossRef]
34. Wang, Y.; Li, B.; Wang, J.; Wang, C.; Wang, X.; Ma, C.; Wang, S.; Sun, Y.; Hao, T. Dietary *Myo*-inositol Requirement of Juvenile Turbot(*Scophthalmus maximus* L.). *Chin. J. Anim. Nutr.* **2019**, *31*, 5122–5132.

35. Jiang, W.; Feng, L.; Liu, Y.; Jiang, J.; Zhou, X. Growth, digestive capacity and intestinal microflora of juvenile Jian carp (*Cyprinus carpio* var. Jian) fed graded levels of dietary inositol. *Aquac. Res.* **2010**, *40*, 955–962. [CrossRef]
36. Yu, Q.; Han, F.; Huang, M.; Wang, X.; Qin, J.G.; Chen, L.; Li, E. The effects of dietary *myo*-inositol on growth and physiological, biochemical, and molecular responses in the Pacific white shrimp (*Litopenaeus vannamei*). *Aquaculture* **2023**, *568*, 739323. [CrossRef]
37. Zhang, S.; Wang, C.; Lu, S.; Liu, S.; Wang, Y.; Liu, H.; Yang, Y.; Xu, Q. Effects of dietary *myo*-Inositol on digestive physiology of *Acipenser Schrenckii*. *Feed Ind.* **2022**, *43*, 53–59.
38. Yone, Y.; Furuichi, M.; Shitanda, K. Vitamin requirement of the red sea bream: Relationship between inositol requirements and glucose levels in diet. *Bull. Jpn. Soc. Sci. Fish.* **1971**, *37*, 149–155. [CrossRef]
39. Lin, Y.; Zhou, X. Dietary glutamine supplementation improves structure and function of intestine of juvenile Jian carp (*Cyprinus carpio* var. Jian). *Aquaculture* **2006**, *256*, 389–394.
40. Fang, Z.; Wang, C.; Wei, H. Effects of mercury and selenium on $Na^+$,$K^+$-ATPase activity in xiphophorus heller heckle. *Chin. J. Appl. Environ. Biol.* **2006**, *12*, 220–223.
41. Xu, Q.; Wang, C.; Xu, H.; Zhu, Q.; Yin, J. Effects of glutamine dipeptide on intestine antioxidant, abilities of digestion and absorption of *Hucho taimen* larvae. *J. Fish. Sci. China* **2010**, *17*, 351–356.
42. Ye, J.; Han, Y.; Zhao, J.; Lu, T.; Liu, H.; Yang, Y. Effects of dietary olaquindox on antioxidant enzymes system in hepatopancreas of *Cyprinus Carpio*. *J. Fish. China* **2004**, *28*, 231–235.
43. Liu, Z.; Song, W.; Yu, B. Effects of inositol on tissue lipid and protein oxidation and antioxidant of crucian. *China Feed* **2019**, *12*, 39–43.
44. Hu, K.; Li, S.; Feng, L.; Jiang, W.; Wu, P.; Liu, Y.; Jiang, J.; Kuang, S.; Tang, L.; Zhou, X. Protective effect of *myo*-inositol on oxidative damage of head kidney and spleen in juvenile grass carp (*Ctenopharyngodon idella*) induced by Aeromonas hydrophila. *J. Fish. China* **2019**, *43*, 2256–2267.

*fishes*

MDPI

*Article*

# The Extract of *Astragalus membranaceus* Inhibits Lipid Oxidation in Fish Feed and Enhances Growth Performance and Antioxidant Capacity in Jian Carp (*Cyprinus carpio* var. Jian)

**Jing Xu [1], Gangfu Chen [1], Min Wu [1], Qihui Yang [2] and Huatao Li [1,*]**

[1] Key Laboratory of Sichuan Province for Conservation and Utilization of Fishes Resources in the Upper Reaches of the Yangtze River, College of Life Sciences, Neijiang Normal University, Neijiang 641100, China; 10001934@njtc.edu.cn (J.X.); 10002034@njtc.edu.cn (G.C.); 40001246@njtc.edu.cn (M.W.)

[2] College of Fisheries, Guangdong Ocean University, Zhanjiang 524088, China; yangqh@gdou.edu.cn

* Correspondence: 10001187@njtc.edu.cn

**Abstract:** In this study, a linoleic and linolenic acid were incubated with petroleum ether extract, ethyl acetate extract, acetone extract (AE) and aqueous extract of *Astragalus membranaceus*. The phenolic content and total antioxidant capacity (T-AOC) were determined in the extracts of *Astragalus membranaceus* (EAms) above. Results showed that EAms decreased the levels of malonaldehyde, conjugated diene, and peroxide value levels in material above. Of all of EAms, AE showed the strongest T-AOC and inhibitory effect on the lipid oxidation. Next, fish feeds were incubated with graded levels of AE. The results showed that AE inhibited lipid oxidation in fish feed. The appropriate dosage for reducing lipid oxidation was 6.74 g AE kg$^{-1}$ feeds. The effect of EAms on the lipid oxidation may be closely associated with their phenolic content. Then, juvenile Jian carp (*Cyprinus carpio* var. Jian, 10.2 ± 0.3 g) were fed with diets containing graded levels of AE (0.0, 1.0, 2.0, 3.0, 4.0, 5.0, 6.0, and 7.0 g kg$^{-1}$) for 60 days. Current data displayed that dietary AE increased the growth performance of fish. The optimum dosage for growth promotion was 5.15 g AE kg$^{-1}$ diet. This result of AE may be ascribed to its enhancing effect on the activity of digestive and absorptive enzymes and antioxidant capacity in digestive organs of fish. Our present study indicated that EAm holds promise as a natural antioxidant for fish and their feed.

**Keywords:** *Astragalus membranaceus*; lipid oxidation; growth performance; digestion; absorption; antioxidant

**Key Contribution:** A. Extracts of *Astragalus membranaceus* can depress lipid oxidation in fish feed. B. Extracts of *Astragalus membranaceus* can improve fish growth performance and antioxidant capacity.

**Citation:** Xu, J.; Chen, G.; Wu, M.; Yang, Q.; Li, H. The Extract of *Astragalus membranaceus* Inhibits Lipid Oxidation in Fish Feed and Enhances Growth Performance and Antioxidant Capacity in Jian Carp (*Cyprinus carpio* var. Jian). *Fishes* **2023**, *8*, 594. https://doi.org/10.3390/fishes8120594

Academic Editor: Francisco Javier Alarcón López

Received: 30 October 2023
Revised: 27 November 2023
Accepted: 28 November 2023
Published: 1 December 2023



## 1. Introduction

Fish feed is distinguished by its high lipid content, which notably includes a substantial proportion of unsaturated fatty acids [1,2]. In aquatic feeds, the presence of an abundance of redox-active metal ions and reactive oxygen species (ROS) leads to the rapid oxidation of these unsaturated fatty acids. This oxidation process results in the formation of various oxidative byproducts, including peroxides (PO), conjugated dienes (CD), and malondialdehyde (MDA) [3]. It is recognized that lipid oxidation can degrade nutritional components, produce toxic metabolites, and reduce the shelf life of the feed [4]. Moreover, oxidative deterioration of feed may impair fish growth, decrease disease resistance, and increase mortality rates [5]. It is well known that the antioxidants can quench lipid oxidation in fish feed [6]. Antioxidants are crucial components in the formulation of animal feeds. Currently, the feed industry predominantly utilizes synthetic antioxidants, such as ethoxyquin (EQ), butylhydroxyanisole (BHA), and dibutylhydroxytoluene (BHT). However, these additives have not been shown to have significant benefits for the growth and health of fish [6]. Recent studies from our laboratory have indicated that EQ and BHT are effective in curbing lipid oxidation in fish feed [7]. Yet, there are concerns as several reports suggest potential carcinogenicity and toxicity of these synthetic antioxidants in animal models [8]. Considering their adverse effects on both animal and human health, it is imperative to enforce stringent regulations on the usage of these synthetic antioxidants in the food industry [9]. Otherwise, it would lead to residues and enrichment in fish products, causing food safety risks and endangering human health by consumption [10–12]. Meanwhile, the disadvantages associated with synthetic antioxidants will lead to increased market demand for antioxidants from natural ingredients, and it is of great significance to develop new natural feed antioxidants.

In *Astragalus membranaceus*, a traditional Chinese medicine herb, the main bioactive chemical components are phenols, flavonoids, saponins, and polysaccharides [13]. Recently, our laboratory reported that supplementation of the extract of *Astragalus membranaceus* (EAm) decreased the generation of ROS in fish erythrocytes induced by the hydroxyl radical ($^{\bullet}OH$) [14]. Moreover, study in vitro showed that *Astragalus mongholicus* extract has the effects of scavenging lipid free radicals detected by electron spin resonance (ESR) [15]. Similarly, in the liver of juvenile *Pangasianodon hypophthalmus*, dietary inclusion of *Astragalus membranaceus* extract decreased MDA levels, which could be a biomarker of lipid peroxidation [16]. Meanwhile, dietary supplementation of *Astragalus membranaceus* extract not only enhanced fish growth performance and feed utilization but also improved digestive ability and antioxidant capacity [16]. So, it can be concluded that EAm may be used as a potential natural antioxidant in fish feed. However, there are limited reports on the use of EAm in fish feed.

Jian carp (*Cyprinus carpio* var. Jian) is one of the most important freshwater farmed and high-yield fish in China because of its fast growth, strong resistance again disease, and strong adaptability [17]. At the same time, its high nutritional value, cheap price, and delicious taste, are favored by consumers [18]. On the basis of former research, EAm might has great application potential as a natural antioxidant element in fish, but the effects of EAm on Jian carp has not been evaluated. Therefore, this study evaluated the mitigation effect of EAm on lipid oxidation in feed, as well as its effects on growth performance, digestion and absorption capability, and antioxidant capacity in Jian carp, providing reference for the application of EAm in aquaculture species.

## 2. Materials and Methods

### 2.1. Chemical Reagent

In this research, key reagents of analytical reagent (AR) grade were carefully selected from reputable sources. Chengdu Kelong Chemical Reagent Factory, located in Chengdu of China, supplied essential chemicals such as petroleum ether, ethyl acetate, Tween 20, and acetone. These reagents were instrumental in our experimental procedures. Specifically, for extracting active compounds from *Astragalus membranaceus*, water were utilized to isolate water-soluble substances, while petroleum ether, ethyl acetate and acetone were effective in extracting water-insoluble substances [19]. Moreover, to ensure high accuracy in our analysis, linoleic acid and linolenic acid of exceptional purity (exceeding 97% and 95%, respectively) were procured from Shanghai Biochemical Reagent Co., Ltd., a renowned supplier based in Shanghai, China. All other chemicals used in the study conformed to the stringent requirements of analytical reagent grade, ensuring the reliability and precision of our experimental results.

### 2.2. Preparation of EAm

The root of *Astragalus membranaceus* utilized in this study was obtained from the Chengdu Pharmaceuticals market located in Chengdu, China. The roots, once dried, were processed into fine powder; ensuring particles did not exceed 0.32 mm in size, using a traditional Chinese medicine mill (Model RHP-2000A, Ronghao, Zhejiang, China). For the extraction process, in line with the methodology outlined by Wojcikowski et al. [20],

50 g of this powdered root underwent extraction. This involved the use of 500 mL each of petroleum ether, ethyl acetate, acetone, and water, respectively. The procedure was carried out at a controlled temperature of 20 °C over 8 h, employing an agitator (Model OS40-S, Dalong, Beijing, China) for each solvent. The extraction process for each solvent was meticulously repeated three times to ensure consistency. Post-extraction, the filtrates were subjected to evaporation under reduced pressure using a rotary evaporator (Model RE-52CS, Jinye, Shanghai, China), continuing until a constant dry mass was achieved, as referenced in studies [14,21]. Consequently, this method ensured the complete volatilization of organic solvents in the ethyl acetate extract (EAE), petroleum ether extract (PEE), and acetone extract (AE). The resulting extracts, obtained using water, ethyl acetate, petroleum ether, and acetone, were then securely stored in airtight containers, shielded from light, and maintained at a temperature of −80 °C until further analysis was conducted.

### 2.3. Measurement of Phenolic Content and Total Antioxidant Capacity (T-AOC)

The quantification of the phenolic compounds in the EAms was conducted based on the protocol established by Yuan et al. [22]. For this analysis, 100 μL of the extract samples were blended with 2 mL of a 2% solution of $Na_2CO_3$. This mixture was then left to stand at ambient temperature for 2 min. Following this incubation period, 100 μL of a 50% solution of Folin–Ciocalteau reagent was introduced to the mixture. Subsequently, the samples were left to settle at ambient temperature for an additional period of 30 min without any disturbance. Following this incubation phase, absorbance values were recorded at a 720 nm wavelength. The phenolic composition of the extracts was quantified in terms of gallic acid equivalents. This quantification was carried out through four individual replicates for each sample to ensure accuracy and reproducibility. In assessing the total antioxidant capacity of EAms, the method described by Serpen et al. [23] was employed based on ABTS rapid oxidation. Trolox, a standard antioxidant, was utilized as a reference for calibrating the antioxidant capacity of the samples. This calibration allowed for the expression of the samples' antioxidant ability in terms of Trolox equivalent antioxidant capacity (TEAC).

### 2.4. Measurement of Lipid Oxidation in Linoleic Acid and Linolenic Acid Emulsion

To investigate the antioxidant properties of EAms on the emulsion of linoleic acid and linolenic acid, our methodology was adapted from Yuan et al. and Li et al. [22,24]. We prepared the emulsion with equal volumes (0.1 mL) of both acids, combined with 9.9 mL of a phosphate buffer (pH 7.0, 0.02 M) and 50 μL Tween 20, followed by homogenization using a FJ200-SH homogenizer (Shanghai, China) for two 10 s periods at 21,000× *g* [24]. EAm concentrations, including a control (0 mg $mL^{-1}$) and 1.0 mg $mL^{-1}$, were integrated into the emulsion. After 8-day incubation at 45 °C, the levels of MDA, CD, and PO were quantified using the Maqsood and Benjakul method [25]. For this, the emulsion was combined with 2.5 mL of TBA solution, heated, cooled, and centrifuged. Absorbance readings were carried out of the supernatant at 532 nm for MDA. The absorbance of CD was determined by UV-1601 spectrophotometer (Shimadzu, Kyoto, Japan) at 234 nm. The PO measurement involved an emulsion sample with ethanol, ammonium thiocyanate, and ferrous chloride solution, followed by an absorbance reading at 500 nm after 3 min.

### 2.5. Measurement of Lipid Oxidation in Fish Feed

In this experimental study, AE demonstrated superior total antioxidant capacity (T-AOC) and effectiveness in counteracting oxidative processes in both linoleic and linolenic acids. Given these findings, AE was chosen as the optimal EAm for evaluating the lipid oxidation markers in fish feed. For the preparation of experimental diets, we adhered to a rigorous protocol established by our research team [26]. A series of ten diets was meticulously formulated, incorporating increasing concentrations of AE, starting from a baseline of 0.0 g $kg^{-1}$ (serving as the control) and escalating up to 9.0 g $kg^{-1}$ in 1.0 g $kg^{-1}$ increments. To maintain nutritional integrity, any augmentation in AE levels was counter-balanced by a proportional reduction in the cellulose content of the diet. As delineated in Table 1, the basic composition of the diet included 34.32% crude protein and 6.84% crude lipid. The processing method for this feed mirrored the emulsification process earlier described for linoleic and linolenic acids. To quantify the lipid oxidation in the fish feed, we employed established analytical methods to measure the levels of malondialdehyde (MDA), conjugated diene (CD), and peroxide (PO), with the results expressed as a percentage relative to the control group.

**Table 1.** Composition and nutrient content of the basal diet.

| Ingredients | % | Proximate Analysis [4] | % |
|---|---|---|---|
| Fish meal | 27.00 | Dry matter | 92.82 |
| Soybean meal | 36.00 | Crude protein | 34.32 |
| Wheat flour | 32.50 | Crude lipid | 6.84 |
| $Ca(H_2PO_4)_2$ [1] | 0.50 | Crude Ash | 6.13 |
| Corn oil | 2.00 | | |
| Vitamin mixture [2] | 1.00 | | |
| Mineral mixture [3] | 1.00 | | |

[1] Content ≥ 99%, food grade. [2] Per kg of vitamin mix: cholecalciferol (500,000 IU $g^{-1}$), 0.48 g; retinyl acetate (500,000 IU $g^{-1}$), 0.80 g; menadione (23%), 0.43 g; DL-α-tocopherol acetate (50%), 20.00 g; riboflavine (80%), 0.63 g; thiamin nitrate (90%), 0.11 g; cyanocobalamin (1%), 0.10 g; pyridoxine HCl (81%), 0.92 g; D-calcium pantothenate (90%), 2.73 g; ascorhyl acetate (93%), 7.16 g; D-biotin (2%), 5.00 g; niacin (99%), 2.82 g; folic acid (96%), 0.52 g; meso-inositol (99%), 52.33 g. [3] Per kg of mineral mix: $CuSO_4 \cdot 5H_2O$ (25% Cu), 1.20 g; $FeSO_4 \cdot 7H_2O$ (20% Fe), 69.70 g; $MnSO_4 \cdot H_2O$ (32% Mn), 4.09 g; $ZnSO_4 \cdot 7H_2O$ (23% Zn), 21.64 g; KI (4% I), 2.90 g; $CaCO_3$, 897.98 g; $Na_2SeO_3 \cdot 5H_2O$ (1% Se), 2.50 g. [4] Proximate analyses were carried out by the program of Chen et al. [27].

### 2.6. Feeding Trial

Juvenile Jian carp were obtained from a fish farm in Neijiang, China. In the laboratory, their habitat was meticulously replicated to mirror their natural environment, maintaining a temperature of (22.0 ± 1) °C, with daily changes of dechlorinated water and a natural light cycle [26,27]. The baseline diet was provided for the fish. A total of 640 juvenile Jian carp, each weighing an average of 10.2 ± 0.3 g, were distributed randomly into eight experimental groups. Each group was allocated four tanks, with each tank housing 20 fish. The experimental diets fed to these carp were carefully prepared with incremental concentrations of AE, namely 0.0, 1.0, 2.0, 3.0, 4.0, 5.0, 6.0, and 7.0 g per kilogram of feed. Over a period of 60 days, each group of fish in the study received eight daily feedings. This feeding regime was meticulously monitored to achieve full satiation without the risk of overfeeding, as per the guidelines in the referenced study [28]. To accurately assess various growth parameters, the fish were weighed and counted both at the onset and conclusion of the trial. These measurements facilitated the calculation of several key indicators: survival rate (SR), feed intake (FI), weight gain (WG), feed efficiency (FE), and specific growth rate (SGR).

The methodology for collecting samples was in line with the standardized procedures developed by our research team [29]. Following the completion of the experimental phase, the fish were gently sedated using a 50 mg $L^{-1}$ benzocaine solution to minimize stress. Once the fish was sedated and killed, and key organs such as the hepatopancreas and intestine were carefully extracted for further study. To preserve the integrity of these tissues for subsequent analyses, they were immediately frozen and stored at an ultra-low temperature, specifically −80 °C.

### 2.7. Biochemical Analysis

Tissue specimens were processed by homogenization in a solution comprising nine parts (*w*/*v*) of chilled physiological saline. This homogenate then underwent centrifugation at 3200× *g* at a temperature of 4 °C for duration of 20 min. The resulting supernatant was preserved for the analysis of various enzymes, including trypsin, lipase, alpha-amylase (referred to as amylase), alkaline phosphatase (AKP), and $Na^+/K^+$-ATPase, as delineated in our preceding study [29]. Additionally, assessments for anti-superoxide anion (ASA) capacity and reduced glutathione (GSH) level were conducted using the same method.

Parallel, enzyme activities such as catalase (CAT), superoxide dismutase (SOD), glutathione reductase (GR), and glutathione peroxidase (GPx) in the intestinal or hepatopancreatic tissues were analyzed following the methodologies established by Jiang et al. [30]. MDA measurement was consistent with those in feed as described by Maqsood and Benjakul [25]. Protein content was quantified according to the method of Bradford [31]. The above indexes read the color intensity at different wavelengths to calculate enzyme activities or contents.

### 2.8. Statistical Analysis

The study's results are displayed as mean values ± standard deviation (S.D), with in vitro studies involving three replicates ($n = 3$) and animal experiments comprising four replicates ($n = 4$). For the statistical evaluation of the data, a one-way analysis of variance (ANOVA) was utilized, employing SPSS software, version 15.0. Prior to conducting the ANOVA, the data underwent preliminary tests for consistency and normality, specifically through Levene's test for homogeneity and the Shapiro–Wilk test for normal distribution. To identify significant variances between the groups post-analysis, Duncan's multiple range tests were applied for detailed post hoc comparison.

## 3. Results

### 3.1. Phenolic Content and Total Antioxidant Capacity in EAms

We determined glycosides, flavonoids and phenolic content in EAms. The correlation analysis results show that the phenolic compounds have the most effective antioxidant activity. Thus, only phenolic compounds were listed in this study. According to Table 2, AE had a significantly higher phenolic content than other EAms ($p < 0.05$). The phenolic content in EAE was substantially lower than in AE ($p < 0.05$). AQE phenolic content was notably lower compared to other extracts ($p < 0.05$), while PEE had a higher phenolic content than AQE. The content of phenolic of PEE was higher than AQE ($p < 0.05$). Regarding T-AOC, AE exhibited higher levels than other extracts ($p < 0.05$), followed by EAE, PEE, and AQE. Figure 1 presents the power regression analysis of phenolic content in EAms against T-AOC, revealing a positive correlation between T-AOC and phenolic content in EAms.

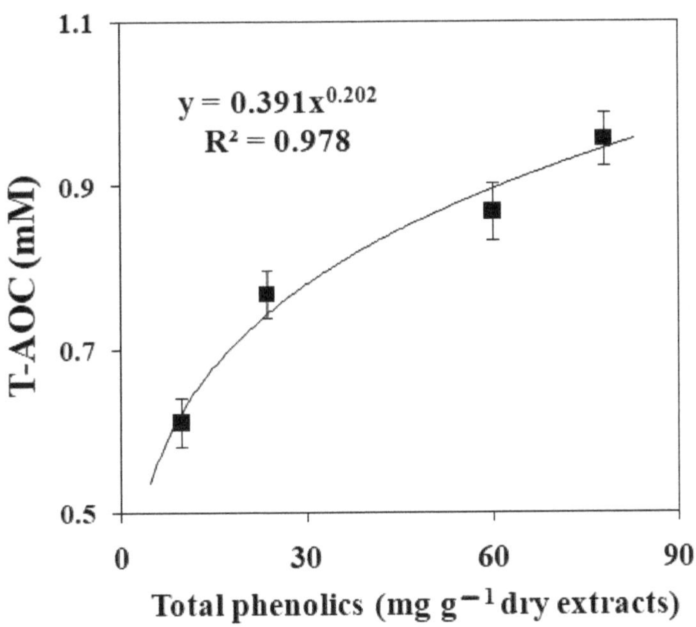


**Figure 1.** Power regression of the total antioxidant capacity (T-AOC) of *Astragalus membranaceus* extracts containing different levels of phenolics. The data were the means ± S.D. of 3 replicates.

**Table 2.** The total antioxidant capacity (T-AOC) and phenolic content of petroleum ether extract (PEE), acetone extract (AE), ethyl acetate extract (EAE), and aqueous extract (AQE) of *Astragalus membranaceus*.

| Extracts | Phenolics (mg $g^{-1}$ Dry Extracts) | T-AOC (mM of Trolox) |
|---|---|---|
| PEE | $23.67 \pm 1.35$ $^{b}$ | $0.77 \pm 0.03$ $^{b}$ |
| EAE | $59.87 \pm 4.12$ $^{c}$ | $0.87 \pm 0.03$ $^{c}$ |
| AE | $77.79 \pm 5.35$ $^{d}$ | $0.96 \pm 0.03$ $^{d}$ |
| AQE | $9.75 \pm 0.47$ $^{a}$ | $0.61 \pm 0.03$ $^{a}$ |

Within the same column, there were significant differences among different superscripts ($p < 0.05$), and the data were the means ± S.D. of 3 replicates.

### 3.2. Effects of EAm on the Lipid Oxidation in Linoleic Acid and Linolenic Acid Emulsion

Table 3 presents a correlation analysis between EAms and the concentrations of MDA, CD, and PO in a linoleic acid emulsion. It was observed that the levels of PO were significantly reduced when treated with AE and EAE, with PEE also demonstrating a reduction, albeit to a lesser extent ($p < 0.05$). A similar trend was noted for CD concentrations under various EAm treatments. In terms of reducing MDA content, AE was the most effective, closely followed by EAE and PEE. Conversely, the application of AQE resulted in the highest recorded levels of PO, CD, and MDA, suggesting its relative ineffectiveness in curtailing the oxidation of linoleic acid ($p < 0.05$).

**Table 3.** The peroxide value (PO), malonaldehyde (MDA), and conjugated diene (CD) levels in a linoleic acid emulsion treated with 1.0 mg $mL^{-1}$ of ethyl acetate extract (EAE), petroleum ether extract (PEE), aqueous extract (AQE), and acetone extract (AE) of *Astragalus membranaceus*.

| Treatment | PO (% of Control) | CD (% of Control) | MDA (% of Control) |
|---|---|---|---|
| PEE | $79.56 \pm 5.07$ $^{b}$ | $78.38 \pm 2.85$ $^{b}$ | $80.36 \pm 3.58$ $^{b}$ |
| EAE | $72.26 \pm 5.17$ $^{a}$ | $70.43 \pm 2.56$ $^{a}$ | $75.42 \pm 3.21$ $^{ab}$ |
| AE | $69.83 \pm 4.35$ $^{a}$ | $66.95 \pm 1.67$ $^{a}$ | $71.86 \pm 1.88$ $^{a}$ |
| AQE | $90.39 \pm 3.47$ $^{c}$ | $84.66 \pm 5.16$ $^{c}$ | $91.12 \pm 4.31$ $^{c}$ |

Within the same column, there were significant differences among different superscripts ($p < 0.05$), and the data were the means ± S.D. of 3 replicates.

Table 4 outlines the effects of EAms on the levels of MDA, CD, and PO in a linolenic acid emulsion. The results indicated a significant reduction in PO concentrations with AE treatment, while EAE and PEE treatments also contributed to decreases, although to a smaller degree ($p < 0.05$). The influence of AE and EAE was notable in lowering CD levels, with PEE showing a similar but more modest impact ($p < 0.05$). This trend was mirrored in the reduction of MDA content across the different EAm treatments. Among the extracts tested, AQE was associated with the highest concentrations of PO, CD, and MDA, suggesting its limited efficacy in reducing the oxidation of linolenic acid ($p < 0.05$).

**Table 4.** The peroxide value (PO), malonaldehyde (MDA), and conjugated diene (CD) levels in a linolenic acid emulsion treated with 1.0 mg $mL^{-1}$ of ethyl acetate extract (EAE), petroleum ether extract (PEE), aqueous extract (AQE), and acetone extract (AE) of *Astragalus membranaceus*.

| Treatment | PO (% of Control) | CD (% of Control) | MDA (% of Control) |
|---|---|---|---|
| PEE | $81.26 \pm 3.30$ $^{c}$ | $79.47 \pm 4.72$ $^{b}$ | $78.23 \pm 4.23$ $^{b}$ |
| EAE | $74.40 \pm 5.62$ $^{b}$ | $72.00 \pm 3.40$ $^{ab}$ | $71.10 \pm 3.69$ $^{a}$ |
| AE | $66.12 \pm 4.35$ $^{a}$ | $67.47 \pm 2.40$ $^{a}$ | $69.46 \pm 5.07$ $^{a}$ |
| AQE | $89.38 \pm 3.23$ $^{d}$ | $88.36 \pm 3.74$ $^{c}$ | $86.99 \pm 3.79$ $^{c}$ |

Within the same column, there were significant differences among different superscripts ($p < 0.05$), and the data were the means ± S.D. of 3 replicates.

Figure 2 depicts the association between the phenolic content present in EAms and the concentrations of PO, MDA, and CD in linolenic and linoleic acid emulsions. The data indicate a direct relationship where increased phenolic content in EAms correlates with higher levels of PO, CD, and MDA.

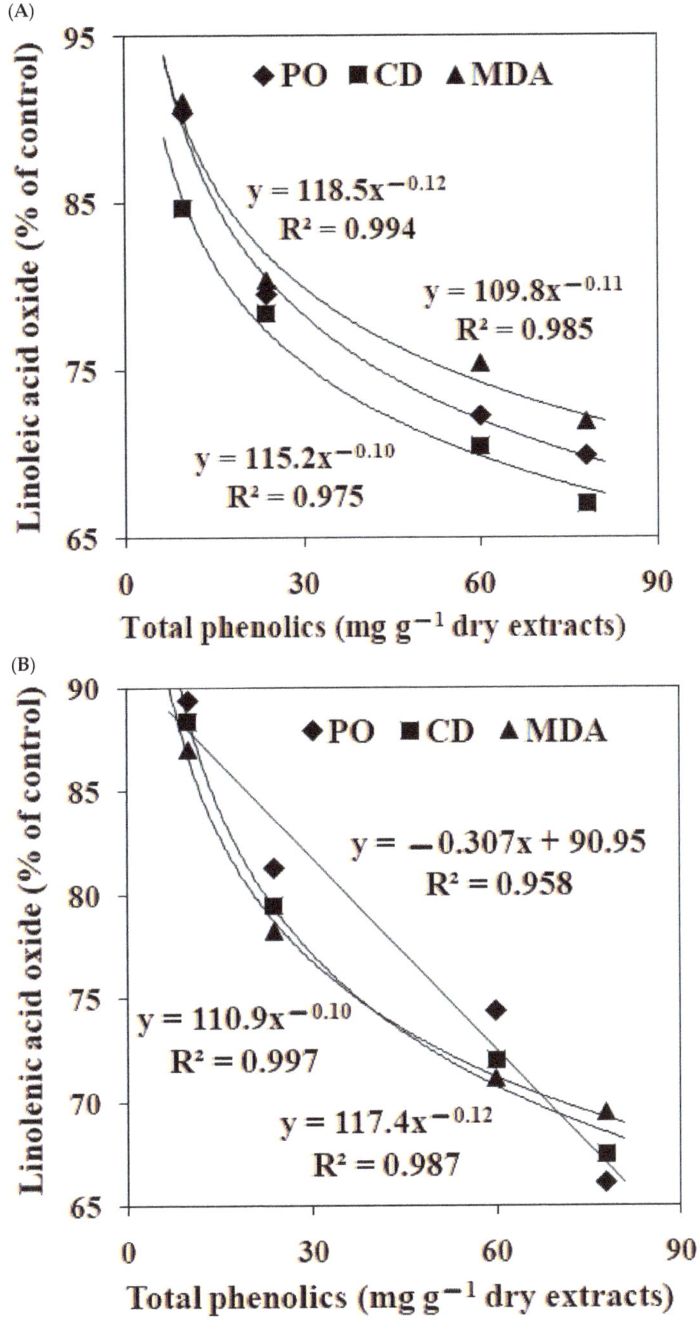


**Figure 2.** The correlations of phenolic content in the extracts of *Astragalus membranaceus* with the levels of peroxide value (PO), conjugated diene (CD), and malondialdehyde (MDA) in linoleic acid emulsion (**A**) and linolenic acid emulsion (**B**). The data were the means ± S.D. of 3 replicates.

*3.3. The Effects of AE on the Lipid Oxidation in Fish Feed*

Figure 3 details the impact of AE on the levels of PO, MDA, and CD in fish feed. It was observed that the inclusion of AE in the diet led to a significant reduction in the concentrations of these compounds. This reduction was most notable with AE amounts of up to 6.0 g $kg^{-1}$, as evidenced by a decrease in PO, CD, and MDA levels ($p < 0.05$). Beyond this concentration of AE in the diet, the downward trend in these compounds' levels reached a plateau, with no significant changes observed at higher concentrations of AE ($p > 0.05$). Broken line analysis indicated that the optimal AE inclusion levels for minimizing PO, CD, and MDA in fish feed were 6.09 g $kg^{-1}$, 6.74 g $kg^{-1}$, and 6.29 g $kg^{-1}$ of the diet, respectively.

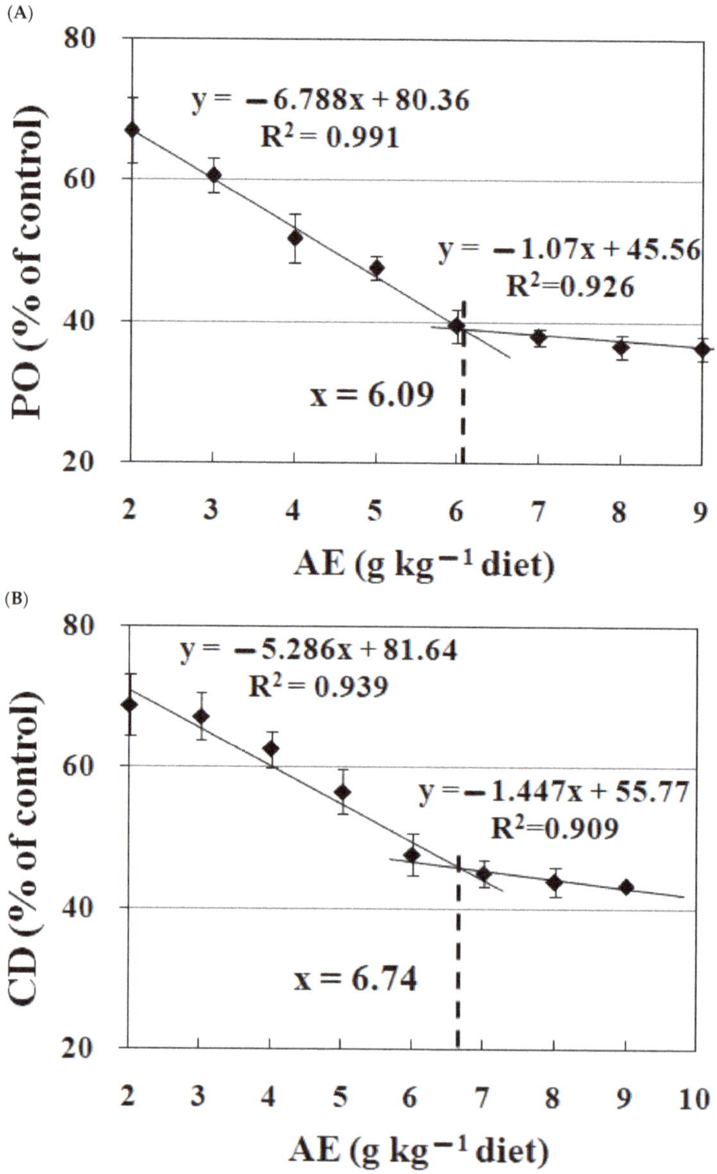


**Figure 3.** *Cont.*

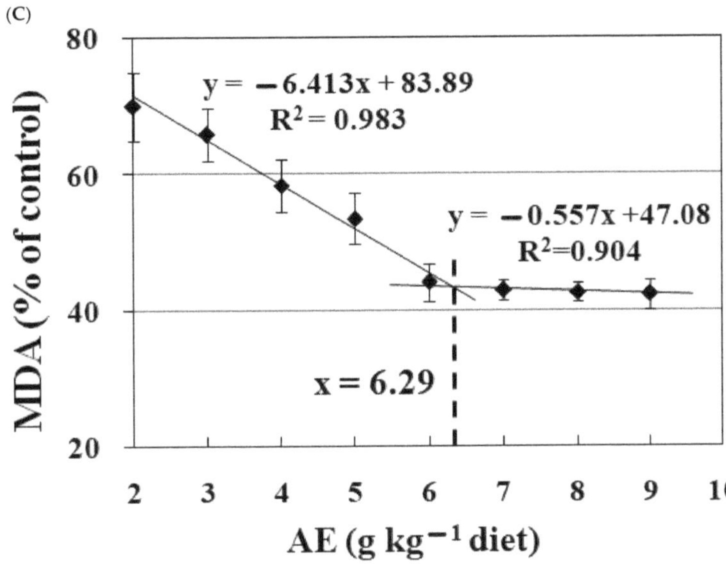


**Figure 3.** Broken-line analysis of peroxide value (PO, (**A**)), conjugated diene (CD, (**B**)), and malondialdehyde (MDA, (**C**)) levels in fish feed treated with graded levels of acetone extract (AE) of *Astragalus membranaceus*. The date were the means ± S.D. of 3 replicates.

### 3.4. Effects of Dietary AE on Fish Growth Performance

Table 5 shows that varying AE levels in the diet did not significantly impact fish FE and SR ($p > 0.05$). Increasing dietary AE to 3.0 g kg$^{-1}$ significantly enhanced WG ($p < 0.05$), with no further significant changes beyond this concentration ($p > 0.05$). Similar trends were observed for FBW, SGR, and FI. Polynomial regression analysis identified 5.15 g kg$^{-1}$ as the optimal AE inclusion level for WG, as depicted in Figure 4. The maximum effective AE level in the diet was determined to be 10.30 g kg$^{-1}$.

**Table 5.** Initial body weight (IBW), final body weight (FBW), weight gain (WG), specific growth rate (SGR), feed intake (FI), feed efficiency (FE), and survival ratio (SR) of Jian carp fed diets containing graded levels of acetone extract (AE) of *Astragalus membranaceus* for 60 days.

| AE (g kg$^{-1}$ Diet) | IBW (g Fish$^{-1}$) | FBW (g Fish$^{-1}$) | WG (g Fish$^{-1}$) | SGR (% d$^{-1}$) | FI (g Fish$^{-1}$) | FE (%) | SR (%) |
|---|---|---|---|---|---|---|---|
| 0.0 | 10.30 ± 0.32 $^{a}$ | 37.62 ± 2.38 $^{a}$ | 27.32 ± 2.39 $^{a}$ | 2.16 ± 0.11 $^{a}$ | 52.95 ± 2.53 $^{a}$ | 51.54 ± 2.43 $^{a}$ | 100.00 ± 0.00 $^{a}$ |
| 1.0 | 10.19 ± 0.29 $^{a}$ | 43.35 ± 1.60 $^{b}$ | 33.16 ± 1.64 $^{b}$ | 2.41 ± 0.08 $^{b}$ | 62.88 ± 2.74 $^{b}$ | 52.77 ± 2.38 $^{a}$ | 100.00 ± 0.00 $^{a}$ |
| 2.0 | 10.22 ± 0.29 $^{a}$ | 48.51 ± 3.24 $^{c}$ | 38.28 ± 3.40 $^{c}$ | 2.59 ± 0.14 $^{c}$ | 68.90 ± 3.86 $^{c}$ | 55.52 ± 2.66 $^{a}$ | 100.00 ± 0.00 $^{a}$ |
| 3.0 | 10.28 ± 0.33 $^{a}$ | 52.59 ± 2.51 $^{d}$ | 42.31 ± 2.18 $^{d}$ | 2.72 ± 0.03 $^{cd}$ | 75.90 ± 3.19 $^{d}$ | 55.76 ± 2.31 $^{a}$ | 100.00 ± 0.00 $^{a}$ |
| 4.0 | 10.23 ± 0.30 $^{a}$ | 53.31 ± 1.96 $^{d}$ | 43.08 ± 2.07 $^{d}$ | 2.75 ± 0.09 $^{d}$ | 78.32 ± 4.11 $^{d}$ | 55.14 ± 4.26 $^{a}$ | 100.00 ± 0.00 $^{a}$ |
| 5.0 | 10.25 ± 0.32 $^{a}$ | 55.52 ± 2.43 $^{d}$ | 45.27 ± 2.62 $^{d}$ | 2.82 ± 0.11 $^{d}$ | 82.30 ± 3.67 $^{d}$ | 55.10 ± 4.16 $^{a}$ | 100.00 ± 0.00 $^{a}$ |
| 6.0 | 10.26 ± 0.31 $^{a}$ | 54.37 ± 2.81 $^{d}$ | 44.11 ± 2.66 $^{d}$ | 2.78 ± 0.07 $^{d}$ | 80.33 ± 4.81 $^{d}$ | 54.96 ± 2.56 $^{a}$ | 100.00 ± 0.00 $^{a}$ |
| 7.0 | 10.27 ± 0.33 $^{a}$ | 52.84 ± 2.05 $^{d}$ | 42.57 ± 2.08 $^{d}$ | 2.73 ± 0.08 $^{cd}$ | 78.10 ± 3.40 $^{d}$ | 54.63 ± 4.28 $^{a}$ | 100.00 ± 0.00 $^{a}$ |

The data were means ± S.D. of 4 replicates, with 20 fish in each replicate. Within the same column, there were significant differences among different superscripts ($p < 0.05$).

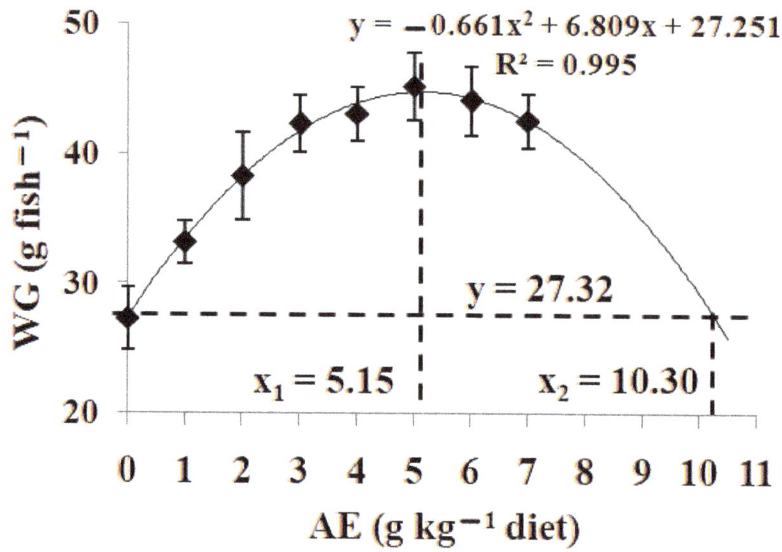


**Figure 4.** Polynomial regression analysis of weight gain (WG) for Jian carp fed diets containing graded levels of acetone extract (AE) of *Astragalus membranaceus* for 60 days. The data were means ± S.D. of 4 replicates, with 20 fish in each replicate.

### 3.5. Effects of Dietary AE on the Biochemical Parameters in Hepatopancreas of Jian Carp

According to the data in Table 6, increasing AE up to 3.0 g $kg^{-1}$ in the diet led to a notable rise in trypsin activity, a change that was statistically significant ($p < 0.05$). Beyond this level of AE, the increase in trypsin activity stabilized, showing no significant improvement with higher AE concentrations ($p > 0.05$). This pattern was also observed in SOD and CAT activities. ASA capacity and lipase activity in hepatopancreas were higher in fish feed with AE supplementation of 4.0 and 2.0 g $kg^{-1}$ diet, respectively. Enzyme amylase and GPx similarly showed significant increases in their activities at AE concentrations of 3.0 g $kg^{-1}$ and 2.0 g $kg^{-1}$ of the diet ($p < 0.05$), respectively, showing no significant improvement with higher AE concentrations ($p > 0.05$). Conversely, MDA content in the intestine was reduced when the diet included 6.0 and 7.0 g $kg^{-1}$ AE.

**Table 6.** The activities of trypsin, lipase, amylase, superoxide dismutase (SOD), anti-superoxide anion (ASA), glutathione peroxidase (GPx), catalase (CAT), and malondialdehyde level (MDA) in hepatopancreas of Jian carp fed diets containing different levels of acetone extract (AE) of *Astragalus membranaceus* for 60 days.

| AE (g $kg^{-1}$ Diet) | Trypsin (U $mg^{-1}$ Protein) | Lipase (U $g^{-1}$ Protein) | Amylase (U $mg^{-1}$ Protein) | ASA (U $g^{-1}$ Protein) | MDA (nmol $mg^{-1}$ Protein) | SOD (U $mg^{-1}$ Protein) | CAT (U $mg^{-1}$ Protein) | GPx (U $mg^{-1}$ Protein) |
|---|---|---|---|---|---|---|---|---|
| 0.0 | $1254 \pm 111^{a}$ | $43.81 \pm 2.66^{a}$ | $1.14 \pm 0.09^{a}$ | $50.19 \pm 4.05^{a}$ | $13.55 \pm 1.13^{b}$ | $87.29 \pm 5.61^{a}$ | $23.00 \pm 1.89^{a}$ | $345 \pm 24.93^{a}$ |
| 1.0 | $1420 \pm 108^{ab}$ | $54.94 \pm 3.74^{b}$ | $1.22 \pm 0.09^{ab}$ | $53.33 \pm 4.47^{ab}$ | $12.93 \pm 0.88^{ab}$ | $89.78 \pm 5.15^{a}$ | $23.18 \pm 0.98^{a}$ | $367 \pm 26.51^{a}$ |
| 2.0 | $1403 \pm 108^{ab}$ | $57.89 \pm 4.52^{bc}$ | $1.28 \pm 0.08^{ab}$ | $61.88 \pm 3.35^{c}$ | $13.01 \pm 0.84^{ab}$ | $98.35 \pm 5.64^{ab}$ | $25.09 \pm 1.36^{ab}$ | $415 \pm 33.59^{b}$ |
| 3.0 | $1512 \pm 98$ $b^{c}$ | $57.04 \pm 4.46^{bc}$ | $1.25 \pm 0.05^{b}$ | $59.79 \pm 4.36^{bc}$ | $12.38 \pm 0.59^{ab}$ | $97.56 \pm 6.76^{ab}$ | $26.92 \pm 1.60^{b}$ | $425 \pm 26.19^{b}$ |
| 4.0 | $1789 \pm 87^{d}$ | $63.24 \pm 4.57^{c}$ | $1.32 \pm 0.08^{b}$ | $60.06 \pm 5.13^{bc}$ | $12.52 \pm 0.84^{ab}$ | $109.09 \pm 9.18^{b}$ | $31.03 \pm 2.27^{c}$ | $421 \pm 23.68^{b}$ |
| 5.0 | $1722 \pm 158^{d}$ | $60.26 \pm 4.82^{bc}$ | $1.27 \pm 0.07^{b}$ | $59.85 \pm 3.90^{bc}$ | $12.31 \pm 0.68^{ab}$ | $105.74 \pm 7.69^{b}$ | $30.52 \pm 2.03^{c}$ | $431 \pm 24.47^{b}$ |
| 6.0 | $1725 \pm 114^{d}$ | $61.28 \pm 3.85^{bc}$ | $1.29 \pm 0.07^{b}$ | $58.92 \pm 4.22^{bc}$ | $11.83 \pm 0.93^{a}$ | $110.48 \pm 9.66^{b}$ | $32.45 \pm 1.49^{c}$ | $430 \pm 33.70^{b}$ |
| 7.0 | $1654 \pm 143^{cd}$ | $58.18 \pm 4.55^{bc}$ | $1.26 \pm 0.06^{b}$ | $56.94 \pm 3.58^{bc}$ | $11.89 \pm 0.74^{a}$ | $105.59 \pm 6.56^{b}$ | $31.86 \pm 1.64^{c}$ | $431 \pm 26.08^{b}$ |

The data were means ± S.D. of 4 replicates, with 5 fish in each replicate. Within the same column, there were significant differences among different superscripts ($p < 0.05$).

### 3.6. Effects of Dietary AE on the Biochemical Parameters in Intestine of Jian Carp

Table 7 presents data showing that the activities of trypsin and $Na^+/K^+$-ATPase, along with the GSH content in the fish intestine, were enhanced as AE levels in the diet increased, up to a maximum of 6.0 g kg$^{-1}$. Amylase activity showed significant improvement at AE concentrations of 3.0 g kg$^{-1}$ ($p < 0.05$), but this enhancement stabilized, with no further significant changes at higher AE levels ($p > 0.05$). ASA capacity and GR activity exhibited similar trends. Notably, lipase activity rose significantly increasing concomitant with AE levels up to 4.0 g kg$^{-1}$ ($p < 0.05$), remained constant at an AE content of 6.0 g kg$^{-1}$ ($p > 0.05$), and then decreased with further increases in AE concentration. AKP activity demonstrated a similar trend. Conversely, MDA content in the intestine was reduced when the diet included 6.0 g kg$^{-1}$ of AE.

**Table 7.** The activities of amylase, lipase, trypsin, alkaline phosphatase (AKP), $Na^+/K^+$-ATPase, glutathione reductase (GR), and anti-superoxide anion (ASA) as well as the contents of reduced glutathione (GSH) and malondialdehyde (MDA) in intestine of Jian carp fed diets containing different levels of acetone extract (AE) of *Astragalus membranaceus* for 60 days.

| AE (g kg$^{-1}$ Diet) | Trypsin (U mg$^{-1}$ Protein) | Lipase (U mg$^{-1}$ Protein) | Amylase (U mg$^{-1}$ Protein) | $Na^+/K^+$-ATPase (U mg$^{-1}$ Protein) | AKP (U mg$^{-1}$ Protein) | ASA (U g$^{-1}$ Protein) | MDA (nmol mg$^{-1}$ Protein) | GSH (mg g$^{-1}$ Protein) | GR (U g$^{-1}$ Protein) |
|---|---|---|---|---|---|---|---|---|---|
| 0.0 | 1068 ± 73 $^{a}$ | 46.41 ± 2.30 $^{a}$ | 1.01 ± 0.07 $^{a}$ | 3.29 ± 0.21 $^{a}$ | 289 ± 22 $^{a}$ | 45.58 ± 2.41 $^{a}$ | 14.21 ± 1.17 $^{c}$ | 14.38 ± 0.90 $^{a}$ | 13.74 ± 1.44 $^{a}$ |
| 1.0 | 1095 ± 92 $^{ab}$ | 47.93 ± 4.30 $^{ab}$ | 1.10 ± 0.08 $^{a}$ | 3.25 ± 0.25 $^{a}$ | 292 ± 24 $^{ab}$ | 50.14 ± 4.70 $^{ab}$ | 14.03 ± 0.84 $^{c}$ | 14.59 ± 0.78 $^{a}$ | 15.38 ± 1.23 $^{ab}$ |
| 2.0 | 1156 ± 113 $^{ab}$ | 50.96 ± 4.20 $^{ab}$ | 1.08 ± 0.08 $^{a}$ | 3.42 ± 0.20 $^{a}$ | 316 ± 21 $^{abc}$ | 49.78 ± 3.13 $^{ab}$ | 13.55 ± 0.86 $^{bc}$ | 14.57 ± 0.95 $^{a}$ | 14.96 ± 1.20 $^{ab}$ |
| 3.0 | 1193 ± 116 $^{ab}$ | 49.78 ± 4.20 $^{ab}$ | 1.23 ± 0.10 $^{b}$ | 4.02 ± 0.31 $^{b}$ | 303 ± 23 $^{abc}$ | 52.77 ± 2.95 $^{b}$ | 13.60 ± 1.02 $^{bc}$ | 15.04 ± 1.20 $^{ab}$ | 15.87 ± 1.41 $^{b}$ |
| 4.0 | 1188 ± 112 $^{ab}$ | 53.00 ± 4.10 $^{b}$ | 1.29 ± 0.07 $^{b}$ | 3.99 ± 0.23 $^{b}$ | 336 ± 22 $^{c}$ | 52.6 ± 2.25 $^{b}$ | 12.84 ± 1.11 $^{abc}$ | 14.96 ± 1.12 $^{ab}$ | 16.12 ± 1.19 $^{b}$ |
| 5.0 | 1264 ± 110 $^{bc}$ | 53.10 ± 3.60 $^{b}$ | 1.27 ± 0.09 $^{b}$ | 4.31 ± 0.31 $^{bc}$ | 332 ± 30 $^{c}$ | 53.56 ± 4.49 $^{b}$ | 13.01 ± 0.77 $^{abc}$ | 15.87 ± 0.80 $^{ab}$ | 16.48 ± 1.22 $^{b}$ |
| 6.0 | 1359 ± 120 $^{c}$ | 54.06 ± 3.70 $^{b}$ | 1.34 ± 0.07 $^{b}$ | 4.53 ± 0.32 $^{c}$ | 332 ± 30 $^{c}$ | 52.11 ± 3.82 $^{b}$ | 12.18 ± 0.87 $^{ab}$ | 16.29 ± 0.98 $^{b}$ | 16.16 ± 1.44 $^{b}$ |
| 7.0 | 1231 ± 94 $^{abc}$ | 52.30 ± 3.60 $^{ab}$ | 1.22 ± 0.08 $^{b}$ | 4.41 ± 0.26 $^{bc}$ | 329 ± 21 $^{bc}$ | 52.45 ± 3.49 $^{b}$ | 11.98 ± 0.90 $^{a}$ | 15.98 ± 0.92 $^{ab}$ | 16.24 ± 1.20 $^{b}$ |

The data were means ± S.D. of 4 replicates, with 5 fish in each replicate. Values with the different superscripts in the same column are significantly different ($p < 0.05$).

## 4. Discussion

### 4.1. EAm Inhibited Lipid Oxidation in Fish Feed

PO, CD, and MDA represent different stages of lipid oxidation in fish feed, occurring in the early, middle, and later phases, respectively, as noted in previous studies [24]. This research found that AE, an extract of *Astragalus membranaceus*, effectively reduced these oxidative products in fish feed. This aligns with previous findings where AE was shown to decrease MDA levels in fish erythrocytes in vitro [14]. However, the interaction between EAm and lipid oxidative products in fish feed has not been extensively studied before. Our findings suggest that the EAm can mitigate lipid oxidation in fish feed, positioning it as a potential natural antioxidant. Utilizing a broken-line model analysis for PO, CD, and MDA, we identified the optimal AE supplementation levels in fish feed as 6.74 g kg$^{-1}$.

In the realm of aquaculture nutrition, the oxidative stability of fish feed is a subject of considerable importance, primarily due to the inherent nature of its fatty acid composition. Predominantly, the presence of unsaturated fatty acids within the feed is a critical aspect in this context. The process of autoxidation of these unsaturated fats stands as the primary mechanism triggering lipid oxidation in fish feeds, a phenomenon well-documented in scholarly research [26]. Among these unsaturated fatty acids, certain types, such as linoleic acid (18:2n-6) and linolenic acid (18:3n-3), are recognized for their essentiality in the dietary requirements of freshwater fish species [32]. Our investigation delved into the role of EAm in mitigating this oxidation process. This study specifically highlights that EAm exhibits a notable efficacy in lowering the concentrations of various oxidative markers, including PO, CD, and MDA, within emulsions composed of linoleic and linolenic acids. This reduction points towards the potential of EAm in impeding the lipid oxidation cycle, particularly in unsaturated fatty acids. The action of EAm appears to extend beyond merely halting the initial phase of oxidation; it seemingly disrupts the subsequent secondary phase as well, thereby suggesting a comprehensive impact on the lipid oxidation pathway. Among the various extracts tested, AE of *Astragalus membranaceus* stood out as the most effective in reducing lipid oxidation in fatty acid emulsions. This finding is consistent with previous research, which demonstrated that the ethanol extract of *Astragalus membranaceus* effectively inhibited lipid oxidation in vitro [15].

ROS are crucial in the lipid oxidation of unsaturated fatty acids [7]. T-AOC measuring free radical scavenging activity, indicates antioxidative capacity in both hydrophilic and lipophilic substances [33,34]. In this study, EAm demonstrated strong T-AOC similar to trolox that was determined by the ABTS assay. Previous research from our laboratory showed that EAm reduced ROS levels in fish erythrocytes, particularly against hydroxyl radicals ($^{\bullet}$OH) [14]. In vitro studies also revealed EAm effectiveness in scavenging lipid free radicals, confirmed by electron spin resonance (ESR) analysis [15]. These results align with previous reports, suggesting that EAm could prevent lipid oxidation in fish feed unsaturated fatty acids by enhancing free radical scavenging activity.

The advantageous influence of EAm on mitigating lipid oxidation in fish feed might be intricately linked to its chemical constituents. In our research, a pronounced correlation was observed between the phenolic content of EAms and the reduced levels of PO, CD, and MDA in unsaturated fatty acids. This is in alignment with existing literature which suggests that phenolic compounds play a protective role against the oxidation of the free fatty acid fraction [35]. The hydroxyl groups present in these phenolic compounds are essential for their antioxidative functionality. They not only contribute hydrogen atoms to neutralize lipid free radicals, but also facilitate the formation of stable lipid derivatives and antioxidative free radicals, enhancing the overall antioxidant defense mechanism [36]. In the processes of the free radical chain reaction of lipid oxidation, phenolic compounds participate in inhibiting two crucial steps by reacting with ROO$^{\bullet}$ and inhibiting formation of hydroperoxides, as well as reacting with RO$^{\bullet}$ and inhibiting formation of aldehyde [37]. This result indicated that phenolic compounds in EAms could depress lipid oxidation in unsaturated fatty acids. Meanwhile, in the present study, the T-AOC of EAms was positively correlated with the total phenol contents. Similarly, strong relationships were observed between phenolic content in the extract of *Astragalus membranaceus* flowers and their antioxidant capacity [38]. This result suggested that phenolic compounds in EAms enhanced free radical scavenging activity. It has been reported that there are three possible antioxidant mechanisms for phenolic compounds at the molecular level, namely, inactivation of free radicals, chelation of free metals, as well as avoiding the reactions of free radical generation [39]. These results confirmed that the antioxidant activity of EAm in fish feed might be caused by the phenolic compounds.

### *4.2. Dietary EAm Supplementation Improved the Growth Performance in Fish*

In this research, the addition of AE to the diet was observed to enhance FI, WG, and SGR in Jian carp, suggesting that EAm supplementation can positively impact the growth of these fish. Through polynomial regression analysis focusing on WG, it was determined that the ideal amount of AE supplementation for juvenile Jian carp is 5.15 g per kg of diet. This finding aligns with previous studies reporting the beneficial effects of dietary *Astragalus membranaceus* on the growth of other fish species, such as bluegill sunfish (*Lepomis macrochirus*) and yellow perch (*Perca flavescens*) [40,41]. Notably, Elabd et al. identified an enhancement in growth performance attributed to the up-regulation of insulin-like growth factor-1, a key molecular biomarker associated with growth [41]. However, it is important to note that there have been relatively few studies exploring the specific growth-promoting effects of EAm on fish.

The growth performance of fish is intricately linked to their ability to digest nutrients, a process largely governed by the efficacy of digestive enzymes [29]. Such enzymes, including amylase, trypsin, chymotrypsin, and lipase are primarily synthesized in the exocrine pancreas of fish and subsequently secreted into the intestine to aid digestion [42,43]. In this research, it was observed that the supplementation of the diet with AE led to an increase in the activities of digestive enzymes such as trypsin, lipase, and amylase. This enhancement suggests that the dietary inclusion of EAm potentially improves the digestive capability of fish, facilitating better nutrient assimilation. Our results were shown to be in conformity with the report that dietary *Astragalus membranaceus* nanoparticles enhanced digestive ability, which was manifested by increased lipase and amylase activities in Nile tilapia (*Oreochromis niloticus*) [44].

The growth performance in fish is not only dependent on nutrient digestion but also significantly influenced by the efficiency of nutrient absorption. The efficacy of absorption is greatly influenced by brush border membrane enzymes [29]. $Na^+/K^+$-ATPase, a key enzyme in this process, significantly contributes to the generation of potential energy from the sodium gradient, which is vital for the cellular uptake of essential nutrients like glucose, phosphate, and amino acids [45]. Additionally, AKP is integral for the absorption of food molecules and is a measure of the absorptive capacity in fish [46]. In our investigation, we found that diet supplementation with AE increased the activity levels of both AKP and $Na^+/K^+$-ATPase, indicating an improvement in nutrient absorption in fish. This enhancement is supported by previous studies, which showed an increase in AKP activity in shrimp following the introduction of *Astragalus membranaceus* in their diet [47].

### 4.3. Dietary EAm Improved the Antioxidant Capacity in Digestive and Absorptive Organs of Fish

In aquatic species, antioxidants play a crucial role in maintaining the health and functionality of tissues and organs [48]. This is particularly evident in fish, where the structural integrity of the intestines is essential for effective digestion and nutrient absorption [29]. ROS, produced during aerobic metabolic activities and potentially sourced externally, contribute to cellular damage, negatively impacting various cellular components like lipids and proteins, thereby potentially impairing cellular functions [49]. Lipid peroxidation, indicated by elevated levels of MDA, is a primary marker of such damage, especially relevant in fish due to their high PUFA content [48]. Our research indicates that incorporating AE into the diet significantly lowers MDA levels in the fish's hepatopancreas and intestine, implying that EAm supplementation in diets can mitigate lipid peroxidation in these organs. This is consistent with earlier studies, which have showned reduced liver MDA levels in juvenile *Pangasianodon hypophthalmus* fed diets enriched with *Astragalus membranaceus* extract [16].

Moreover, an increase in ROS production, particularly $^{\bullet}OH$ and superoxide anion ($O_2^{\bullet-}$), was identified as a catalyst for initiating lipid oxidation [30]. This escalation also compromises the structural stability of proteins and lipids due to the excess generation of various free radicals, such as $O_2^{\bullet-}$ and $^{\bullet}OH$. Such oxidative stress leads to molecular damage, impacting cellular functions and health [50]. This research found that supplementation of AE in the diet enhanced ASA capacity in the hepatopancreas and intestine of fish, suggesting an increase in $O_2^{\bullet-}$ scavenging ability in fish digestive and absorptive organs. This aligns with previous findings indicating that EAm decreased the $O_2^{\bullet-}$ levels in fish erythrocytes under $^{\bullet}OH$ stress [14]. An in vitro study revealed that *Astragalus membranaceus* exhibits antioxidant effects by scavenging $O_2^{\bullet-}$ [51]. Similarly, extract of *Astragalus mongholicus* demonstrated a dose-dependent effect in scavenging $^{\bullet}OH$ in vitro [15]. Additionally, oral administration of *Astragalus mongholicus* was shown to decrease ROS levels in rat brains [15]. Abuelsaad's research indicated that oral *Astragalus* polysaccharides reduced ROS production in the intestinal tissues of infected mice [52]. These effects of EAm could be related to the chemical structure of its phenolic compounds, which are known for their free radical-scavenging abilities, as reported by Hidalgo and Almajano [53]. It has also been demonstrated that phenol can protect against ROS in rainbow trout gill cells [54].

The antioxidative mechanisms within fish, encompassing both non-enzymatic and enzymatic components, have evolved to effectively counteract or repair the damage caused by ROS to fish tissues and overall organism health [55]. Within these antioxidative enzyme systems, SOD plays a pivotal role as the primary enzyme that interacts with $O_2^{\bullet-}$, converting it into hydrogen peroxide ($H_2O_2$) and dioxygen [48]. Subsequently, $H_2O_2$ is detoxified by enzymes such as GPx and CAT [56]. In this particular study, the addition of AE to the diet was found to elevate the activities of GPx, CAT, and SOD in the hepatopancreas, suggesting that dietary EAm supplementation enhances the antioxidative capacity in fish. This observation aligns with prior research which demonstrated that diets enriched with

*Astragalus membranaceus* extract boosted the activities of CAT, GPx, and SOD in the liver of juvenile *Pangasianodon hypophthalmus* [16]. The beneficial effects of EAm may be correlated with the phenolic compound. In Cat fish (*Clarias batrachus*), the increased activities of CAT and SOD in liver were response to dietary supplementation of phenolic compounds [57]. Moreover, a mixture of *Crataegus hupehensis*, *Angelica sinensis*, and *Astragalus membranaceus* supplementation showed significant up-regulation of SOD and CAT genes in the intestine of Nile Tilapia [58].

Maintaining robust glutathione antioxidant status is key in controlling cell damage which induced by free radicals [59]. The glutathione-dependent antioxidant system, crucial in preventing radical-induced cellular damage, includes GR, glutathione-*S*-transferase (GST), and GSH [60]. GR plays a vital role in regenerating reduced glutathione from its oxidized form, using NADPH as the electron donor [61]. In the present study, dietary AE supplementation increased activity of GR in fish intestine. These results were in conformity with our laboratory report that EAm restored the GR activity in fish erythrocytes under $^{\bullet}OH$ exposure [14]. In this study, dietary AE supplementation improved GSH content in fish intestine. These results were in conformity with the report that phenolic compounds increased the GSH level in the liver of Cat fish [57]. Therefore, our present study indicated that EAm improved enzymatic and non-enzymatic antioxidant capacity so as to enhance the radical-scavenging ability in fish intestine. However, the mechanism of EAm regulated fish antioxidant capacity needs to be further explored.

## 5. Conclusions

Hence, it can be deduced that EAm plays an important role in minimizing lipid oxidation in fish feed by safeguarding unsaturated fatty acids. Through broken-line analysis, the ideal dietary levels of AE supplementation in fish feed were determined to be 6.74 g $kg^{-1}$. The antioxidant properties of EAm are intrinsically linked to their phenolic content. Additionally, incorporating EAm into the diet was found to enhance fish growth. Polynomial regression analysis revealed that the optimal concentration of AE supplementation for juvenile Jian carp was 5.15 g $kg^{-1}$ diet. Moreover, dietary EAm supplementation boosted the digestive, absorptive and antioxidant capabilities of the fish's digestion and absorption organs, thereby preserving their structure and function, enhancing nutrient absorption, and consequently improving growth performance. The antioxidant efficiency of EAm is intimately associated with its phenolic constituents. This study suggests that EAm holds promise as a natural antioxidant for fish and their feed.

**Author Contributions:** Conceptualization, J.X. and H.L.; methodology, J.X., G.C., M.W. and Q.Y.; formal analysis, G.C. and M.W.; validation, J.X., G.C., Q.Y. and H.L.; data curation, J.X. and M.W.; resources, H.L.; writing—original draft preparation, J.X., G.C., Q.Y. and H.L. All authors have read and agreed to the published version of the manuscript.

**Funding:** This research was funded by the Sichuan Science and Technology Program [grant number 2018JY0214], the Patent Project of Neijiang Normal University [grant number Z2019027], and the Talent Program of Neijiang Normal University [grant number R2019015].

**Institutional Review Board Statement:** All procedures were approved by the Neijiang Normal University Institutional Animal Care and Use Committee (Approval Code: JM2016-11; Approval Date: 1 November 2023).

**Data Availability Statement:** The data used to generate the results in this manuscript can be made available if requested from the corresponding author.

**Conflicts of Interest:** The authors declare no conflict of interest.

## References

1. Godwin, A.; Prabhu, H.R. Lipid peroxidation of fish oils. *Indian J. Clin. Biochem.* **2006**, *21*, 202–204. [CrossRef]
2. Zeb, A. Concept, mechanism, and applications of phenolic antioxidants in foods. *J. Food Biochem.* **2020**, *44*, e13394. [CrossRef]

3. Guo, X.Y.; Chen, S.N.; Cao, J.Y.; Zhou, J.Y.; Chen, Y.Z.; Jamali, M.A.; Zhang, Y.W. Hydrolysis and oxidation of protein and lipids in dry-salted grass carp (*Ctenopharyngodon idella*) as affected by partial substitution of NaCl with KCl and amino acids. *RSC Adv.* **2019**, *9*, 39545–39560. [CrossRef]
4. Filipe, D.; Gonçalves, M.; Fernandes, H.; Oliva-Teles, A.; Peres, H.; Belo, I.; Salgado, J.M. Shelf-Life performance of fish feed supplemented with bioactive extracts from fermented olive mill and winery by-products. *Foods* **2023**, *12*, 305. [CrossRef]
5. Andrews, J.T.; Giesen, A.F.; Scott, F.R. Antioxidants manage effects of oxidation on feeds, feed ingredients. *Glob. Aquac. Advocate* **2004**, *12*, 66–68.
6. Hamre, K.; Kolås, K.; Sandnes, K. Protection of fish feed, made directly from marine raw materials, with natural antioxidants. *Food Chem.* **2010**, *119*, 270–278. [CrossRef]
7. Li, H.; Zhou, X.; Gao, P.; Li, Q.; Li, H.; Huang, R.; Wu, M. Inhibition of lipid oxidation in foods and feeds and hydroxyl radical-treated fish erythrocytes: A comparative study of *Ginkgo biloba* leaves extracts and synthetic antioxidants. *Anim. Nutr.* **2016**, *2*, 234–241. [CrossRef]
8. Ito, N.; Fukushima, S.; Tsuda, H. Carcinogenicity and modification of the carcinogenic response by BHA, BHT, and other antioxidants. *Crit. Rev. Toxicol.* **1985**, *15*, 109–150. [CrossRef]
9. EFSA. Safety and efficacy of a feed additive consisting of ethoxyquin (6-ethoxy-1, 2-dihydro-2, 2, 4-trimethylquinoline) for all animal species (FEFANA asbl). *EFSA J.* **2022**, *20*, e07166.
10. Harnedy, P.A.; FitzGerald, R.J. Bioactive peptides from marine processing waste and shellfish: A review. *J. Funct. Foods* **2012**, *4*, 6–24. [CrossRef]
11. Ribeiro, J.S.; Santos, M.J.M.C.; Silva, L.K.R.; Pereira, L.C.L.; Santos, I.A.; Silva Lannes, S.C.; Silva, M.V. Natural antioxidants used in meat products: A brief review. *Meat Sci.* **2019**, *148*, 181–188. [CrossRef]
12. Singh, B.K.; Tiwari, S.; Dubey, N.K. Essential oils and their nanoformulations as green preservatives to boost food safety against mycotoxin contamination of food commodities: A review. *J. Sci. Food Agric.* **2021**, *101*, 4879–4890. [CrossRef]
13. Agyemang, K.; Han, L.; Liu, E.; Zhang, Y.; Wang, T.; Gao, X. Recent advances in *Astragalus membranaceus* anti-diabetic research: Pharmacological effects of its phytochemical constituents. *Evid.-Based Compl. Alt.* **2013**, *2013*, 654643. [CrossRef]
14. Li, H.; Zhou, X.; Wu, M.; Deng, M.; Wang, C.; Hou, J.; Mou, P. The cytotoxicity and protective effects of *Astragalus membranaceus* extracts and butylated hydroxyanisole on hydroxyl radical-induced apoptosis in fish erythrocytes. *Anim. Nutr.* **2016**, *2*, 376–382. [CrossRef]
15. Aldarmaa, J.; Liu, Z.; Long, J.; Mo, X.; Ma, J.; Liu, J. Anti-convulsant effect and mechanism of *Astragalus mongholicus* extract in vitro and in vivo: Protection against oxidative damage and mitochondrial dysfunction. *Neurochem. Res.* **2010**, *35*, 33–41. [CrossRef]
16. Abdel-Latif, H.M.; Ahmed, H.A.; Shukry, M.; Chaklader, M.R.; Saleh, R.M.; Khallaf, M.A. *Astragalus membranaceus* extract (AME) enhances growth, digestive enzymes, antioxidant capacity, and immunity of *Pangasianodon hypophthalmus* juveniles. *Fishes* **2022**, *7*, 319. [CrossRef]
17. Mirghaed, A.T.; Fayaz, S.; Hoseini, S.M. Effects of dietary 1, 8-cineole supplementation on serum stress and antioxidant markers of common carp (*Cyprinus carpio*) acutely exposed to ambient ammonia. *Aquaculture* **2019**, *509*, 8–15. [CrossRef]
18. Li, S.; Ji, H.; Zhang, B.; Zhou, J.; Yu, H. Defatted black soldier fly (*Hermetia illucens*) larvae meal in diets for juvenile Jian carp (*Cyprinus carpio* var. Jian): Growth performance, antioxidant enzyme activities, digestive enzyme activities, intestine and hepatopancreas histological structure. *Aquaculture* **2017**, *477*, 62–70. [CrossRef]
19. Ning, R.; Li, Y.; Wang, C.; Ji, M.; Shi, Z. Research on antimicrobial activities of crude extracts from *Artemesia apiacea* Hance. *Jiangsu Agric. Sci.* **2007**, *3*, 61–63.
20. Wojcikowski, K.; Stevenson, L.; Leach, D.; Wohlmuth, H.; Gobe, G. Antioxidant capacity of 55 medicinal herbs traditionally used to treat the urinary system: A comparison using a sequential three-solvent extraction process. *J. Altern. Complem. Med.* **2007**, *13*, 103–110. [CrossRef]
21. Dalar, A.; Türker, M.; Zabaras, D.; Konczak, I. Phenolic composition, antioxidant and enzyme inhibitory activities of Eryngium bornmuelleri leaf. *Plant Food. Hum. Nutr.* **2014**, *69*, 30–36. [CrossRef] [PubMed]
22. Yuan, Y.V.; Bone, D.E.; Carrington, M.F. Antioxidant activity of dulse (*Palmaria palmata*) extract evaluated in vitro. *Food Chem.* **2005**, *91*, 485e94. [CrossRef]
23. Serpen, A.; Gökmen, V.; Fogliano, V. Total antioxidant capacities of raw and cooked meats. *Meat Sci.* **2012**, *90*, 60–65. [CrossRef] [PubMed]
24. Li, H.; Tang, S.; Du, W.; Jiang, J.; Peng, P.; Yuan, P.; Liao, Y.; Long, J.; Zhou, S. The effects of ethoxyquin and *Angelica sinensis* extracts on lipid oxidation in fish feeds and growth, digestive and absorptive capacities and antioxidant status in juvenile red carp (*Cyprinus carpio* var. xingguonensis): A comparative study. *Fish Physiol. Biochem.* **2019**, *45*, 43–61. [CrossRef] [PubMed]
25. Maqsood, S.; Benjakul, S. Comparative studies of four different phenolic compounds on in vitro antioxidative activity and the preventive effect on lipid oxidation of fish oil emulsion and fish mince. *Food Chem.* **2010**, *119*, 123–132. [CrossRef]
26. Li, H.; Yang, D.; Li, Z.; He, M.; Li, F.; Jiang, J.; Tang, S.; Peng, P.; Du, W.; Ma, Y.; et al. Effects of *Angelica sinensis* extracts on lipid oxidation in fish feeds and growth performance of juvenile Jian carp (*Cyprinus carpio* var. Jian). *Anim. Nutr.* **2019**, *5*, 109–114. [CrossRef]

27. Chen, G.F.; Feng, L.; Kuang, S.Y.; Liu, Y.; Jiang, J.; Hu, K.; Jiang, W.D.; Li, S.H.; Tang, L.; Zhou, X.Q. Effect of dietary arginine on growth, intestinal enzyme activities and gene expression in muscle, hepatopancreas and intestine of juvenile Jian carp (*Cyprinus carpio* var. Jian). *Brit. J. Nutr.* **2012**, *108*, 195–207. [CrossRef]
28. Li, H.; Lu, L.; Wu, M.; Xiong, X.; Luo, L.; Ma, Y.; Liu, Y. The effects of dietary extract of mulberry leaf on growth performance, hypoxia-reoxygenation stress and biochemical parameters in various organs of fish. *Aquacult. Rep.* **2020**, *18*, 100494. [CrossRef]
29. Saxena, T.B.; Zachariassen, K.E.; Jørgensen, L. Effects of ethoxyquin on the blood composition of turbot, *Scophthalmus maximus* L. *Comp. Biochem. Physiol. Part C Pharmacol. Toxicol. Endocrinol.* **2000**, *127*, 1–9. [CrossRef]
30. Jiang, W.D.; Wu, P.; Kuang, S.Y.; Liu, Y.; Jiang, J.; Hu, K.; Li, S.H.; Tang, L.; Feng, L.; Zhou, X.Q. *Myo*-inositol prevents copper-induced oxidative damage and changes in antioxidant capacity in various organs and the enterocytes of juvenile Jian carp (*Cyprinus carpio* var. *Jian*). *Aquat. Toxicol.* **2011**, *105*, 543–551. [CrossRef]
31. Bradford, M.M. A rapid and sensitive method for the quantitation of microgram quantities of protein utilizing the principle of protein-dye binding. *Anal. Biochem.* **1976**, *72*, 248–254. [CrossRef]
32. El-Husseiny, O.M.; Abdul-Aziz, G.M.; Goda, A.S.; Suloma, A. Effect of altering linoleic acid and linolenic acid dietary levels and ratios on the performance and tissue fatty acid profiles of Nile tilapia *Oreochromis niloticus* fry. *Aquacult. Int.* **2010**, *18*, 1105–1119. [CrossRef]
33. Wei, X.B.; Liu, H.Q.; Sun, X.; Fu, F.; Zhang, X.; Wang, J.; An, J.; Ding, H. Hydroxysafflor yellow A protects rat brains against ischemia-reperfusion injury by antioxidant action. *Neurosci. Lett.* **2005**, *386*, 58–62. [CrossRef]
34. Zhang, H.; Wang, S.; Duan, X.; Feng, X.; Yang, Y. The interaction effects of coke oven emissions exposure and metabolic enzyme Gene variants on total antioxidant capacity of workers. *Environ. Toxicol. Pharmacol.* **2019**, *70*, 103197. [CrossRef]
35. Servili, M.; Montedoro, G. Contribution of phenolic compounds to virgin olive oil quality. *Eur. J. Lipid Sci. Technol.* **2002**, *104*, 602–613. [CrossRef]
36. Shahidi, F.; Ambigaipalan, P. Phenolics and polyphenolics in foods, beverages and spices: Antioxidant activity and health effects. *J. Funct. Foods* **2015**, *18*, 820–897. [CrossRef]
37. Frankel, E.N. Natural phenolic antioxidants and their impact on health. In *Antioxidant Food Supplements in Human Health*; Academic Press: Cambridge, MA, USA, 1999; pp. 385–392.
38. Li, Y.; Guo, S.; Zhu, Y.; Yan, H.; Qian, D.; Wang, H.; Yu, J.; Duan, J. Flowers of *Astragalus membranaceus* var. *mongholicus* as a novel high potential by-product: Phytochemical characterization and antioxidant activity. *Molecules* **2019**, *24*, 434.
39. Leopoldini, M.; Russo, N.; Toscano, M. The molecular basis of working mechanism of natural polyphenolic antioxidants. *Food Chem.* **2011**, *125*, 288–306. [CrossRef]
40. Elabd, H.; Wang, H.P.; Shaheen, A.; Yao, H.; Abbass, A. *Astragalus membranaceus* (AM) enhances growth performance and antioxidant stress profiles in bluegill sunfish (*Lepomis macrochirus*). *Fish Physiol. Biochem.* **2016**, *42*, 955–966. [CrossRef]
41. Elabd, H.; Wang, H.P.; Shaheen, A.; Yao, H.; Abbass, A. Feeding *Glycyrrhiza glabra* (liquorice) and *Astragalus membranaceus* (AM) alters innate immune and physiological responses in yellow perch (*Perca flavescens*). *Fish Shellfish Immun.* **2016**, *54*, 374–384. [CrossRef]
42. Gilloteaux, J.; Kashouty, R.; Yono, N. The perinuclear space of pancreatic acinar cells and the synthetic pathway of zymogen in *Scorpaena scrofa* L.: Ultrastructural aspects. *Tissue Cell* **2008**, *40*, 7–20. [CrossRef] [PubMed]
43. Zambonino Infante, J.L.; Cahu, C.L. Ontogeny of the gastrointestinal tract of marine fish larvae. *Comp. Biochem. Physiol. Part C Toxicol. Pharmacol.* **2001**, *130*, 477–487. [CrossRef] [PubMed]
44. Elabd, H.; Wang, H.P.; Shaheen, A.; Matter, A. *Astragalus membranaceus* nanoparticles markedly improve immune and anti-oxidative responses; and protection against *Aeromonas veronii* in Nile tilapia *Oreochromis niloticus*. *Fish Shellfish Immun.* **2020**, *97*, 248–256. [CrossRef] [PubMed]
45. Geering, K. Subunit assembly and functional maturation of Na,K-ATPase. *J. Membr. Biol.* **1990**, *115*, 109–121. [CrossRef]
46. Suzer, C.; Aktülün, S.; Çoban, D.; Okan Kamacı, H.; Saka, Ş.; Fırat, K.; Alpbaz, A. Digestive enzyme activities in larvae of sharpsnout seabream (*Diplodus puntazzo*). *Comp. Biochem. Physiol. Part A Mol. Integr. Physiol.* **2007**, *148*, 470–477. [CrossRef]
47. Angela, C.; Wang, W.; Lyu, H.; Zhou, Y.; Huang, X. The effect of dietary supplementation of *Astragalus membranaceus* and *Bupleurum chinense* on the growth performance, immune-related enzyme activities and genes expression in white shrimp, *Litopenaeus vannamei*. *Fish Shellfish Immun.* **2020**, *107*, 379–384. [CrossRef]
48. Feng, L.; Zhao, S.; Chen, G.; Jiang, W.; Liu, Y.; Jiang, J.; Hu, K.; Li, S.; Zhou, X. Antioxidant status of serum, muscle, intestine and hepatopancreas for fish fed graded levels of biotin. *Fish Physiol. Biochem.* **2013**, *40*, 499–510. [CrossRef]
49. Gliszczyńska-Świgło, A. Antioxidant activity of water soluble vitamins in the TEAC (trolox equivalent antioxidant capacity) and the FRAP (ferric reducing antioxidant power) assays. *Food Chem.* **2006**, *96*, 131–136. [CrossRef]
50. Woo, S.J.; Chung, J.K. Effects of trichlorfon on oxidative stress, neurotoxicity, and cortisol levels in common carp, *Cyprinus carpio* L., at different temperatures. *Comp. Biochem. Physiol. Part C Toxicol. Pharmacol.* **2020**, *229*, 108698. [CrossRef]
51. Yang, W.J.; Li, D.P.; Li, J.K.; Li, M.H.; Chen, Y.L.; Zhang, P.Z. Synergistic antioxidant activities of eight traditional Chinese herb pairs. *Biol. Pharm. Bull.* **2009**, *32*, 1021–1026. [CrossRef]
52. Abuelsaad, A.S. Supplementation with *Astragalus polysaccharides* alters *Aeromonas*-induced tissue-specific cellular immune response. *Microb. Pathog.* **2014**, *66*, 48–56. [CrossRef] [PubMed]
53. Hidalgo, G.I.; Almajano, M.P. Red fruits: Extraction of antioxidants, phenolic content, and radical scavenging determination: A review. *Antioxidants* **2017**, *6*, 7. [CrossRef] [PubMed]

54. Chung, M.J.; Walker, P.A.; Hogstrand, C. Dietary phenolic antioxidants, caffeic acid and Trolox, protect rainbow trout gill cells from nitric oxide-induced apoptosis. *Aquat. Toxicol.* **2006**, *80*, 321–328. [CrossRef] [PubMed]
55. Martínez-Álvarez, R.M.; Morales, A.E.; Sanz, A. Antioxidant defenses in fish: Biotic and abiotic factors. *Rev. Fish Biol. Fish.* **2005**, *15*, 75–88. [CrossRef]
56. Ni, M.; Liu, M.; Lou, J.; Mi, G.; Gu, Z. Stocking density alters growth performance, serum biochemistry, digestive enzymes, immune response, and muscle quality of largemouth bass (*Micropterus salmoides*) in in-pond raceway system. *Fish Physiol. Biochem.* **2021**, *47*, 1243–1255. [CrossRef] [PubMed]
57. Charan, A.A.; Charan, A.I.; Verma, O.P.; Naushad, S.S. Profiling of Antioxidant enzymes in Cat fish (*Clarias batrachus*) exposed to phenolic compounds. *Asian J. Bio Sci.* **2015**, *10*, 6–14. [CrossRef]
58. Abarike, E.D.; Jian, J.C.; Tang, J.F.; Cai, J.; Yu, H.; Chen, L.H. Traditional Chinese medicine enhances growth, immune response, and resistance to *Streptococcus agalactiae* in Nile tilapia. *J. Aquat. Anim. Health* **2019**, *31*, 46–55. [CrossRef]
59. Kirshenbaum, L.A.; Singal, P.K. Increase in endogenous antioxidant enzymes protects hearts against reperfusion injury. *Am. J. Physiol.* **1993**, *265*, H484–H492. [CrossRef]
60. Yim, T.K.; Wu, W.K.; Pak, W.F.; Mak, D.H.F.; Ko, K.M. Myocardial protection against ischaemia-reperfusion injury by a *Polygonum multiflorum* extract supplemented 'Dang-Gui decoction for enriching blood', a compound formulation, ex vivo. *Phytother. Res.* **2000**, *14*, 195–199. [CrossRef]
61. Reed, D.J. Glutathione: Toxicological implications. *Annu. Rev. Pharmacol. Toxicol.* **1990**, *30*, 603–631. [CrossRef]

fishes

MDPI

Article

# Extract of *Ginkgo biloba* Leaves (EGb) Decrease Lipid Oxidation in Fish Feed and Meat and Enhance Growth and Antioxidant Capacity in Jian Carp (*Cyprinus carpio* var. Jian)

**Gangfu Chen [1], Jing Xu [1], Min Wu [1], Huatao Li [1,*], Qihui Yang [2] and Lin Feng [3]**

[1] Key Laboratory of Sichuan Province for Conservation and Utilization of Fishes Resources in the Upper Reaches of the Yangtze River, College of Life Sciences, Neijiang Normal University, Neijiang 641100, China; fanyahr@163.com (G.C.)
[2] College of Fisheries, Guangdong Ocean University, Zhanjiang 524088, China; qihuiyang03@163.com
[3] Animal Nutrition Institute, Sichuan Agricultural University, Chengdu 611130, China
* Correspondence: huataoli217@126.com

**Abstract:** Firstly, a linoleic and linolenic acid emulsion were incubated with petroleum ether extract, ethyl acetate extract (EAE), ethanol extract and aqueous extract of *Ginkgo biloba* leaves. The flavonoids content, total antioxidant capacity (T-AOC) and metal-chelating ability (MCA) were determined in EGbs above. Results showed that the extracts of *Ginkgo biloba* leaves (EGbs) inhibited the lipid oxidation in material above. Of all of EGbs, EAE showed the strongest T-AOC, MCA and protective effects against the lipid oxidation. Next, fish feeds were incubated with graded levels of EAE. The results showed that EAE inhibited lipid oxidation in fish feeds. The optimal inclusion levels of EAE for minimizing lipid oxidation were 4.26 g $kg^{-1}$ feeds. The effect of EGbs on the lipid oxidation may be closely associated with their flavonoid content. Finally, juvenile carp (14.8 $\pm$ 0.4 g) were fed with EAE at concentrations ranging from 0.0 to 6.0 g $kg^{-1}$ for 60 days. Current data displayed that dietary EAE increased the growth performance of fish. This result of EAE may be ascribed to its enhancing effect on the activity of digestive and absorptive enzymes and antioxidant capacity in digestive organs of fish. Furthermore, dietary EAE decreased the hot-drying-induced lipid oxidation in fish meat through inhibiting the induction effect of hemoglobin in erythrocytes. Our study suggests that EGb can be considered as a potential natural antioxidant for fish and fish feed.

**Keywords:** ginkgo leaf; lipid oxidation; fish feedstuff; growth performance; fish meat; natural antioxidant

**Key Contribution:** A. Extract of ginkgo leaves decreases lipid oxidation in fish feed and meat. B. Extract of ginkgo leaves improves fish growth performance and antioxidant capacity.

**Citation:** Chen, G.; Xu, J.; Wu, M.; Li, H.; Yang, Q.; Feng, L. Extract of *Ginkgo biloba* Leaves (EGb) Decrease Lipid Oxidation in Fish Feed and Meat and Enhance Growth and Antioxidant Capacity in Jian Carp (*Cyprinus carpio* var. Jian). *Fishes* **2023**, *8*, 564. https://doi.org/10.3390/fishes8110564

Academic Editors: Qiyou Xu, Jianhua Ming, Fei Song, Changle Qi, Chuanpeng Zhou and Marina Paolucci

Received: 27 September 2023
Revised: 8 November 2023
Accepted: 9 November 2023
Published: 20 November 2023



## 1. Introduction

Fish feed contains high amounts of lipids as well as unsaturated fatty acids [1]. In fish feed, under the action of reactive oxygen species (ROS) and excessive redox metal ions, unsaturated fatty acids are easily oxidized to produce oxidation products such as peroxide, malondialdehyde and conjugated diene [2]. On the one hand, fish meat is prone to lipid oxidation due to the presence of high amounts of polyunsaturated fatty acids (PUFAs) [3]. On the other hand, in fish processing, blood is not always removed, retaining a high level of hemoglobin (Hb) in fish meat [4]. ROS can be generated by the auto-oxidation of Hb, which induces heme iron release in erythrocytes and lipid oxidation in fish meat [5]. Meanwhile, lipid oxidation may be harmful to fish meat quality, such as breakdown of nutritional components, generation of toxic metabolites, and degradation of color, aroma and taste, as well as reducing the shelf life [6]. Furthermore, dietary oxidized lipid had a negative effect on fish growth performance as manifested by the decreased weight gain ratio and special growth ratio [7].

Synthetic antioxidants (ethoxyquin, butylated hydroxytoluene and butylated hydroxyanisole) are commonly used in the foods, cosmetics and therapeutic industries for lipid oxidation prevention [6]. Recently, our laboratory reported that ethoxyquin and butylated hydroxytoluene inhibited lipid oxidation of fish feed [8]. However, these synthetic antioxidants can accumulate in tissues and would be toxic and carcinogenic to animals [9], and then are transmitted to humans via the consumption of these animals [6,8]. Based on the detrimental effects on animals and humans, the application of these synthetic antioxidants in the food industry should be under strict regulation [10]. In this sense, natural ingredients may be the promising substitute for synthetic antioxidants [11].

*Ginkgo biloba* L. (Gb), a Chinese-origin tree, has been cultivated as a gardening plant and used in medicine in China, with leaves that are rich in the main bioactive chemical components, such as terpenes and flavonoids [12]. Tan et al. (2018) reported that extract of *Ginkgo biloba* leaves (EGb) increased hepatic antioxidant enzyme (superoxide dismutase and catalase) activities and decreased hepatic malonaldehyde content in hybrid grouper (*Epinephelus lanceolatus*♂ × *Epinephelus fuscoguttatus*♀) [13]. EGb761, a standardized extract of *Ginkgo biloba* leaves, is an antioxidative natural product [14]. Previous investigations have established that EGb effectively inhibits lipid peroxidation, a capacity demonstrated in both edible and animal feed matrices under in vitro conditions [8]. Given these findings, EGb presents itself as a promising candidate for a natural antioxidant within aquaculture systems.

In China, Jian carp (*Cyprinus carpio* var. Jian) is the most widely cultivated fish species because it grows faster and has higher meat quality than common carp [15]. This investigation assessed the impact of EGb on the oxidative stability of linoleic acid and linolenic acid emulsions and its effects on the growth performance and antioxidant capacity of Jian carp when incorporated into their diet at varying concentrations over a period of 60 days. The findings offer a foundation for the potential inclusion of EGb as an antioxidant agent in the diets of aquatic species.

## 2. Materials and Methods

### 2.1. Chemical Reagent

Reagents such as Tween 20, petroleum ether, ethyl acetate, and ethanol of analytical reagent (AR) grade were sourced from Chengdu Kelong Chemical Reagent Factory, located in Chengdu, China. High-purity linoleic acid and linolenic acid, with purities exceeding 97% and 95%, respectively, were procured from Shanghai Biochemical Reagent Co., Ltd., based in Shanghai, China. The remainder of the chemicals utilized in this study conformed to analytical reagent (AR)-grade standards.

### 2.2. EGbs Preparation and Composition Analyses

Ginkgo leaves were harvested from Gb in the vicinity of Neijiang Normal University, Neijiang, China, during the month of November. The extraction of EGbs followed methodologies previously established by our research collective [8]. Extracts derived from the leaves using dry petroleum ether, ethyl acetate (EAE), ethanol, and water were stored in darkness within airtight containers at a temperature maintained at −80 °C pending subsequent analysis. The characterization of the EAE was conducted in adherence to analytical protocols delineated in prior publications by our team [16].

### 2.3. Measurement of Flavonoids Content, Total Antioxidant Capacity and Metal-Chelating Ability

The flavonoids content, total antioxidant capacity and metal-chelating ability of EGbs were determined according to the methods of Zou et al. [17], Cao et al. [18] and Zhao et al. [19], respectively.

### 2.4. Measurement of Lipid Oxidation in Linoleic Acid and Linolenic Acid Emulsion

The antioxidative impact of EGb on an emulsified mixture of linoleic acids or linolenic acids was evaluated following a slightly adapted procedure from Yuan et al. [20]. An emulsion was constituted by blending 0.1 mL each of linoleic acids or linolenic acids with 9.9 mL of chilled phosphate buffer (with a pH of 7.0 and concentration of 0.02 M) and 50 μL of Tween 20, followed by homogenization on ice using a FJ200-SH homogenizer (Shanghai, China) for two intervals of 10 s at a centrifugal force of 21,000× *g*. Different concentrations of EGbs, namely 0 (as a control) and 1.0 mg $mL^{-1}$, were then solubilized into the emulsion. Post an 8-day incubation period at 45 °C, the resultant levels of malondialdehyde, conjugated diene, and peroxide were quantified employing the methodologies previously described by Maqsood and Benjakul [21].

## 2.5. Measurement of Lipid Oxidation in Fish Feed

Our diet preparation protocol adhered to the established methods within our group [22]. As indicated in Table 1, the fundamental diet's composition included 34.30% of crude protein and 6.25% crude lipid. This feed underwent treatment employing a method analogous to the emulsification process used for linoleic acids or linolenic acids as previously detailed.

**Table 1.** Composition and nutrients content of the basal diet

| Ingredients | % | Proximate Analysis [3] | % |
|---|---|---|---|
| Fish meal | 27.00 | Dry matter | 92.46 |
| Soybean meal | 36.00 | Crude protein | 34.30 |
| Wheat flour | 32.50 | Crude lipid | 6.25 |
| $Ca(H_2PO_4)_2$ | 0.50 | Crude Ash | 5.78 |
| Corn oil | 2.00 | | |
| Vitamin mixture [1] | 1.00 | | |
| Mineral mixture [2] | 1.00 | | |

[1] Per kg of vitamin mix: retinyl acetate (500,000 IU $g^{-1}$), 0.80 g; cholecalciferol (500,000 IU $g^{-1}$), 0.48 g; DL-α-tocopherol acetate (50%), 20.00 g; menadione (23%), 0.43 g; thiamin nitrate (90%), 0.11 g; riboflavine (80%), 0.63 g; pyridoxine HCl (81%), 0.92 g; cyanocobalamin (1%), 0.10 g; ascorhyl acetate (93%), 7.16 g; D-calcium pantothenate (90%), 2.73 g; niacin (99%), 2.82 g; D-biotin (2%), 5.00 g; meso-inositol (99%), 52.33 g; folic acid (96%), 0.52 g. [2] Per kg of mineral mix: $FeSO_4 \cdot 7H_2O$ (20% Fe), 69.70 g; $CuSO_4 \cdot 5H_2O$ (25% Cu), 1.20 g; $ZnSO_4 \cdot 7H_2O$ (23% Zn), 21.64 g; $MnSO_4 \cdot H_2O$ (32% Mn), 4.09 g; $Na_2SeO_3 \cdot 5H_2O$ (1% Se), 2.50 g; KI (4% I), 2.90 g; $CaCO_3$, 897.98 g. [3] Proximate analyses were carried out by the procedures by Chen et al. [15].

## 2.6. Feeding Trial

Juvenile Jian carp were acquired from an aquaculture facility in Neijiang, China, and were subsequently housed under conditions that replicated their natural light environment in our laboratory [22]. Groups of juvenile Jian carp, averaging 14.8 ± 0.4 g, were assigned randomly across 7 treatment sets. Each set comprised 4 tanks, with 20 fish in each tank. Following the formulation techniques delineated by Chen et al. [15], seven distinct experimental diets were produced. Over a course of sixty days, the carp were fed diets augmented with varying concentrations of EAE: specifically, 0.0, 1.0, 2.0, 3.0, 4.0, 5.0, and 6.0 g per kilogram of feed. Vital growth parameters such as survival rate, feed intake, body mass increment, feed conversion efficiency, and specific growth rate were computed in alignment with methodologies established in our preceding research [22].

Consistent with the protocol outlined in our previous work [16], at the experiment's conclusion, a quintet of fish from each aquarium was subjected to desiccation at 105 °C for 48 h, adapting the technique prescribed by Li et al. [8] with minor adjustments. Subsequent to this drying phase, these specimens were utilized to assess malondialdehyde, conjugated diene, and peroxide levels, employing the analytical methods established by Maqsood and Benjakul [21]. For further investigation, five fish from each tank were exsanguinated and sectioned, followed by preservation at −20 °C for a span of two months. Additionally, an equivalent number of fish were sedated using a benzocaine solution at 50 mg $L^{-1}$ before being humanely sacrificed via cervical dissection. Post-mortem, vital organs such as the hepatopancreas, intestine and gills were meticulously harvested and subsequently stored at an ultra-low temperature of −80 °C for impending analyses.

In accordance with our established procedures, the activities of several enzymes, including $Na^+/K^+$-ATPase, amylase, lipase, trypsin, and alkaline phosphatase, were evaluated. Additionally, we measured the concentrations of hydrogen peroxide, reduced glutathione, and malondialdehyde, as well as the enzymatic activities of glutathione S-transferase, catalase, glutathione peroxidase, superoxide dismutase, and the capacity for anti-hydroxyl radical and anti-superoxide anion action. Blood samples were collected from 15 fish in each experimental group using a heparinized syringe via caudal puncture. Subsequently, these samples were centrifuged at 1000× *g* for 3 min at 4 °C within one hour of collection. The plasma was then separated from the erythrocytes to assay the aforementioned parameters, aligning with the methods detailed by Li et al. [23]. Within the erythrocyte fraction, we determined the activities of $Na^+/K^+$-ATPase, glutamate-pyruvate transaminase, superoxide dismutase, catalase, glutathione reductase, lactate dehydrogenase, and the content of malondialdehyde, reduced glutathione, met-hemoglobin, superoxide anion, and hydrogen peroxide. Furthermore, we quantified the activities of glutamate-oxaloacetate transaminase and glutamate-pyruvate transaminase in the plasma, and the plasma's ammonia levels.

### 2.7. Biochemical Analysis

Enzymatic assays were conducted to evaluate the activities of $Na^+/K^+$-ATPase, lipase, alkaline phosphatase, amylase, and trypsin, as well as the transaminases glutamate-oxaloacetate and glutamate-pyruvate, replicating methods from our preceding research [15]. Determinations of the anti-superoxide anion capacity, hydrogen peroxide concentration, plasma ammonia levels, and the presence of glutathione and malondialdehyde, in conjunction with assays for superoxide dismutase, catalase, glutathione peroxidase, lactate dehydrogenase, and glutathione reductase, were carried out in alignment with procedures reported by Chen et al. [22]. Measurement protocols for superoxide anion and met-hemoglobin levels adhered to those established by Li et al. [24]. Moreover, the activities of glutathione S-transferase and the anti-hydroxyl radical capacity were quantified following the techniques detailed by Jiang et al. [25].

### 2.8. Statistical Analysis

Results are presented as the means ± S.D. Statistical analyses were performed using a one-way analysis of variance (ANOVA) with the software SPSS, version 15.0. Post hoc comparisons to determine significant differences among groups were made using Duncan's multiple range tests.

## 3. Results

### 3.1. Flavonoids Content, Total Antioxidant Capacity and Metal-Chelating Ability in EGbs

As shown in Table 2, the flavonoids content was significantly higher in EAE than other extracts ($p < 0.05$). Flavonoids content of petroleum ether extract of *Ginkgo biloba* leaves was significantly lower than other extracts ($p < 0.05$). Flavonoids content of ethanol extract of *Ginkgo biloba* leaves was higher than aqueous extract ($p < 0.05$). The total antioxidant capacity was significantly higher in EAE of *Ginkgo biloba* leaves than other extracts ($p < 0.05$), followed by ethanol, aqueous and petroleum ether extract. A similar trend was found in metal-chelating ability when treated with different EGbs. Power regression of flavonoids content in EGbs on total antioxidant capacity and metal-chelating ability were presented in Figure 1. The total antioxidant capacity and metal-chelating ability were positively correlated with the flavonoid contents of EGbs.

**Table 2.** The flavonoids content, total antioxidant capacity (T-AOC) and metal-chelating ability (MCA) of petroleum ether extract (PEE), ethyl acetate extract (EAE), ethanol extract (EE) and aqueous extract (AQE) of *Ginkgo biloba* leaves.

| Extracts | Flavonoids (mg/g Dry Extracts) | T-AOC (mM of Trolox) | MCA (% of Control) |
|---|---|---|---|
| PEE | 11.49 ± 0.64 [a] | 0.45 ± 0.03 [a] | 16.20 ± 0.93 [a] |
| EAE | 70.60 ± 5.05 [d] | 1.02 ± 0.05 [d] | 70.42 ± 2.44 [d] |
| EE | 32.79 ± 2.29 [c] | 0.88 ± 0.06 [c] | 40.38 ± 2.34 [c] |
| AQE | 19.94 ± 1.10 [b] | 0.69 ± 0.03 [b] | 24.30 ± 1.53 [b] |

Within a same column, values with different superscripts are significantly different ($p < 0.05$), and the data represent the means ± S.D. of three replicates.

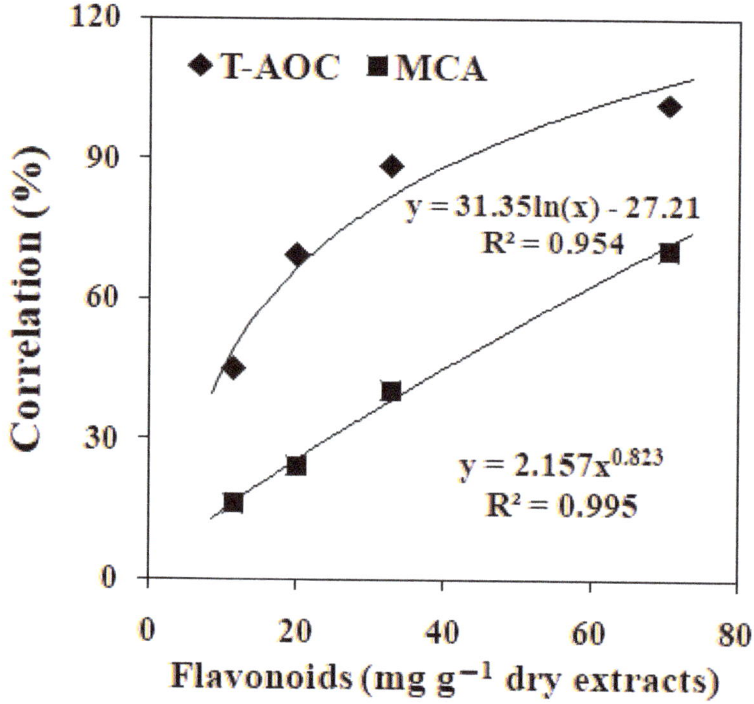


**Figure 1.** The correlations of flavonoids content with the total antioxidant capacity (T-AOC) and metal-chelating ability (MCA) in the extracts of *Ginkgo biloba* leaves. Values are the means of three replicates.

*3.2. Effects of EGbs on the Lipid Oxidation in Linoleic Acid and Linolenic Acid*

The relationship between EGbs and the peroxide, conjugated diene and malonaldehyde levels in a linoleic acid emulsion was shown in Table 3. The peroxide level was lower when treated with EAE, followed by ethanol, aqueous and petroleum ether extract ($p < 0.05$). The conjugated diene content of EAE was significantly lower than other extracts ($p < 0.05$). The conjugated diene content of ethanol extract was lower than aqueous and petroleum ether extract ($p < 0.05$). A similar trend was found in malonaldehyde level when treated with different EGb. Of the examined compounds, the levels of peroxide, conjugated diene and malonaldehyde were higher when treated with aqueous and petroleum ether extract, suggesting that these two extracts had the lowest inhibitory effects on the oxidation of linoleic acid ($p < 0.05$).

**Table 3.** The peroxide (PO), conjugated diene (CD) and malonaldehyde (MDA) levels in a linoleic acid emulsion treated with 1.0 mg $mL^{-1}$ of petroleum ether extract (PEE), ethyl acetate extract (EAE), ethanol extract (EE) and aqueous extract (AQE) of *Ginkgo biloba* leaves.

| Treatment | PO (% of Control) | CD (% of Control) | MDA (% of Control) |
|---|---|---|---|
| PEE | $72.38 \pm 4.50^{d}$ | $67.69 \pm 3.23^{c}$ | $69.09 \pm 5.18^{c}$ |
| EAE | $39.90 \pm 2.43^{a}$ | $40.79 \pm 3.07^{a}$ | $44.85 \pm 2.42^{a}$ |
| EE | $55.84 \pm 2.39^{b}$ | $54.97 \pm 2.49^{b}$ | $58.99 \pm 2.52^{b}$ |
| AQE | $64.23 \pm 2.22^{c}$ | $65.20 \pm 2.49^{c}$ | $69.29 \pm 2.99^{c}$ |

Values within the same column with different superscripts are significantly different ($p < 0.05$). The data represent the means ± S.D. of three replicates.

The relationship between EGbs and the peroxide, conjugated diene and malonaldehyde levels in a linolenic acid emulsion was shown in Table 4. The peroxide level was significantly lower when treated with EAE, followed by ethanol, aqueous and petroleum ether extract ($p < 0.05$). The conjugated diene level was lower when treated with EAE ($p < 0.05$), followed by ethanol and aqueous extract. The level of conjugated diene was higher when treated with petroleum ether extract. The malonaldehyde level of EAE was significantly lower than other extracts ($p < 0.05$). The malonaldehyde level of ethanol extract was lower than aqueous and petroleum ether extract ($p < 0.05$). Of the examined compounds, the levels of peroxide, conjugated diene and malonaldehyde were higher when treated with petroleum ether extract, suggesting the petroleum ether extract had the lowest inhibitory effects on the oxidation of linolenic acid ($p < 0.05$).

**Table 4.** The peroxide (PO), conjugated diene (CD) and malonaldehyde (MDA) levels in a linolenic acid emulsion treated with 1.0 mg $mL^{-1}$ of petroleum ether extracts (PEE), ethyl acetate extracts (EAE), ethanol extract (EE) and aqueous extract (AQE) of *Ginkgo biloba* leaves.

| Treatment | PO (% of Control) | CD (% of Control) | MDA (% of Control) |
|---|---|---|---|
| PEE | $59.35 \pm 2.44^{d}$ | $62.22 \pm 3.13^{c}$ | $55.10 \pm 2.63^{c}$ |
| EAE | $33.97 \pm 1.99^{a}$ | $37.50 \pm 2.08^{a}$ | $34.61 \pm 1.88^{a}$ |
| EE | $39.06 \pm 2.80^{b}$ | $50.22 \pm 2.52^{b}$ | $45.54 \pm 2.62^{b}$ |
| AQE | $49.18 \pm 1.67^{c}$ | $53.00 \pm 2.53^{b}$ | $54.90 \pm 2.04^{c}$ |

Values within the same column with different superscripts are significantly different ($p < 0.05$). The data represent the means ± S.D. of three replicates.

The correlations of flavonoids content in EGb on peroxide, conjugated diene and malonaldehyde levels in linoleic acid and linolenic acid emulsion were presented in Figure 2. The peroxide, conjugated diene and malonaldehyde levels were negatively correlated with the flavonoids contents of EGbs.

### 3.3. The EAE Composition

As showed in Figure 3 and Table 5, there are some components in EAE, including 1,5,6,7-tetramethylbicyclo[3.2.0]hept-6-en-3-one, 1,4-cyclohexadiene, 3,3,6,6-tetramethyl-, acetic acid 3,7,11,15-tetramethyl-2-hexadecenyl and 1,2-dihexylcyclopropene.

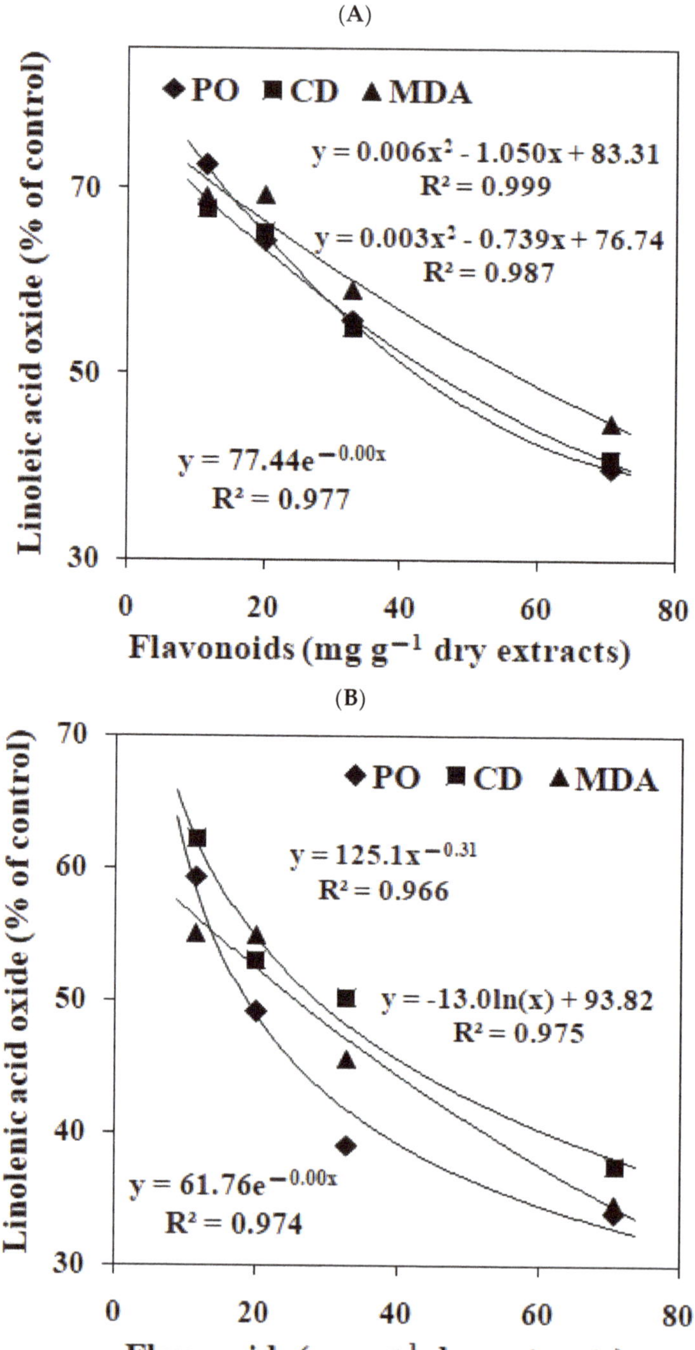


**Figure 2.** The correlations of flavonoids content in the extracts of *Ginkgo biloba* leaves with the levels of peroxide (PO), conjugated diene (CD) and malonaldehyde (MDA) in linoleic acid emulsion (**A**) and linolenic acid emulsion (**B**). The data represent the means of three replicates.

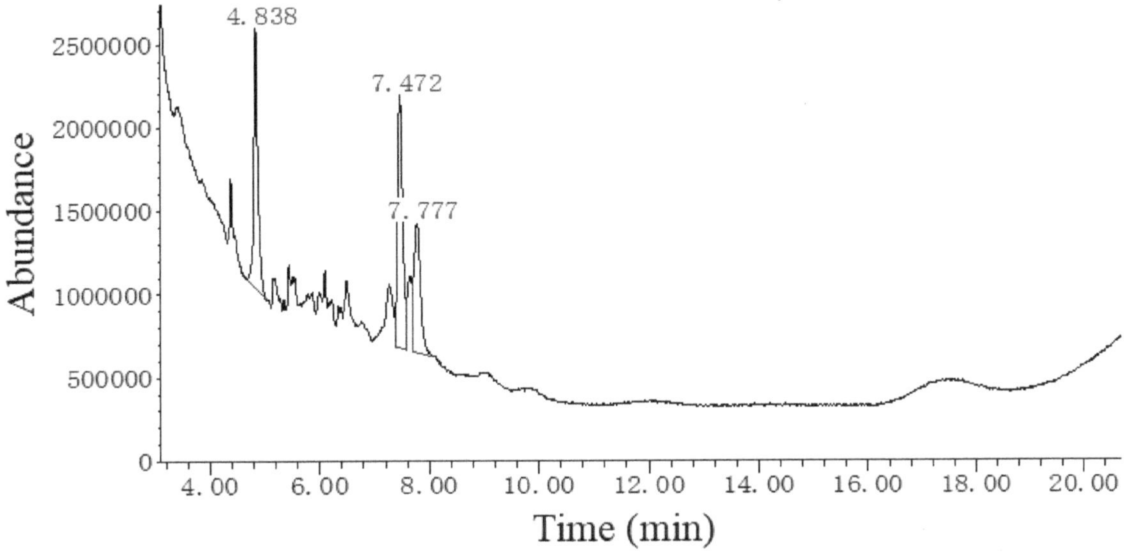


**Figure 3.** The gas chromatogram of ethyl acetate extract of *Ginkgo biloba* leaves (EAE). This experiment was repeated three times with similar results achieved.

**Table 5.** The composition analyses of ethyl acetate extract of *Ginkgo biloba* leaves (EAE) by gas chromatograph–mass spectrometry (GC–MS).

| Retention Time (min) | Compound Name | Molecular Weight (amu) | Molecular Formula | Matching Degree (%) |
|---|---|---|---|---|
| 4.838 | 1,5,6,7-Tetramethylbicyclo[3.2.0]hept-6-en-3-one | 164 | $C_{11}H_{16}O$ | 80 |
| 7.472 | 1,4-Cyclohexadiene, 3,3,6,6-tetramethyl- | 136 | $C_{10}H_{16}$ | 80 |
| | 1,5,6,7-Tetramethylbicyclo[3.2.0]hept-6-en-3-one | 164 | $C_{11}H_{16}O$ | 76 |
| 7.777 | Phytol, acetate | 338 | $C_{22}H_{42}O_2$ | 82 |
| | 1,2-Dihexylcyclopropene | 208 | $C_{15}H_{28}$ | 78 |

This experiment was repeated three times with similar results achieved.

### *3.4. The Effects of EAE on the Lipid Oxidation in Fish Feed*

Figure 4 illustrates the impact of escalating levels of EAE in the diet on the quantities of peroxides, conjugated dienes, and malonaldehyde within the fish feed. A significant reduction in these compounds was observed concomitant with increased EAE inclusion in the diet up to 4.0 g $kg^{-1}$, as evidenced by $p$-values less than 0.05. This declining trend plateaued when EAE concentrations in the diet exceeded this amount, indicated by $p$-values greater than 0.05. The broken line analysis suggested that the optimal inclusion levels of EAE for minimizing peroxides, conjugated dienes, and malonaldehyde were 4.26, 4.18, and 4.22 g $kg^{-1}$ of the diet, respectively.

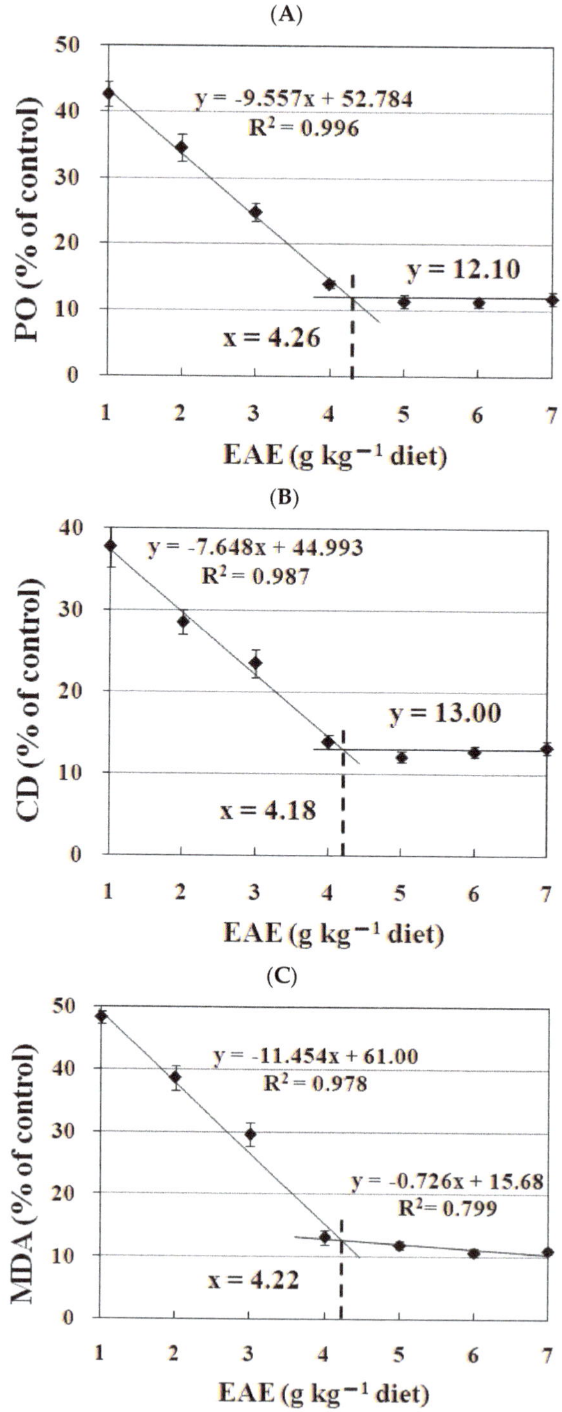


**Figure 4.** Broken-line analysis of peroxide (PO, (**A**)), conjugated diene (CD, (**B**)) and malondialdehyde (MDA, (**C**)) levels in fish feeds treated with graded levels of ethyl acetate extract (EAE) of *Ginkgo biloba* leaves. Values are the means ± S.D. of three replicates.

### 3.5. Effects of Dietary EAE on Fish Growth Performance

Table 6 delineates the lack of significant alterations in the fish feed efficiency and survival rates upon varying the levels of EAE in the diet (with $p$-values exceeding 0.05). Incremental increases in dietary EAE up to 2.0 g $kg^{-1}$ were associated with a notable enhancement in weight gain, which reached a threshold as $p$-values exceeded 0.05 beyond this concentration. Subsequent increases to 4.0 g $kg^{-1}$ did not yield further benefits and, in fact, led to a diminution in weight gain with higher EAE levels. Analogous patterns were observed regarding final body weight and specific growth rate, with both metrics experiencing similar trends. Feed intake exhibited an increase with rising dietary EAE levels up to 1.0 g $kg^{-1}$, reached a steady state with continued elevation to 4.0 g $kg^{-1}$ ($p$-values again surpassing 0.05), and subsequently decreased significantly when dietary EAE levels were augmented further ($p$-values falling below 0.05).

**Table 6.** Initial body weight (IBW), final body weight (FBW), weight gain (WG), specific growth rate (SGR), feed intake (FI), feed efficiency (FE) and survival ratio (SR) of Jian carp fed diets containing graded levels of ethyl acetate extract of *Ginkgo biloba* leaves (EAE) for 60 days.

| EAE (g $kg^{-1}$ Diet) | 0.0 | 1.0 | 2.0 | 3.0 | 4.0 | 5.0 | 6.0 |
|---|---|---|---|---|---|---|---|
| IBW (g $fish^{-1}$) | $14.78 \pm 0.37^{a}$ | $14.75 \pm 0.36^{a}$ | $14.80 \pm 0.40^{a}$ | $14.79 \pm 0.29^{a}$ | $14.81 \pm 0.38^{a}$ | $14.69 \pm 0.34^{a}$ | $14.57 \pm 0.40^{a}$ |
| FBW (g $fish^{-1}$) | $56.06 \pm 2.18^{b}$ | $59.14 \pm 2.32^{cd}$ | $61.61 \pm 2.09^{e}$ | $62.57 \pm 2.33^{e}$ | $60.92 \pm 2.07^{de}$ | $57.93 \pm 2.46^{bc}$ | $52.92 \pm 2.07^{a}$ |
| WG (g $fish^{-1}$) | $41.27 \pm 2.16^{b}$ | $44.38 \pm 1.39^{cd}$ | $46.82 \pm 1.15^{e}$ | $47.78 \pm 1.46^{e}$ | $46.14 \pm 1.14^{de}$ | $43.19 \pm 1.60^{bc}$ | $38.35 \pm 1.74^{a}$ |
| SGR (% $d^{-1}$) | $2.22 \pm 0.07^{b}$ | $2.31 \pm 0.03^{cd}$ | $2.38 \pm 0.03^{de}$ | $2.40 \pm 0.03^{e}$ | $2.36 \pm 0.03^{de}$ | $2.28 \pm 0.04^{bc}$ | $2.15 \pm 0.08^{a}$ |
| FI (g $fish^{-1}$) | $67.38 \pm 2.83^{b}$ | $71.15 \pm 2.96^{bc}$ | $72.40 \pm 1.99^{c}$ | $73.82 \pm 2.32^{c}$ | $71.87 \pm 2.45^{c}$ | $67.83 \pm 2.55^{b}$ | $62.88 \pm 2.54^{a}$ |
| FE (%) | $61.28 \pm 2.61^{a}$ | $62.47 \pm 3.08^{a}$ | $64.72 \pm 2.55^{a}$ | $64.80 \pm 3.31^{a}$ | $64.23 \pm 1.99^{a}$ | $63.78 \pm 3.83^{a}$ | $61.08 \pm 3.81^{a}$ |
| SR (%) | $100.00 \pm 0.00^{a}$ | $100.00 \pm 0.00^{a}$ | $100.00 \pm 0.00^{a}$ | $100.00 \pm 0.00^{a}$ | $100.00 \pm 0.00^{a}$ | $100.00 \pm 0.00^{a}$ | $100.00 \pm 0.00^{a}$ |

Values are the means ± S.D. of four replicates, with 20 fish in each replicate. Values in the same row with the different superscripts are significantly different ($p < 0.05$).

Polynomial regression analysis pinpointed the optimal dietary EAE supplementation for Jian carp at 2.75 g $kg^{-1}$ of the diet, as extrapolated from weight gain data (refer to Figure 5).

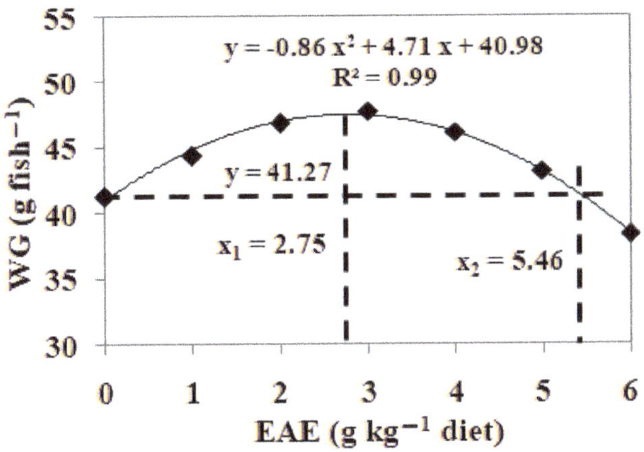


**Figure 5.** Polynomial regression analysis of weight gain (WG) for Jian carp fed diets containing graded levels of ethyl acetate extract (EAE) of *Ginkgo biloba* leaves for 60 days. Values are the means of four replicates, with 20 fish in each replicate.

### 3.6. Effects of Dietary EAE on the Biochemical Parameters in Hepatopancreas of Jian Carp

Table 7 presents a correlation between the elevation of EAE in the diet and enzymatic activities. It was noted that trypsin activity witnessed a significant rise with increments in EAE up to 2.0 g $kg^{-1}$ of the diet, indicated by a $p$-value less than 0.05, followed by a decline upon further elevation in EAE. Enzymes lipase and amylase similarly showed significant increases in their activities at EAE concentrations of 4.0 and 2.0 g $kg^{-1}$ of the diet, respectively, with $p$-values below 0.05. No significant changes in their activities were observed once the concentrations of EAE exceeded these levels, as the $p$-values went above 0.05. Enhanced activity levels of antioxidant enzymes such as anti-superoxide anion, anti-hydroxyl radical, superoxide dismutase, catalase, and glutathione peroxidase were recorded at 4.0, 3.0, 3.0, 4.0, and 4.0 g $kg^{-1}$ EAE in the diet, respectively. In contrast, the malonaldehyde concentration in the hepatopancreas was lower when the diet was supplemented with 4.0 g $kg^{-1}$ EAE. Additionally, a decrease in hydrogen peroxide was observed with increasing dietary EAE up to 1.0 g $kg^{-1}$ ($p < 0.05$), with levels stabilizing with further increments ($p > 0.05$). The reduced glutathione level showed a significant upsurge with increased dietary EAE up to 3.0 g $kg^{-1}$ ($p < 0.05$), but further increases in EAE resulted in a significant reduction in its concentration ($p < 0.05$).

**Table 7.** The activities of trypsin, lipase, amylase, anti-superoxide anion (ASA), anti-hydroxyl radical (AHR), superoxide dismutase (SOD), catalase (CAT), glutathione peroxidase (GPx) and glutathione S-transferase (GST) and the levels of hydrogen peroxide ($H_2O_2$), malondialdehyde (MDA) and reduced glutathione (GSH) in hepatopancreas of Jian carp fed diets containing different levels of ethyl acetate extract of *Ginkgo biloba* leaves (EAE) for 60 days.

| EAE (g $kg^{-1}$ Diet) | 0.0 | 1.0 | 2.0 | 3.0 | 4.0 | 5.0 | 6.0 |
|---|---|---|---|---|---|---|---|
| Trypsin (U $mg^{-1}$ protein) | 769.22 ± 159.11 $^{b}$ | 1645.93 ± 213.49 $^{d}$ | 1694.70 ± 108.87 $^{d}$ | 1253.24 ± 113.93 $^{c}$ | 813.20 ± 167.95 $^{b}$ | 668.35 ± 120.94 $^{ab}$ | 487.08 ± 93.94 $^{a}$ |
| Lipase (U $g^{-1}$ protein) | 35.80 ± 7.16 $^{a}$ | 36.36 ± 3.94 $^{a}$ | 38.12 ± 4.13 $^{a}$ | 53.00 ± 8.35 $^{b}$ | 63.15 ± 7.02 $^{bc}$ | 66.61 ± 10.53 $^{c}$ | 62.35 ± 6.93 $^{bc}$ |
| Amylase (U $mg^{-1}$ protein) | 1.00 ± 0.09 $^{a}$ | 1.04 ± 0.05 $^{ab}$ | 1.14 ± 0.14 $^{abc}$ | 1.27 ± 0.09 $^{bc}$ | 1.32 ± 0.13 $^{c}$ | 1.20 ± 0.08 $^{abc}$ | 1.11 ± 0.08 $^{abc}$ |
| ASA (U $g^{-1}$ protein) | 61.86 ± 2.93 $^{ab}$ | 61.37 ± 3.44 $^{ab}$ | 64.83 ± 1.07 $^{bc}$ | 65.19 ± 1.75 $^{bc}$ | 67.37 ± 4.62 $^{c}$ | 63.28 ± 0.37 $^{b}$ | 58.87 ± 0.75 $^{a}$ |
| AHR (U $mg^{-1}$ protein) | 147.72 ± 23.70 $^{a}$ | 150.65 ± 7.34 $^{ab}$ | 151.98 ± 9.38 $^{ab}$ | 193.46 ± 8.69 $^{c}$ | 179.62 ± 18.40 $^{bc}$ | 180.38 ± 22.10 $^{bc}$ | 159.08 ± 10.15 $^{ab}$ |
| SOD (U $mg^{-1}$ protein) | 83.27 ± 2.71 $^{ab}$ | 83.35 ± 5.23 $^{ab}$ | 90.66 ± 3.09 $^{bc}$ | 95.95 ± 4.44 $^{c}$ | 90.12 ± 2.86 $^{bc}$ | 80.63 ± 9.06 $^{a}$ | 78.95 ± 3.03 $^{a}$ |
| CAT (U $mg^{-1}$ protein) | 20.40 ± 3.66 $^{a}$ | 22.99 ± 4.17 $^{ab}$ | 25.47 ± 3.48 $^{abc}$ | 31.19 ± 2.13 $^{cd}$ | 38.34 ± 3.47 $^{e}$ | 35.77 ± 4.99 $^{de}$ | 26.93 ± 4.68 $^{bc}$ |
| GPx (U $mg^{-1}$ protein) | 278.08 ± 17.38 $^{a}$ | 299.74 ± 34.13 $^{ab}$ | 311.95 ± 36.06 $^{ab}$ | 335.49 ± 49.67 $^{ab}$ | 341.01 ± 26.82 $^{b}$ | 297.89 ± 26.77 $^{ab}$ | 274.40 ± 18.00 $^{a}$ |
| GST (U $mg^{-1}$ protein) | 87.35 ± 8.44 $^{a}$ | 86.43 ± 7.52 $^{a}$ | 93.24 ± 6.71 $^{a}$ | 95.06 ± 4.03 $^{a}$ | 91.25 ± 9.55 $^{a}$ | 91.41 ± 12.90 $^{a}$ | 92.37 ± 2.51 $^{a}$ |
| MDA (nmol $mg^{-1}$ protein) | 12.21 ± 2.31 $^{d}$ | 9.81 ± 1.84 $^{c}$ | 5.90 ± 0.59 $^{b}$ | 4.41 ± 0.54 $^{ab}$ | 3.79 ± 0.28 $^{a}$ | 4.65 ± 0.56 $^{ab}$ | 4.94 ± 0.76 $^{ab}$ |
| $H_2O_2$ (mmol $g^{-1}$ protein) | 31.52 ± 0.55 $^{b}$ | 29.43 ± 2.35 $^{ab}$ | 29.53 ± 1.28 $^{ab}$ | 29.29 ± 1.41 $^{ab}$ | 28.27 ± 0.04 $^{a}$ | 27.76 ± 0.04 $^{a}$ | 27.91 ± 0.04 $^{a}$ |
| GSH (mg $g^{-1}$ protein) | 7.16 ± 1.24 $^{a}$ | 8.53 ± 1.16 $^{ab}$ | 8.87 ± 0.82 $^{bc}$ | 11.77 ± 1.07 $^{d}$ | 10.00 ± 0.66 $^{c}$ | 7.76 ± 0.79 $^{ab}$ | 7.38 ± 0.29 $^{a}$ |

Values are the means ± S.D. of three replicates, with five fish in each replicate. Values in the same row with the different superscripts are significantly different ($p < 0.05$).

### 3.7. Effects of Dietary EAE on the Biochemical Parameters in Intestine of Jian Carp

According to the data in Table 8, it was observed that dietary incorporation of EAE influenced several enzymatic activities in the intestinal tissue of fish. Specifically, trypsin and amylase activity levels were augmented in correlation with EAE increments in the diet up to a threshold of 1.0 g $kg^{-1}$, as indicated by a $p$-value less than 0.05. These enzymatic activities reached a stable state when the EAE concentration in the diet was enhanced to between 5.0 g $kg^{-1}$ and 3.0 g $kg^{-1}$, after which a $p$-value greater than 0.05 was noted and subsequent declines in activity were observed with an increased EAE concentration ($p < 0.05$). The anti-superoxide anion capacity and glutathione S-transferase activity followed a similar pattern, showing elevated activity up to EAE concentrations of 1.0 g $kg^{-1}$ and 2.0 g $kg^{-1}$, respectively, before stabilizing. Superoxide dismutase activity increased with EAE up to 1.0 g $kg^{-1}$, plateaued up to a dietary content of 3.0 g $kg^{-1}$, and then decreased. Catalase activity showed significant enhancement with EAE levels reaching 2.0 g $kg^{-1}$, which then plateaued with no significant changes upon further increase in EAE ($p > 0.05$). Additionally, $Na^+/K^+$-ATPase, alkaline phosphatase, and glutathione peroxidase activities were positively influenced at EAE supplementation levels of 2.0, 3.0, and 4.0 g $kg^{-1}$ of the diet, respectively. A notable increase in glutathione content was observed with EAE supplementation at 6.0 g $kg^{-1}$ of the diet. Hydrogen peroxide concentrations in the fish intestine decreased concomitant with EAE levels up to 1.0 g $kg^{-1}$ of the diet ($p < 0.05$) and remained constant up to an EAE content of 5.0 g $kg^{-1}$ ($p > 0.05$), beyond which an increase was recorded.

**Table 8.** The activities of trypsin, amylase, $Na^+/K^+$-ATPase, alkaline phosphatase (AKP), antisuperoxide anion (ASA), glutathione *S*-transferase (GST), superoxide dismutase (SOD), catalase (CAT) and glutathione peroxidase (GPx) and the content of hydrogen peroxide ($H_2O_2$), malondialdehyde (MDA) and reduced glutathione (GSH) in intestine of Jian carp fed diets containing different levels of ethyl acetate extract of *Ginkgo biloba* leaves (EAE) for 60 days.

| EAE (g $kg^{-1}$ Diet) | 0.0 | 1.0 | 2.0 | 3.0 | 4.0 | 5.0 | 6.0 |
|---|---|---|---|---|---|---|---|
| Trypsin (U $mg^{-1}$ protein) | 1092.62 ± 33.17 $^{ab}$ | 1131.91 ± 75.60 $^{b}$ | 1158.46 ± 97.24 $^{b}$ | 1131.90 ± 42.17 $^{b}$ | 1133.46 ± 48.90 $^{b}$ | 1145.19 ± 37.27 $^{b}$ | 1007.35 ± 61.41 $^{a}$ |
| Amylase (U $mg^{-1}$ protein) | 0.80 ± 0.10 $^{a}$ | 1.23 ± 0.05 $^{b}$ | 1.22 ± 0.19 $^{b}$ | 1.16 ± 0.03 $^{b}$ | 0.86 ± 0.08 $^{a}$ | 0.87 ± 0.04 $^{a}$ | 0.84 ± 0.06 $^{a}$ |
| $Na^+/K^+$-ATPase (U $mg^{-1}$ protein) | 3.70 ± 0.39 $^{b}$ | 3.87 ± 0.13 $^{b}$ | 6.75 ± 0.77 $^{d}$ | 4.89 ± 0.80 $^{c}$ | 4.66 ± 0.65 $^{c}$ | 3.40 ± 0.26 $^{b}$ | 2.48 ± 0.19 $^{a}$ |
| AKP (U $mg^{-1}$ protein) | 280.87 ± 22.65 $^{a}$ | 294.03 ± 5.11 $^{ab}$ | 302.80 ± 9.95 $^{ab}$ | 306.46 ± 12.65 $^{b}$ | 290.77 ± 11.13 $^{ab}$ | 294.11 ± 6.50 $^{ab}$ | 280.37 ± 25.01 $^{a}$ |
| ASA (U $g^{-1}$ protein) | 46.65 ± 3.41 $^{a}$ | 49.27 ± 1.92 $^{ab}$ | 47.89 ± 2.51 $^{ab}$ | 51.57 ± 1.88 $^{b}$ | 51.45 ± 2.68 $^{b}$ | 46.39 ± 3.81 $^{a}$ | 46.52 ± 2.12 $^{a}$ |
| GST (U $mg^{-1}$ protein) | 75.74 ± 8.43 $^{a}$ | 113.66 ± 16.32 $^{b}$ | 141.82 ± 26.15 $^{c}$ | 155.27 ± 24.03 $^{c}$ | 105.04 ± 10.99 $^{b}$ | 103.62 ± 16.26 $^{b}$ | 67.26 ± 6.78 $^{a}$ |
| SOD (U $mg^{-1}$ protein) | 74.64 ± 15.90 $^{a}$ | 135.32 ± 15.79 $^{c}$ | 151.70 ± 5.24 $^{c}$ | 148.36 ± 12.97 $^{c}$ | 97.37 ± 13.14 $^{b}$ | 96.45 ± 15.99 $^{b}$ | 83.75 ± 6.83 $^{b}$ |
| CAT (U $mg^{-1}$ protein) | 13.62 ± 1.32 $^{a}$ | 16.27 ± 3.37 $^{ab}$ | 18.58 ± 1.60 $^{bc}$ | 18.64 ± 1.52 $^{bc}$ | 18.44 ± 1.59 $^{bc}$ | 19.14 ± 2.52 $^{bc}$ | 23.16 ± 4.21 $^{c}$ |
| GPx (U $mg^{-1}$ protein) | 421.62 ± 59.02 $^{b}$ | 422.56 ± 56.11 $^{b}$ | 419.92 ± 67.01 $^{b}$ | 418.34 ± 65.99 $^{b}$ | 597.12 ± 6.12 $^{c}$ | 332.33 ± 37.49 $^{ab}$ | 258.80 ± 9.29 $^{a}$ |
| GSH (mg $g^{-1}$ protein) | 3.06 ± 0.29 $^{a}$ | 3.06 ± 0.25 $^{a}$ | 3.42 ± 0.11 $^{a}$ | 3.45 ± 0.20 $^{a}$ | 4.25 ± 0.57 $^{a}$ | 8.86 ± 0.80 $^{b}$ | 12.15 ± 1.75 $^{c}$ |
| $H_2O_2$ (mmol $g^{-1}$ protein) | 17.51 ± 1.70 $^{c}$ | 12.97 ± 1.48 $^{ab}$ | 11.91 ± 2.29 $^{ab}$ | 11.12 ± 1.79 $^{a}$ | 10.01 ± 2.15 $^{a}$ | 12.68 ± 1.98 $^{ab}$ | 14.54 ± 2.09 $^{b}$ |

Values are the means ± S.D. of three replicates, with five fish in each replicate. Values in the same row with the different superscripts are significantly different ($p < 0.05$).

### 3.8. Effects of EAE on the Biochemical Parameters in Gills of Jian Carp

In the investigation detailed by Table 9, the anti-superoxide anion capability in the gill tissue of fish was enhanced when their feed was supplemented with 4.0 g $kg^{-1}$ of EAE. Concurrently, there was a noted decrease in malondialdehyde levels with the elevation of EAE in the diet up to 2.0 g $kg^{-1}$, marked by a statistically significant threshold ($p < 0.05$). This downward trend in malondialdehyde content reached a plateau with no notable differences at EAE concentrations of up to 4.0 g $kg^{-1}$, and an upsurge in levels occurred with further increments in EAE. Regarding antioxidative enzymatic activities and associated compounds in fish gills, this study found an uptick in the presence of superoxide dismutase, catalase, and glutathione peroxidase, with their activities peaking when the diet contained 3.0, 4.0, and 3.0 g $kg^{-1}$ of EAE, respectively. The reduced glutathione content was observed to increase in diets containing 2.0 g $kg^{-1}$ of EAE. Furthermore, glutathione S-transferase activity rose in tandem with higher dietary EAE supplementation, with the most substantial activities recorded at EAE concentrations of 5.0 and 6.0 g $kg^{-1}$.

**Table 9.** The level of malondialdehyde (MDA) and reduced glutathione (GSH) and the activities of antisuperoxide anion (ASA), superoxide dismutase (SOD), catalase (CAT), glutathione peroxidase (GPx) and glutathione *S*-transferase (GST) in gills of Jian carp fed diets containing different levels of ethyl acetate extract of *Ginkgo biloba* leaves (EAE) for 60 days.

| EAE (g kg$^{-1}$ Diet) | 0.0 | 1.0 | 2.0 | 3.0 | 4.0 | 5.0 | 6.0 |
|---|---|---|---|---|---|---|---|
| ASA (U g$^{-1}$ protein) | 53.53 ± 8.94 $^{ab}$ | 56.23 ± 12.07 $^{ab}$ | 57.46 ± 13.83 $^{ab}$ | 67.29 ± 3.93 $^{bc}$ | 74.73 ± 9.17 $^{c}$ | 61.39 ± 7.02 $^{abc}$ | 51.05 ± 5.46 $^{a}$ |
| MDA (nmol mg$^{-1}$ protein) | 6.48 ± 0.48 $^{c}$ | 6.19 ± 0.63 $^{bc}$ | 5.30 ± 0.73 $^{ab}$ | 4.40 ± 0.72 $^{a}$ | 4.95 ± 0.68 $^{a}$ | 6.89 ± 0.22 $^{c}$ | 9.65 ± 1.04 $^{d}$ |
| SOD (U mg$^{-1}$ protein) | 18.27 ± 2.06 $^{ab}$ | 19.26 ± 1.40 $^{abc}$ | 21.15 ± 2.09 $^{bc}$ | 21.57 ± 1.59 $^{c}$ | 18.63 ± 1.76 $^{ab}$ | 18.72 ± 1.80 $^{a}$ | 17.76 ± 0.83 $^{a}$ |
| CAT (U mg$^{-1}$ protein) | 5.30 ± 0.69 $^{a}$ | 5.23 ± 0.73 $^{a}$ | 5.38 ± 0.78 $^{a}$ | 6.26 ± 1.34 $^{ab}$ | 9.49 ± 2.07 $^{c}$ | 8.01 ± 1.93 $^{bc}$ | 5.56 ± 0.94 $^{a}$ |
| GPx (U mg$^{-1}$ protein) | 64.17 ± 9.60 $^{a}$ | 69.38 ± 11.51 $^{a}$ | 77.02 ± 9.11 $^{ab}$ | 117.36 ± 19.23 $^{d}$ | 107.71 ± 7.37 $^{cd}$ | 98.10 ± 9.13 $^{bcd}$ | 92.48 ± 13.13 $^{bc}$ |
| GSH (mg g$^{-1}$ protein) | 6.86 ± 0.72 $^{a}$ | 8.48 ± 0.56 $^{b}$ | 11.40 ± 1.66 $^{c}$ | 8.31 ± 0.54 $^{ab}$ | 7.52 ± 0.61 $^{ab}$ | 7.53 ± 0.96 $^{ab}$ | 7.70 ± 0.88 $^{ab}$ |
| GST (U mg$^{-1}$ protein) | 25.49 ± 1.40 $^{a}$ | 35.84 ± 4.63 $^{a}$ | 48.72 ± 7.91 $^{b}$ | 50.47 ± 8.46 $^{b}$ | 48.15 ± 8.25 $^{b}$ | 62.09 ± 9.21 $^{c}$ | 67.43 ± 9.18 $^{c}$ |

Values are the means ± SD of four replicates, with five fish in each replicate. Values with the different superscripts in the same row are significantly different ($p < 0.05$).

### 3.9. Effects of EAE on the Biochemical Parameters in Plasma and Erythrocytes of Jian Carp

Table 10 reveals that the activities of both glutamate-oxaloacetate transaminase and glutamate-pyruvate transaminase in plasma showed a significant reduction as the levels of EAE in the diet were increased ($p < 0.05$). This trend persisted until reaching a dietary EAE concentration of 3.0 g kg$^{-1}$, beyond which no significant changes were observed, with activities rising upon further EAE enrichment. Additionally, plasma ammonia concentrations were observed to be lower in fish whose diet included an EAE supplement of 6.0 g kg$^{-1}$.

**Table 10.** The activities glutamate-oxaloacetate transaminase (GOT) and glutamate-pyruvate transaminase (GPT) in plasma and plasma ammonia content (PAC) of Jian carp fed diets containing different levels of ethyl acetate extract of *Ginkgo biloba* leaves (EAE) for 60 days.

| EAE (g kg$^{-1}$ Diet) | 0.0 | 1.0 | 2.0 | 3.0 | 4.0 | 5.0 | 6.0 |
|---|---|---|---|---|---|---|---|
| PAC (μmol mL$^{-1}$) | 352.4 ± 25.46 $^{b}$ | 348.46 ± 33.09 $^{ab}$ | 338.56 ± 27.31 $^{ab}$ | 331.5 ± 23.4 $^{ab}$ | 330.13 ± 22.56 $^{ab}$ | 333.73 ± 16.93 $^{ab}$ | 310.25 ± 14.57 $^{a}$ |
| GOT (U L$^{-1}$) | 121.8 ± 18.41 $^{b}$ | 76.2 ± 14.42 $^{a}$ | 85.0 ± 9.07 $^{a}$ | 88.9 ± 15.10 $^{a}$ | 115.3 ± 6.88 $^{b}$ | 115.4 ± 9.78 $^{b}$ | 124.9 ± 19.68 $^{b}$ |
| GPT (U L$^{-1}$) | 159.80 ± 27.02 $^{d}$ | 120.83 ± 18.28 $^{c}$ | 62.36 ± 8.47 $^{ab}$ | 34.63 ± 3.37 $^{a}$ | 95.37 ± 1.33 $^{bc}$ | 227.92 ± 9.46 $^{e}$ | 237.34 ± 36.36 $^{e}$ |

Values are the means ± S.D. of three replicates, with five fish in each replicate. Values with the different superscripts in the same row are significantly different ($p < 0.05$).

Table 11 illustrates an increase in the activities of $Na^+/K^+$-ATPase in fish erythrocytes correlating with a rise in dietary EAE up to 2.0 g kg$^{-1}$ ($p < 0.05$), with a subsequent decline noted with additional EAE inclusion. The activities of glutamate-pyruvate transaminase escalated with an increase in EAE up to 5.0 g kg$^{-1}$, and continued to rise with further EAE supplementation. Erythrocytes exhibited lower lactate dehydrogenase activity when the fish diet was supplemented with 2.0 g kg$^{-1}$ of EAE. Moreover, the levels of met-hemoglobin, superoxide anion, and hydrogen peroxide in fish erythrocytes were reduced in diets supplemented with EAE at 4.0, 4.0, and 3.0 g kg$^{-1}$, respectively. Malondialdehyde content showed a significant decrease in conjunction with elevated dietary EAE up to 2.0 g kg$^{-1}$ ($p < 0.05$), with a stabilization of levels observed upon reaching an EAE concentration of 3.0 g kg$^{-1}$, followed by an increase with higher EAE amounts. The activities of superoxide dismutase and catalase, along with the reduced glutathione levels in fish erythrocytes, were higher in diets containing EAE at 3.0, 4.0, and 6.0 g kg$^{-1}$, respectively. Glutathione reductase activity was enhanced with increased dietary EAE up to 3.0 g kg$^{-1}$ ($p < 0.05$), plateauing with subsequent EAE increments.

**Table 11.** The activities of $Na^+/K^+$-ATPase, glutamate-pyruvate transaminase (GPT) and lactate dehydrogenase (LDH) in erythrocytes, superoxide dismutase (SOD), catalase (CAT) and glutathione reductase (GR) and the content of met-hemoglobin (Met-Hb), superoxide anion ($O_2^{\bullet-}$), hydrogen peroxide ($H_2O_2$), malondialdehyde (MDA) and reduced glutathione (GSH) in erythrocytes of Jian carp fed diets containing different levels of ethyl acetate extract of *Ginkgo biloba* leaves (EAE) for 60 days.

| EAE ($g\ kg^{-1}$ Diet) | 0.0 | 1.0 | 2.0 | 3.0 | 4.0 | 5.0 | 6.0 |
|---|---|---|---|---|---|---|---|
| $Na^+/K^+$-ATPase ($U\ mg^{-1}$ Hb) | $1.61 \pm 0.02^{ab}$ | $1.63 \pm 0.09^{ab}$ | $1.67 \pm 0.1^{b}$ | $1.54 \pm 0.17^{ab}$ | $1.52 \pm 0.08^{ab}$ | $1.53 \pm 0.05^{ab}$ | $1.48 \pm 0.12^{a}$ |
| GPT ($U\ g^{-1}$ Hb) | $5.24 \pm 0.29^{a}$ | $5.82 \pm 0.64^{ab}$ | $6.59 \pm 0.79^{b}$ | $6.63 \pm 0.55^{b}$ | $6.77 \pm 0.6^{bc}$ | $7.68 \pm 0.89^{c}$ | $5.75 \pm 0.55^{ab}$ |
| LDH ($U\ g^{-1}$ Hb) | $571.43 \pm 33.46^{d}$ | $565.58 \pm 21.61^{d}$ | $192.88 \pm 30.55^{a}$ | $259.4 \pm 32.33^{b}$ | $269.48 \pm 46.38^{b}$ | $281.31 \pm 64.67^{b}$ | $378.16 \pm 39.09^{c}$ |
| Met-Hb ($g\ L^{-1}$) | $1.84 \pm 0.03^{d}$ | $1.81 \pm 0.05^{d}$ | $1.79 \pm 0.04^{cd}$ | $1.64 \pm 0.08^{b}$ | $1.42 \pm 0.08^{a}$ | $1.66 \pm 0.14^{bc}$ | $1.77 \pm 0.13^{bcd}$ |
| $O_2^{\bullet-}$ ($U\ g^{-1}$ Hb) | $42.3 \pm 2.97^{bc}$ | $41.42 \pm 2.94^{bc}$ | $39.53 \pm 1.20^{ab}$ | $38.76 \pm 3.76^{ab}$ | $35.61 \pm 2.52^{a}$ | $40.14 \pm 2.34^{ab}$ | $45.06 \pm 3.99^{c}$ |
| $H_2O_2$ ($mmol\ g^{-1}$ Hb) | $35.03 \pm 3.83^{bc}$ | $34.46 \pm 3.53^{bc}$ | $31.28 \pm 1.27^{b}$ | $26.39 \pm 0.69^{a}$ | $36.64 \pm 3.02^{bc}$ | $39.60 \pm 2.91^{c}$ | $46.06 \pm 2.88^{d}$ |
| MDA ($nmol\ mg^{-1}$ Hb) | $4.82 \pm 0.44^{c}$ | $3.50 \pm 0.51^{b}$ | $1.46 \pm 0.14^{a}$ | $1.94 \pm 0.31^{a}$ | $3.94 \pm 0.36^{b}$ | $3.85 \pm 0.72^{b}$ | $4.68 \pm 0.40^{c}$ |
| SOD ($U\ mg^{-1}$ Hb) | $89.68 \pm 12.27^{a}$ | $101.21 \pm 3.72^{ab}$ | $130.25 \pm 13.88^{d}$ | $131.87 \pm 9.72^{d}$ | $121.88 \pm 5.47^{cd}$ | $114.50 \pm 16.77^{bcd}$ | $104.73 \pm 18.35^{abc}$ |
| CAT ($U\ mg^{-1}$ Hb) | $6.55 \pm 0.98^{a}$ | $6.93 \pm 1.13^{ab}$ | $6.85 \pm 1.02^{ab}$ | $7.11 \pm 0.86^{ab}$ | $8.18 \pm 0.90^{b}$ | $7.58 \pm 0.89^{ab}$ | $7.69 \pm 0.92^{ab}$ |
| GSH ($mg\ g^{-1}$ Hb) | $7.41 \pm 0.35^{a}$ | $7.44 \pm 0.23^{a}$ | $8.25 \pm 1.14^{a}$ | $8.12 \pm 0.79^{a}$ | $9.44 \pm 0.71^{b}$ | $9.62 \pm 0.55^{b}$ | $15.35 \pm 1.06^{c}$ |
| GR ($U\ g^{-1}$ Hb) | $33.89 \pm 2.39^{a}$ | $37.24 \pm 1.79^{b}$ | $42.98 \pm 2.91^{c}$ | $45.14 \pm 2.36^{cd}$ | $46.44 \pm 1.49^{cd}$ | $46.05 \pm 2.02^{d}$ | $48.25 \pm 1.53^{d}$ |

Values are the means ± SD of three replicates, with 10 fish in each replicate. Values in the same row with different superscripts are significantly different ($p < 0.05$).

### 3.10. Effects of EAE on the Biochemical Parameters in Meat of Jian Carp

Upon examination of Table 12, it was observed that the concentration of hydrogen peroxide and malondialdehyde in fish tissues diminished as dietary EAE was augmented to 4.0 g $kg^{-1}$ and 3.0 g $kg^{-1}$, respectively. Concurrently, there was a significant elevation in the capability to neutralize hydroxy radicals, the activity of glutathione S-transferase, and the reduced glutathione concentration when the diet was supplemented with EAE up to 4.0, 4.0, and 3.0 g $kg^{-1}$, respectively ($p < 0.05$). However, these metrics saw a decline when the levels of dietary EAE were increased beyond these points. Additionally, the catalase activity in fish meat showed a progressive increment correlating with higher dietary EAE concentrations.

**Table 12.** The level of hydrogen peroxide ($H_2O_2$), malondialdehyde (MDA) and reduced glutathione (GSH) and the activities of anti-hydroxy radical (AHR), catalase (CAT) and glutathione *S*-transferase (GST) in the meat of Jian carp fed diets containing different levels of ethyl acetate extract of *Ginkgo biloba* leaves (EAE) for 60 days.

| EAE ($g\ kg^{-1}$ Diet) | 0.0 | 1.0 | 2.0 | 3.0 | 4.0 | 5.0 | 6.0 |
|---|---|---|---|---|---|---|---|
| $H_2O_2$ ($mmol\ g^{-1}$ protein) | $28.47 \pm 3.59^{c}$ | $29.29 \pm 4.30^{c}$ | $15.46 \pm 3.19^{b}$ | $13.12 \pm 1.36^{ab}$ | $11.35 \pm 0.42^{a}$ | $10.2 \pm 0.80^{a}$ | $9.79 \pm 0.29^{a}$ |
| AHR ($U\ mg^{-1}$ protein) | $173.24 \pm 23.39^{a}$ | $177.68 \pm 9.20^{a}$ | $175.41 \pm 8.73^{a}$ | $180.41 \pm 2.98^{ab}$ | $211.97 \pm 13.18^{c}$ | $201.27 \pm 7.67^{bc}$ | $195.15 \pm 12.74^{abc}$ |
| MDA ($nmol\ mg^{-1}$ protein) | $9.06 \pm 1.79^{c}$ | $5.49 \pm 0.44^{b}$ | $5.13 \pm 0.53^{ab}$ | $3.90 \pm 0.58^{a}$ | $5.10 \pm 0.32^{ab}$ | $4.84 \pm 0.54^{ab}$ | $6.15 \pm 0.70^{b}$ |
| CAT ($U\ mg^{-1}$ protein) | $10.47 \pm 0.71^{a}$ | $14.27 \pm 2.63^{b}$ | $14.19 \pm 1.87^{b}$ | $15.98 \pm 0.58^{b}$ | $16.52 \pm 0.80^{b}$ | $16.47 \pm 1.58^{b}$ | $23.83 \pm 2.83^{c}$ |
| GSH ($mg\ g^{-1}$ protein) | $6.93 \pm 0.70^{a}$ | $11.31 \pm 1.40^{b}$ | $14.03 \pm 0.98^{bcd}$ | $15.76 \pm 2.33^{cd}$ | $16.48 \pm 3.12^{d}$ | $13.32 \pm 0.24^{bc}$ | $6.06 \pm 0.54^{a}$ |
| GST ($U\ mg^{-1}$ protein) | $66.02 \pm 12.48^{a}$ | $102.49 \pm 7.35^{b}$ | $140.54 \pm 28.63^{c}$ | $185.63 \pm 36.47^{d}$ | $174.72 \pm 10.52^{d}$ | $133.81 \pm 24.64^{bc}$ | $109.24 \pm 17.77^{bc}$ |

Values are the means ± SD of four replicates, with five fish in each replicate. Values with the different superscripts in the same row are significantly different ($p < 0.05$).

### 3.11. Effects of Dietary EAE on the Hot-Drying-Induced Lipid Oxidation in Meat of Jian Carp

Figure 6 illustrates the impact of varying EAE supplementation on peroxide, conjugated diene, and malondialdehyde levels within piscine muscle tissues. A reduction in peroxide concentration was observed with EAE inclusion in the diet up to 4.0 g $kg^{-1}$, followed by an ascent upon further elevation of EAE levels. This pattern was also mirrored in the measurements of conjugated diene and malondialdehyde. Regressive analysis deduced that the optimal inclusion rates of dietary EAE, calculated from the concentrations of peroxide, conjugated diene, and malondialdehyde in fish tissues, were 3.72 g $kg^{-1}$, 3.05 g $kg^{-1}$, and 3.73 g $kg^{-1}$, respectively.

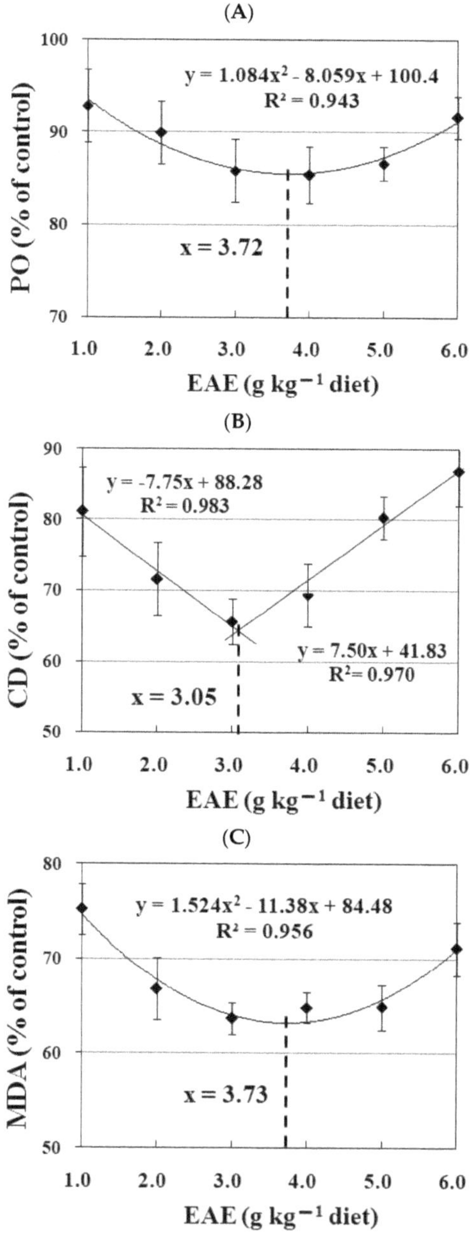


**Figure 6.** Regressive analysis of peroxide (PO, (**A**)), conjugated diene (CD, (**B**)) and malondialdehyde (MDA, (**C**)) levels in fish fed diets containing graded levels of ethyl acetate extract (EAE) of *Ginkgo biloba* leaves, following drying at 105 °C for 48 h. Values are the means ± SD of four replicates.

## 4. Discussion

### 4.1. EGb Inhibited the Lipid Oxidation in Fish Feed

Lipid oxidation is a common problem in foods and feeds [8]. In this study, EAE reduced peroxide, conjugated diene and malondialdehyde levels in fish feed. Similarly, malondialdehyde level was relieved in fish erythrocytes by EAE in vitro [8]. The investigation revealed that EGb could potentially serve as an inhibitor of lipid oxidation within aqua feed. By employing a broken-line model for peroxide, conjugated diene, and malondialdehyde, it was ascertained that the optimal levels of EAE supplementation in aqua feed are 4.26 g $kg^{-1}$. Lipid oxidation within aqua feed is likely linked to the presence of unsaturated fatty acids. Peroxides and hydro-peroxides are known as the principal products in the initial phase of lipid oxidation in unsaturated fatty acids [26]. The molecular architecture of hydro-peroxides, characterized by conjugated dienes, arises via the resonance of methylene-interrupted double bonds in unsaturated fatty acids, upon the assault of free radicals [27]. Consequently, peroxide and conjugated diene are acknowledged as precise indicators of early oxidation [28]. In the subsequent phase, peroxides and hydro-peroxides undergo reactions with oxygen or redox-active metals, culminating in the formation of malondialdehyde [29]. Essential unsaturated fatty acids like linolenic acid (18:3n − 3) and linoleic acid (18:2n − 6) are vital in freshwater fish nutrition [30]. Our study noted that EGbs diminishes the levels of peroxide, conjugated diene, and malondialdehyde within linoleic and linolenic acid emulsions, indicating its capability to mitigate lipid oxidation in aqua feed by disrupting both the initiation and propagation stages of the lipid oxidation process. Out of all extracts tested, EAE was identified as the most efficacious in attenuating lipid oxidation in these fatty acid emulsions. This aligns with prior observations where EAE inhibited lipid oxidation, verified via thiocyanate and thiobarbituric acid reactive substance assays [8].

ROS and redox metal ions play an important role in lipid oxidation of unsaturated fatty acids [8]. The total antioxidant capacity is an indicator of ROS scavenging activity, which reflects the overall antioxidative capability in hydrophilic or lipophilic substances [31,32]. Lipid oxidation is propagated under the catalytic actions of iron and other redox metal ions [8]. In this study, EGb displayed the strong metal chelation capabilities and total antioxidant capacity that was determined by ABTS assay and similar to trolox. Previous studies in our laboratory reported that EGb decreased ROS production in fish erythrocytes caused by hydroxyl radical [8]. Moreover, study in vitro showed that EGb can act as primary and/or secondary antioxidants, and free radical scavengers in prevention of lipid oxidation [33]. The current finding is consistent with the reports mentioned above. These results suggested that EGb may prevent unsaturated fatty acids of fish feed from undergoing lipid oxidation through enhancing metal inactivation and free radical scavenging activity.

The beneficial effects of EGbs on lipid oxidation in fish feed may be closely associated with their chemical constituents. The flavonoids content of *Ginkgo biloba* leaves is concerned with total antioxidant activity, which was determined by Near-infrared spectroscopy method [34]. In vitro, the ethyl acetate extract and methanol extract from *Ginkgo biloba* leaves evidenced a marked radical scavenging activity, which was based on the metal-chelating activity of flavonoids [35]. Within the ambit of radical-scavenging mechanisms, flavonoids are equipped with intramolecular hydrogen bonds that serve to enhance the antioxidative efficacy of hydroxyl functionalities, wherein the hydroxyl groups act as receptors of hydrogen bonds, notably at the 4′-OH position [36]. In addition, flavonoids compounds were showed to protect the polyunsaturated fatty acid from oxidation [37]. The hydroxyl groups in flavonoids compound are the crucial structure for antioxidant capacity, which could not only provide hydrogen atoms to lipid radicals, but also generate stable lipid derivatives and antioxidant radicals [38]. In the present study, not only the close correlations of flavonoids content in EGbs with the levels of peroxide, conjugated diene and malonaldehyde in linoleic acid and linolenic acid, but also with their total antioxidant capacity and the metal-chelating activity were observed. This is in good agreement with the report that there are four possible antioxidant mechanisms for flavonoids, namely inhibiting free radical generation of related enzymes, chelating metals, quenching free radical elements as well as stimulating the internal antioxidant enzymes [39]. These results confirmed that the antioxidant activity of EGb in fish feed might be caused by the flavonoids compounds.

### 4.2. Dietary EAE Supplementation Improved Fish Growth Performance

In piscine mitochondria, the presence of aminotransferases such as glutamate-oxaloacetate transaminase and glutamate-pyruvate transaminase is notable, given their extensive distribution across various tissues including the liver, muscle, heart, and kidney [40]. Tissue cell damage triggers the release of these enzymes into the bloodstream [41], rendering the levels of glutamate-oxaloacetate transaminase and glutamate-pyruvate transaminase in plasma reliable biomarkers for assessing fish health [42]. Findings from the current investigation indicate that inclusion of dietary EAE has led to a reduction in the activity of both aminotransferases in fish plasma, implying that EAE in the diet could potentially bolster fish health by mitigating liver damage. The positive impacts observed from EGb supplementation on piscine well-being may be attributable to its flavonoid content [43]. Nonetheless, elucidation of the precise mechanisms by which EGb influences fish health warrants additional research.

In the present study, dietary EAE increased feed intake, weight gain and specific growth rate, which indicates that dietary EGb improves Jian carp growth performance. Based on polynomial regression analysis of weight gain, the optimum dietary EAE for Jian carp was 2.75 g $kg^{-1}$ diet. Similarly, these results were consistent with the report on Nile tilapia (*Oreochromis niloticus*) [44]. Similar observations have been reported in *Penaeus vannamei* and hybrid grouper [13,45]. Furthermore, EGb can improve energy metabolism in rats, partly because it might increase the expression of ghrelin, which was an endogenous ligand of the growth hormone [46,47]. However, to date few studies have investigated the growth-promoting effect of EGb on fish.

The growth performance of fish is closely related to the digestion of nutrients, which has been attributed to the levels of digestive enzymes activities [15]. Digestive enzymes (amylase, trypsin, chymotrypsin and lipase) are synthesized in fish exocrine pancreas and secreted into the intestine [48,49]. In this study, dietary EAE increased activities of trypsin, lipase and amylase, indicates that dietary EAE supplementation improved fish digestive ability. Meanwhile, the growth performance of fish is also closely related to the absorption of nutrients, which has been attributed to the levels of brush-border membrane enzymes activities [22]. $Na^+/K^+$-ATPase is able to create potential energy of Na gradient to transport nutrients into cells, such as glucose, phosphate and amino acids [50]. Alkaline phosphatase, employed to be a general marker of fish absorption capacity, is an important enzyme in absorption of food molecules [51]. In this study, dietary EAE increased activities of alkaline phosphatase and $Na^+/K^+$-ATPase, which indicates that dietary EGb improved fish absorptive ability. This is in line with the reports that the intestinal length is increased in Nile tilapia after feeding EGb [44].

The structure and function of tissues and organs in aquatic animals are largely determined by their antioxidant status [52]. The integrity of intestinal morphology is the foundation for the digestive and absorptive process for fish [15]. ROS, which are produced as a result of aerobic metabolism and exposure to exogenous agents, have been shown to be damaging to lipids and proteins [53]. Malonaldehyde is used as a sensitive biomarker for lipid peroxidation [52]. In this study, dietary EAE decreased malonaldehyde content in hepatopancreas and intestine, which indicates that dietary EGb decreases lipid peroxidation in fish digestive and absorptive organs. These results were in conformity to the report that short-term oral administration of *Ginkgo biloba* extract reduces malonaldehyde level in washed platelets of type 2 diabetic patients [54]. Oral ingestion of *Ginkgo biloba* extract protected the liver injury by blunting the rises of malonaldehyde in rats [55].

In addition, the increased generation of ROS, including superoxide anion and hydroxyl radical, can induce lipid oxidation [25]. In this study, dietary EAE supplementation increased anti-superoxide anion capacity in hepatopancreas and intestine, which indicates that dietary EAE enhanced superoxide anion-scavenging ability in fish digestive and absorptive organs. These results were in conformity the report that EGb decreased the superoxide anion levels of in carp erythrocytes under hydroxyl radical stress [8]. In vitro study showed that *Ginkgo biloba* leaves exerted antioxidant properties by scavenging of superoxide anions [56]. Furthermore, it was reported that *Ginkgo biloba* extract showed a good capability of scavenging ROS in intestine of rats [57]. The effects of EAE may be linked to the chemical structure of their flavonoid compounds. Seyoum et al. reported that the flavonoids compound had free radical-scavenging ability [58]. In the meantime, it was demonstrated that uptake of flavonoids could decrease the activity of ROS in human colon adenocarcinoma Caco-2 cells [59].

The antioxidative systems in fish, inclusive of non-enzymatic and enzymatic components, have been refined through evolution to mitigate or rectify the damage inflicted by ROS on piscine tissues and organisms [60]. Within the enzymatic segment of this defense system, superoxide dismutase emerges as the initial enzyme to interact with the superoxide anion, facilitating its conversion into $H_2O_2$ and dioxygen [52]. Subsequently, catalase and glutathione peroxidase act on $H_2O_2$ to decompose it further [61]. Current research demonstrates that inclusion of EAE in the diet elevates the activity levels of superoxide dismutase, catalase, and glutathione peroxidase within the hepatopancreas, signifying an enhancement in the antioxidant capabilities of fish owing to EGb inclusion in feed. These findings align with studies showing that EGb-supplemented diets increase the activities of hepatic antioxidant enzymes such as superoxide dismutase and catalase in hybrid grouper [13]. The advantageous influence of EAE could be attributed to its flavonoid constituents. For instance, in *Channa punctata*, quercetin, a flavonoid supplement, was reported to alleviate stress induced by deltamethrin, as evidenced by restored activities of liver enzymes including superoxide dismutase, glutathione peroxidase, and catalase [62]. Furthermore, the significance of fish antioxidative defense mechanisms is paramount in counteracting oxidative stress [60]. The glutathione-dependent antioxidant system, pivotal in cell damage prevention caused by radicals, encompasses glutathione S-transferase and glutathione reductase, in concert with glutathione itself, a tripeptide of low molecular weight [63,64]. Glutathione S-transferase represents a family of multifunctional enzymes that facilitate the binding of electrophilic metabolites to glutathione's thiol group [13]. Meanwhile, glutathione reductase is responsible for the regeneration of reduced glutathione from its oxidized form, utilizing NADPH as the electron source [65]. In the present study, augmenting the diet with EAE was found to boost both glutathione reductase activity and reduced glutathione levels in the intestinal tissue, indicative of amelioration in enzymatic antioxidative function brought on by EGb. This is corroborated by laboratory evidence showing that EGb reinstated glutathione reductase activity in piscine erythrocytes when subjected to hydroxyl radical stress [8]. Consequently, this study posits that EAE augments the antioxidative defense of digestive organs, thereby preserving their structure and functionality, which in turn supports digestive efficiency and, consequently, fish growth performance. Nonetheless, the mechanisms through which EAE exerts its regulatory effect on fish growth remain to be elucidated further.

### *4.3. Dietary EAE Decreased Lipid Oxidation in Fish Meat through Inhibiting the Induction Effect of Hemoglobin in Erythrocyte*

Lipid oxidation is a key factor that contributes to the quality deterioration in muscle foods [4]. In fish, the content of polyunsaturated fatty acids was commonly higher than in mammalian [3]. In addition, fish is prone to lipid oxidation not only due to the amounts but also the types of polyunsaturated fatty acids. The most common polyunsaturated fatty acids present in mammalian are arachidonic acid, and in fish is eicosa pentanoic acid, which is more susceptible to lipid oxidation [66]. Furthermore, the drying process significantly accelerated lipid oxidation in fish [67]. In this study, dietary EAE supplementation decreased peroxide, conjugated diene and malonaldehyde levels, which indicates that EGb could decrease the lipid oxidation in fish meat induced by hot-drying process. Similar observations have been reported in hybrid grouper showed that dietary supplementation with EGb alleviated lipid peroxidation in fish liver as manifested the decrease in malonaldehyde content [13]. However, the mechanism by which EGb alleviated lipid oxidation in fish needs to be further explored. Based on regressive analysis of peroxide, conjugated diene and malonaldehyde, optimum dietary EAE supplementation were 3.72, 3.05 and 3.73 g $kg^{-1}$ diet, respectively.

Lipid oxidation in fish foods may be similar to fish muscle. Fish muscle is susceptible to oxidative stress, which led to reduce flesh quality [68]. In this study, dietary EAE supplementation decreased hydrogen peroxide levels and enhanced capacity of anti-hydroxyl radical (a maker of hydroxyl radical-scavenging ability), which indicates that EGb could reduce ROS production in fish meat. Similarly, EGb reduced ROS levels in mice lymphocytes [69]. ROS have been shown to be damaging to fish lipids [53]. In this study, dietary EAE supplementation decreased malonaldehyde levels in fish meat, which indicates that EGb could decrease lipid oxidation in fish meat. The beneficial effects of EAE on antioxidant status in fish meat may be linked to its flavonoid compound. Study in Nile tilapia showed that levels of hydrogen peroxide and malonaldehyde in fish flesh decreased by dietary inclusion of quercetin, a kind of flavonol glycoside in *Ginkgo biloba* leaves [70,71]. In this study, dietary supplementation of EGb increased reduced glutathione content and activities of catalase and glutathione *S*-transferase, indicates that dietary EGb could regulate antioxidant defense systems in fish meat, therefore decreasing lipid oxidation and improving the capacity to cope with ROS. Study in broiler chickens showed that dietary inclusion of fermented *Ginkgo biloba* leaves could improve antioxidant properties in thigh meat by enhancing the antioxidant enzyme system [72].

Lipid oxidation of fish meat may be related to the auto-oxidation of hemoglobin. Fish erythrocytes are important for all tissues and organs because they carried hemoglobin to them for oxygen supply [73]. Erythrocytes exhibit a heightened sensitivity to reactive oxygen species (ROS) due to the elevated tension of oxygen, significant iron content from heme groups, and the presence of polyunsaturated fatty acids within their membranes, which are inherently prone to oxidative damage [74,75]. While hemoglobin principally functions as an oxygen-transporting protein, its susceptibility to oxidative stress leads to its conversion into met-hemoglobin, a form incapable of carrying oxygen [74]. Studies have identified hemoglobin as a significant pro-oxidant, catalyzing lipid peroxidation in piscine muscle [76]. The propensity for lipid oxidation in fish tissue is exacerbated by two factors: the substantial hemoglobin content in fish meat, which remains largely unextracted before processing, and the autoxidation of hemoglobin, a continuous process yielding ROS and releasing iron ions within erythrocytes, facilitating further oxidation of lipids in fish muscle [4,5]. The present investigation revealed that incorporating of EAE in fish diets led to reduced levels of met-hemoglobin, as well as diminished concentrations of malondialdehyde, superoxide radicals, and hydrogen peroxide in piscine erythrocytes. This suggests that EAE intake curtails ROS generation and lipid peroxidation within erythrocytes, subsequently mitigating the lipid oxidation process in fish muscle tissue. Similar observations have been reported that *Ginkgo biloba* extract reduced $H_2O_2$-induced oxidative stress in human erythrocytes [77]. Furthermore, dietary supplementation of EAE increased content of reduced glutathione, and activities of glutathione reductase, superoxide dismutase and catalase, indicates that dietary EAE could regulate antioxidant defense systems in fish erythrocytes, therefore decreasing lipid oxidation and ROS production. So, the effects of dietary EGb on the lipid oxidation in fish meat may be attributed to its inhibition on the induction of hemoglobin in erythrocytes of fish.

Fish gills were easier attacked by the ROS than other organs not only because they always exposed to environmental stress but also as a dynamic osmoregulatory and respiratory organ, which gathered high amounts of erythrocytes as well as hemoglobin [78]. ROS have been shown to be damaging to lipids in fish [53]. In this study, dietary EAE supplementation decreased malonaldehyde levels, which indicates that EGb could decrease lipid oxidation in fish gills. In addition, dietary EAE supplementation enhanced anti-superoxide anion capacity, which suggested EGb could reduce ROS production in fish gills. Fishes have developed non-enzymatic antioxidants (such as reduced glutathione) and ROS scavenging enzymes to prevent or repair the harm triggered by ROS [60]. In this study, dietary EAE supplementation increased the reduced glutathione content and the activities of superoxide dismutase, glutathione peroxidase, catalase and glutathione *S*-transferase, indicates that dietary EGb could regulate antioxidant defense systems in fish gills, therefore decreasing lipid oxidation and ROS production. Study in tilapia (*Oreochromis niloticus*) showed that dietary inclusion of *Ginkgo biloba* extract could alleviate oxidative stress by enhancing activities of superoxide dismutase and catalase, and reduced glutathione levels in fish gills induced by glyphosate exposure [79]. The effects of dietary EGb on fish gills confirmed our hypothesis of its effects on fish meat about the role of hemoglobin.

## 5. Conclusions

Therefore, we conclude that EGb could decrease lipid oxidation in fish feed through protecting unsaturated fatty acids from oxidation. The effect of EGbs on the unsaturated fatty acids may be closely associated with their metal chelation capabilities and total antioxidant capacity. Broken-line analysis of peroxide, conjugated diene and malonaldehyde showed that the optimum dietary EAE supplementation was 4.26 g $kg^{-1}$ in fish feed. Meanwhile, current results displayed that dietary EGb increased the growth performance of fish, which may be ascribed to its enhancing effect on the activity of digestive and absorptive enzymes and antioxidant capacity in digestive organs of fish. Based on polynomial regression analysis of weight gain, optimum dietary EAE supplementation for Jian carp was 2.75 g $kg^{-1}$ diet. In addition, dietary EAE supplementation inhibits the induction effect of hemoglobin in erythrocyte, therefore decreasing the hot-drying-induced lipid oxidation in fish meat. The antioxidant capacity of EGb was closely related to their flavonoid content. Our present study indicates that EGb can be considered as a potential natural antioxidant for fish and fish feed.

**Author Contributions:** Conceptualization, G.C. and H.L.; methodology, G.C., J.X., Q.Y., L.F. and M.W.; formal analysis, J.X. and M.W.; validation, G.C., J.X., Q.Y., L.F. and H.L.; data curation, G.C. and M.W.; resources, H.L. and L.F.; writing—original draft preparation, G.C., J.X., Q.Y. and H.L. All authors have read and agreed to the published version of the manuscript.

**Funding:** This research was funded by the Sichuan Science and Technology Program [grant number 2018JY0214]; the Patent Project of Neijiang Normal University [grant number Z2019027]; and the Talent Program of Neijiang Normal University [grant number R2019015].

**Institutional Review Board Statement:** All procedures were approved by the Neijiang Normal University Institutional Animal Care and Use Committee(Approval Code: JM2016-11).

**Informed Consent Statement:** Not applicable.

**Data Availability Statement:** The data used to generate the results in this manuscript can be made available if requested from the corresponding author.

**Conflicts of Interest:** The authors declare no conflict of interest.

## References

1. Godwin, A.; Prabhu, H.R. Lipid peroxidation of fish oils. Indian. *J. Clin. Biochem.* **2006**, *21*, 202–204. [CrossRef] [PubMed]
2. Guo, X.Y.; Chen, S.N.; Cao, J.Y.; Zhou, J.Y.; Chen, Y.Z.; Jamali, M.A.; Zhang, Y.W. Hydrolysis and oxidation of protein and lipids in dry-salted grass carp (*Ctenopharyngodon idella*) as affected by partial substitution of NaCl with KCl and amino acids. *RSC Adv.* **2019**, *9*, 39545–39560. [CrossRef] [PubMed]
3. Hernández, A.; García García, B.; Jordán, M.J.; Hernández, M.D. Natural antioxidants in extruded fish feed: Protection at different storage temperatures. *Anim. Feed. Sci. Technol.* **2014**, *195*, 112–119. [CrossRef]
4. Maqsood, S.; Benjakul, S. Effect of bleeding on lipid oxidation and quality changes of Asian seabass (*Lates calcarifer*) muscle during iced storage. *Food. Chem.* **2011**, *124*, 459–467. [CrossRef]
5. Maqsood, S.; Benjakul, S.; Kamal-Eldin, A. Haemoglobin-mediated lipid oxidation in the fish muscle: A review. *Trends Food Sci. Technol.* **2012**, *28*, 33–43. [CrossRef]

6. Błaszczyk, A.; Augustyniak, A.; Skolimowski, J. Ethoxyquin: An Antioxidant Used in Animal Feed. *Int. J. Food Sci.* **2013**, 1–12. [CrossRef]
7. Song, C.; Liu, B.; Xu, P.; Xie, J.; Ge, X.; Zhou, Q.; Sun, C.; Zhang, H.; Shan, F.; Yang, Z. Oxidized fish oil injury stress in Megalobrama amblycephala: Evaluated by growth, intestinal physiology, and transcriptome-based PI3K-Akt/NF-κB/TCR inflammatory signaling. *Fish Shellfish Immunol.* **2018**, *81*, 446–455.
8. Li, H.; Zhou, X.; Gao, P.; Li, Q.; Li, H.; Huang, R.; Wu, M. Inhibition of lipid oxidation in foods and feeds and hydroxyl radical-treated fish erythrocytes: A comparative study of *Ginkgo biloba* leaves extracts and synthetic antioxidants. *Anim. Nutr.* **2016**, *2*, 234–241. [CrossRef]
9. Ito, N.; Fukushima, S.; Tsuda, H. Carcinogenicity and modification of the carcinogenic response by BHA, BHT, and other antioxidants. *Crit. Rev. Toxicol.* **1985**, *15*, 109–150.
10. EFSA. Safety and efficacy of ethoxyquin (6-ethoxy-1,2-dihydro-2,2,4-trimethylquinoline) for all animal species. *EFSA J.* **2015**, *13*, 4272.
11. Aksoy, L.; Kolay, E.; Agilonu, Y.; Aslan, Z.; Kargioglu, M. Free radical scavenging activity, total phenolic content, total antioxidant status, and total oxidant status of endemic *Thermopsis turcica*. *Saudi J. Biol. Sci.* **2013**, *20*, 235–239. [CrossRef] [PubMed]
12. Lin, Y.; Lou, K.; Wu, G.; Wu, X.; Zhou, X.; Feng, Y.; Zhang, H.; Yu, P. Bioactive metabolites in of *Ginkgo biloba* leaves: Variations by seasonal, meteorological and soil. *Braz. J. Biol.* **2019**, *80*, 790–797. [CrossRef] [PubMed]
13. Tan, X.; Sun, Z.; Liu, Q.; Ye, H.; Zou, C.; Ye, C.; Wang, A.; Lin, H. Effects of dietary *Ginkgo biloba* leaf extract on growth performance, plasma biochemical parameters, fish composition, immune responses, liver histology, and immune and apoptosis-related genes expression of hybrid grouper (*Epinephelus lanceolatus*♂ × *Epinephelus fuscoguttatus*♀) fed high lipid diets. *Fish Shellfish Immun.* **2018**, *72*, 399–409.
14. Mahadevan, S.; Park, Y. Multifaceted therapeutic benefifits of *Ginkgo biloba* L.: Chemistry, effificacy, safety, and uses. *J. Food. Sci.* **2008**, *73*, R14e9. [CrossRef] [PubMed]
15. Chen, G.F.; Feng, L.; Kuang, S.Y.; Liu, Y.; Jiang, J.; Hu, K.; Jiang, W.D.; Li, S.H.; Tang, L.; Zhou, X.Q. Effect of dietary arginine on growth, intestinal enzyme activities and gene expression in muscle, hepatopancreas and intestine of juvenile Jian carp (*Cyprinus carpio* var. Jian). *Br. J. Nutr.* **2012**, *108*, 195–207. [CrossRef] [PubMed]
16. Chen, G.; Long, J.; Li, H.; Xu, J.; Yuan, J.; Yang, Q.; Feng, L.; Wu, M.; Jiang, J. The Protective Effect of a Dietary Extract of Mulberry (*Morus alba* L.) Leaves against a High Stocking Density, Copper and Trichlorfon in Crucian Carp (*Carassius auratus*). *Animals* **2023**, *13*, 2652. [CrossRef] [PubMed]
17. Zou, Y.P.; Lu, Y.H.; Wei, D.Z. Antioxidant activity of a flavonoid-rich extract of *Hypericum perforatum* L. in vitro. *J. Agric. Food. Chem.* **2004**, *52*, 5032–5039. [CrossRef]
18. Cao, G.; Alessio, H.M.; Cutler, R.G. Oxygen-radical absorbance capacity assay for antioxidants. *Free Radic. Biol. Med.* **1993**, *14*, 303–311. [CrossRef]
19. Zhao, J.; Posa, D.K.; Kumar, V.; Hoetker, D.; Kumar, A.; Ganesan, S.; Riggs, D.W.; Bhatnagar, A.; Wempe, M.F.; Baba, S.P. Carnosine protects cardiac myocytes against lipid peroxidation products. *Amino Acids* **2018**, *51*, 123–128. [CrossRef]
20. Yuan, Y.V.; Bone, D.E.; Carrington, M.F. Antioxidant activity of dulse (*Palmaria palmata*) extract evaluated in vitro. *Food Chem.* **2005**, *91*, 485–494. [CrossRef]
21. Maqsood, S.; Benjakul, S. Comparative studies of four different phenolic compounds on in vitro antioxidative activity and the preventive effect on lipid oxidation of fish oil emulsion and fish mince. *Food Chem.* **2010**, *119*, 123–132. [CrossRef]
22. Chen, G.; Wu, M.; Li, H.; Xu, J.; Liu, H.; Du, W.; Yang, Q.; Feng, L.; Jiang, J. *Scoparia dulcis* L. Extract Relieved High Stocking Density-Induced Stress in Crucian Carp (*Carassius auratus*). *Animals* **2023**, *13*, 2522. [CrossRef] [PubMed]
23. Li, H.T.; Wu, M.; Wang, J.; Qin, C.J.; Long, J.; Zhou, S.S.; Yuan, P.; Jing, X.Q. Protective role of *Angelica sinensis* extract on trichlorfon-induced oxidative damage and apoptosis in gills and erythrocytes of fish. *Aquaculture* **2020**, *519*, 734895. [CrossRef]
24. Li, H.T.; Feng, L.; Jiang, W.D.; Liu, Y.; Jiang, J.; Li, S.H.; Zhou, X.Q. Oxidative stress parameters and anti-apoptotic response to hydroxyl radicals in fish erythrocytes: Protective effects of glutamine, alanine, citrulline and proline. *Aquat. Toxicol.* **2013**, *126*, 169–179. [CrossRef]
25. Jiang, W.D.; Wu, P.; Kuang, S.Y.; Liu, Y.; Jiang, J.; Hu, K.; Li, S.H.; Tang, L.; Feng, L.; Zhou, X.Q. *Myo*-inositol prevents copper-induced oxidative damage and changes in antioxidant capacity in various organs and the enterocytes of juvenile Jian carp (*Cyprinus carpio* var. Jian). *Aquat. Toxicol.* **2011**, *105*, 543–551. [CrossRef] [PubMed]
26. Wang, D.; Meng, Y.; Zhao, X.; Fan, W.; Yi, T.; Wang, X. Sunflower oil flavored by essential oil from *Punica granatum cv. Heyinshiliu* peels improved its oxidative stability and sensory properties. *LWT* **2019**, *111*, 55–61. [CrossRef]
27. Recknagel, R.O.; Glende, E.A. Spectrophotometric detection of lipid conjugated dienes. *Method Enzymol.* **1984**, *105*, 331–337.
28. Martínez-Yusta, A.; Goicoechea, E.; Guillén, M.D. A review of thermo-oxidative degradation of food lipids studied by 1HNMR spectroscopy: Influence of degradative conditions and food lipid nature. *Compr. Rev. Food Sci. F* **2014**, *13*, 838–859. [CrossRef]
29. Chen, X.; Zhang, Y.; Zu, Y.G.; Yang, L.; Lu, Q.; Wang, W. Antioxidant effects of rosemary extracts on sunflower oil compared with synthetic antioxidants. *Int. J. Food Sci. Technol.* **2014**, *49*, 385–391. [CrossRef]
30. El-Husseiny, O.M.; Abdul-Aziz, G.M.; Goda, A.S.; Suloma, A. Effect of altering linoleic acid and linolenic acid dietary levels and ratios on the performance and tissue fatty acid profiles of Nile tilapia *Oreochromis niloticus* fry. *Aquac. Int.* **2010**, *18*, 1105–1119. [CrossRef]

31. Wei, X.B.; Liu, H.Q.; Sun, X.; Fu, F.; Zhang, X.; Wang, J.; An, J.; Ding, H. Hydroxysafflor yellow A protects rat brains against ischemia-reperfusion injury by antioxidant action. *Neurosci. Lett.* **2005**, *386*, 58–62. [CrossRef] [PubMed]
32. Zhang, H.; Wang, S.; Duan, X.; Feng, X.; Yang, Y. The interaction effects of coke oven emissions exposure and metabolic enzyme Gene variants on total antioxidant capacity of workers. *Environ. Toxicol. Pharmacol.* **2019**, *70*, 103197. [CrossRef] [PubMed]
33. Lugasi, A.; Dworschak, E.; Horvatovich, P. Additional information to the in vitro antioxidant activity of *Ginkgo biloba* L. *Phytother. Res.* **1999**, *13*, 160–162. [CrossRef]
34. Zhao, L.J.; Liu, W.; Xiong, S.H.; Tang, J.; Lou, Z.H.; Xie, M.X.; Xia, B.H.; Lin, L.M.; Liao, D.F. Determination of total flavonoids contents and antioxidant activity of *Ginkgo biloba* leaf by near-infrared reflectance method. *Int. J. Anal. Chem.* **2018**, 8195784. [CrossRef] [PubMed]
35. Hyun, S.K.; Jung, H.A.; Chung, H.Y.; Choi, J.S. In vitro peroxynitrite scavenging activity of 6-hydroxykynurenic acid and other flavonoids from *Gingko biloba* yellow leaves. *Arch. Pharm. Res.* **2006**, *29*, 1074–1079. [CrossRef]
36. Ni, Y.; Duan, Z.; Zhou, D.; Liu, S.; Wan, H.; Gui, C.; Zhang, H. Identification of Structural Features for the Inhibition of OAT3-Mediated Uptake of Enalaprilat by Selected Drugs and Flavonoids. *Front. Pharmacol.* **2020**, *11*, 802. [CrossRef]
37. Fuhrman, B.; Aviram, M. Flavonoids protect LDL from oxidation and attenuate atherosclerosis. *Curr. Opin. Lipidol.* **2001**, *12*, 41–48. [CrossRef]
38. Chen, Z.Y.; Chan, P.T.; Ho, K.Y.; Fung, K.P.; Wang, J. Antioxidant activity of natural flavonoids is governed by number and location of their aromatic hydroxyl groups. *Chem. Phys. Lipids* **1996**, *79*, 157–163. [CrossRef]
39. Procházková, D.; Boušová, I.; Wilhelmová, N. Antioxidant and prooxidant properties of flavonoids. *Fitoterapia* **2011**, *82*, 513–523. [CrossRef]
40. Rather, A.A. Biochemical responses induced by sub lethal concentrations of carbaryl and parathion on certain enzymes of fresh water catfish *Clarias batrachus*. *Int. J. Biol. Sci.* **2015**, *4*, 52–56.
41. Dikshith, T.S.; Datta, K.; Kushwah, H.; Raizada, R. Histopathological and biochemical changes in guinea pigs after repeated dermal exposure to benzene hexachloride. *Toxicology* **1978**, *10*, 55–66. [CrossRef] [PubMed]
42. Johari, S.A.; Sarkheil, M.; Asghari, S.; Haghighat, F.; Dekani, L.; Keyvanshokooh, S. Comparative toxicity of nanoparticulate and ionic copper following dietary exposure to common carp (*Cyprinus carpio*). *Comp. Biochem. Physiol. C Toxicol. Pharmacol.* **2019**, *229*, 108680. [CrossRef] [PubMed]
43. Ahmadifar, E.; Yousefi, M.; Karimi, M.; Fadaei Raieni, R.; Dadar, M.; Yilmaz, S.; Dawood, M.; Abdel-Latif, H. Benefits of Dietary Polyphenols and Polyphenol-Rich Additives to Aquatic Animal Health: An Overview. *Revi. Fish. Sci. Aquac.* **2021**, *29*, 478–511. [CrossRef]
44. Abdel-Latif, H.M.; Hendam, B.M.; Nofal, M.I.; El-Son, M.A. *Ginkgo biloba* leaf extract improves growth, intestinal histomorphometry, immunity, antioxidant status and modulates transcription of cytokine genes in hapa-reared *Oreochromis niloticus*. *Fish Shellfish Immunol.* **2021**, *117*, 339–349. [CrossRef] [PubMed]
45. Liao, M.; Wang, F.; Huang, L.; Liu, C.; Dong, W.; Zhuang, X.; Yin, X.; Liu, Y.; Wang, W. Effects of dietary *Ginkgo biloba* leaf extract on growth performance, immunity and environmental stress tolerance of *Penaeus vannamei*. *Fish Shellfish Immunol.* **2023**, *132*, 108500. [CrossRef] [PubMed]
46. Xu, Z.; Wu, W.; Lan, T.; Zhang, X. Protective effects of extract of *Ginkgo biloba* on adriamycin-induced heart failure and its mechanism: Role of ghrelin peptide. *Zhongguo Zhong Yao Za Zhi* **2009**, *34*, 2786–2789.
47. Shi, L.; Du, X.; Jiang, H.; Xie, J. Ghrelin and neurodegenerative disorders—A review. *Mol. Neurobiol.* **2017**, *54*, 1144–1155. [CrossRef]
48. Gilloteaux, J.; Kashouty, R.; Yono, N. The perinuclear space of pancreatic acinar cells and the synthetic pathway of zymogen in *Scorpaena scrofa* L.: Ultrastructural aspects. *Tissue Cell* **2008**, *40*, 7–20. [CrossRef]
49. Zambonino Infante, J.L.; Cahu, C.L. Ontogeny of the gastrointestinal tract of marine fish larvae. *Comp. Biochem. Physiol. C Toxicol. Pharmacol.* **2001**, *130*, 477–487. [CrossRef]
50. Geering, K. Subunit assembly and functional maturation of Na, K-ATPase. *J. Membr. Biol.* **1990**, *115*, 109–121. [CrossRef]
51. Suzer, C.; Aktülün, S.; Çoban, D.; Okan Kamacı, H.; Saka, Ş.; Fırat, K.; Alpbaz, A. Digestive enzyme activities in larvae of sharpsnout seabream (*Diplodus puntazzo*). *Comp. Biochem. Physiol. A Mol. Integr. Physiol.* **2007**, *148*, 470–477. [CrossRef] [PubMed]
52. Feng, L.; Zhao, S.; Chen, G.; Jiang, W.; Liu, Y.; Jiang, J.; Hu, K.; Li, S.; Zhou, X. Antioxidant status of serum, muscle, intestine and hepatopancreas for fish fed graded levels of biotin. *Fish Physiol. Biochem.* **2013**, *40*, 499–510. [CrossRef] [PubMed]
53. Gliszczyńska-Świgło, A. Antioxidant activity of water soluble vitamins in the TEAC (trolox equivalent antioxidant capacity) and the FRAP (ferric reducing antioxidant power) assays. *Food Chem.* **2006**, *96*, 131–136. [CrossRef]
54. Kudolo, G.B.; Delaney, D.; Blodgett, J. Short-term oral ingestion of *Ginkgo biloba* extract (EGb 761) reduces malondialdehyde levels in washed platelets of type 2 diabetic subjects. *Diabetes Res. Clin. Pract.* **2005**, *68*, 29–38. [CrossRef] [PubMed]
55. Yuan, G.; Gong, Z.; Li, J.; Li, X. *Ginkgo biloba* extract protects against alcohol-induced liver injury in rats. *Phytother. Res.* **2007**, *21*, 234–238. [CrossRef] [PubMed]
56. Ren, Q.; Chen, J.; Ding, Y.; Cheng, J.; Yang, S.; Ding, Z.; Dai, Q.; Ding, Z. In vitro antioxidant and immunostimulating activities of polysaccharides from *Ginkgo biloba* leaves. *Int. J. Biol. Macromol.* **2019**, *124*, 972–980. [CrossRef] [PubMed]
57. Liu, K.X.; He, W.; Rinne, T.; Liu, Y.; Zhao, M.Q.; Wu, W.K. The effect of *Ginkgo biloba* extract (EGb 761) pretreatment on intestinal epithelial apoptosis induced by intestinal ischemia/reperfusion in rats: Role of ceramide. *Am. J. Chin. Med.* **2007**, *35*, 805–819. [CrossRef]

58. Seyoum, A.; Asres, K.; El-Fiky, F.K. Structure–radical scavenging activity relationships of flavonoids. *Phytochemistry* **2006**, *67*, 2058–2070. [CrossRef]
59. Yokomizo, A.; Moriwaki, M. Effects of uptake of flavonoids on oxidative stress induced by hydrogen peroxide in human intestinal Caco-2 cells. *Biosci. Biotechnol. Biochem.* **2006**, *70*, 1317–1324. [CrossRef]
60. Martínez-Álvarez, R.M.; Morales, A.E.; Sanz, A. Antioxidant defenses in fish: Biotic and abiotic factors. *Rev. Fish Biol. Fisher.* **2005**, *15*, 75–88. [CrossRef]
61. Ni, M.; Liu, M.; Lou, J.; Mi, G.; Gu, Z. Stocking density alters growth performance, serum biochemistry, digestive enzymes, immune response, and muscle quality of largemouth bass (*Micropterus salmoides*) in in-pond raceway system. *Fish Physiol. Biochem.* **2021**, *47*, 1243–1255. [CrossRef] [PubMed]
62. Bhattacharjee, P.; Das, S. Deltamethrin Induced Alteration of Biochemical Parameters in Channa punctata, Bloch and its Amelioration by Quercetin. *Bull. Environ. Contam. Toxicol.* **2017**, *98*, 763–769. [CrossRef] [PubMed]
63. Kirshenbaum, L.A.; Singal, P.K. Increase in endogenous antioxidant enzymes protects hearts against reperfusion injury. *Am. J. Physiol.* **1993**, *265*, H484–H492. [CrossRef]
64. Yim, T.K.; Wu, W.K.; Pak, W.F.; Mak, D.H.F.; Ko, K.M. Myocardial protection against ischaemia- reperfusion injury by a *Polygonum multiflorum* extract supplemented 'Dang-Gui decoction for enriching blood', a compound formulation, ex vivo. *Phytother. Res.* **2000**, *14*, 195–199. [CrossRef]
65. Reed, D.J. Glutathione: Toxicological implications. *Annu. Rev. Pharm. Toxicol.* **1990**, *30*, 603–631. [CrossRef]
66. Sugihara, N.; Tsuruta, Y.; Date, Y.; Furuno, K.; Kohashi, K. High peroxidative susceptibility of fish oil polyunsaturated fatty acid in cultured rat hepatocytes. *Toxicol. Appl. Pharmacol.* **1994**, *126*, 124–128. [CrossRef] [PubMed]
67. Cui, K.; Liu, N.; Sun, Y.; Sun, G.; Wang, S.; Yang, M.; Yang, M.; Wang, X.; Zhou, D.; Ge, Y.; et al. Effect of drying processes on the occurrence of lipid oxidation-derived 4-hydroxy-2-hexenal and 4-hydroxy-2-nonenal in Spanish mackerel (*Scomberomorus niphonius*). *Food Sci. Nutri.* **2022**, *11*, 1013–1023. [CrossRef]
68. Morachis-Valdez, G.; Dublán-García, O.; López-Martínez, L.X.; Galar-Martínez, M.; Saucedo-Vence, K.; Gómez-Oliván, L.M. Chronic exposure to pollutants in Madín Reservoir (Mexico) alters oxidative stress status and flesh quality in the common carp *Cyprinus carpio*. *Environ. Sci. Pollut. Res.* **2015**, *22*, 9159–9172. [CrossRef]
69. Eckert, A.; Keil, U.; Kressmann, S.; Schindowski, K.; Leutner, S.; Leutz, S.; Muller, W.E. Effects of EGb 761 *Ginkgo biloba* extract on mitochondrial function and oxidative stress. *Pharmacopsychiatry* **2003**, *36*, s15–s23.
70. Ibrahim, D.; Kishawy, A.T.Y.; Khater, S.I.; Khalifa, E.; Ismail, T.A.; Mohammed, H.A.; Elnahriry, S.S.; Tolba, H.A.; Sherief, W.; Farag, M.; et al. Interactive effects of dietary quercetin nanoparticles on growth, flesh antioxidant capacity and transcription of cytokines and *Aeromonas hydrophila* quorum sensing orchestrating genes in Nile tilapia (*Oreochromis niloticus*). *Fish Shellfish Immunol.* **2021**, *119*, 478–489. [CrossRef]
71. Eisvand, F.; Razavi, B.M.; Hosseinzadeh, H. The effects of *Ginkgo biloba* on metabolic syndrome: A review. *Phytother. Res.* **2020**, *34*, 1798–1811. [CrossRef] [PubMed]
72. Niu, Y.; Wan, X.L.; Zhang, X.H.; Zhao, L.G.; He, J.T.; Zhang, J.F.; Zhang, L.L.; Wang, T. Effect of supplemental fermented *Ginkgo biloba* leaves at different levels on growth performance, meat quality, and antioxidant status of breast and thigh muscles in broiler chickens. *Poult. Sci.* **2017**, *96*, 869–877. [CrossRef]
73. Kulkeaw, K.; Sugiyama, D. Zebrafish erythropoiesis and the utility of fish as models of anemia. *Stem Cell Res. Ther.* **2012**, *3*, 55. [PubMed]
74. Çimen, M.Y.B. Free radical metabolism in human erythrocytes. *Clin. Chim. Acta* **2008**, *390*, 1–11. [CrossRef]
75. Clemens, M.R.; Ruess, M.; Bursa, Z.; Waller, H.D. The relationship between lipid composition of red blood cells and their susceptibility to lipid peroxidation. *Free Radic. Res. Commun.* **1987**, *3*, 265–271. [CrossRef] [PubMed]
76. Richards, M.P.; Hultin, H.O. Contributions of Blood and Blood Components to Lipid Oxidation in Fish Muscle. *J. Agric. Food Chem.* **2002**, *50*, 555–564. [CrossRef]
77. Köse, K.; Dogan, P. Lipoperoxidation Induced by Hydrogen Peroxide in Human Erythrocyte Membranes. *J. Int. Med. Res.* **1995**, *23*, 9–18. [CrossRef]
78. Sinha, A.K.; Zinta, G.; AbdElgawad, H.; Asard, H.; Blust, R.; De Boeck, G. High environmental ammonia elicits differential oxidative stress and antioxidant responses in five different organs of a model estuarine teleost (*Dicentrarchus labrax*). *Comp. Biochem. Physiol. C Toxicol. Pharmacol.* **2015**, *174–175*, 21–31. [CrossRef]
79. Zheng, T.; Jia, R.; Cao, L.; Du, J.; Gu, Z.; He, Q.; Xu, P.; Yin, G. Alleviative effects of *Ginkgo biloba* extract on oxidative stress, inflammatory response and immune suppression induced by long-term glyphosate exposure in tilapia (*Oreochromis niloticus*). *Aquaculture* **2022**, *546*, 737325. [CrossRef]

*fishes*

MDPI

*Article*

# Immunoprotective Effect of *Coptis chinensis*-Supplemented Diet on *Streptococcus agalactiae* Infection in Tilapia

**Ruijie Guo [1], Kai Yu [1,2], Kai Huang [1,*], Qiang Lin [3,*] and Ting Liu [1]**

1 College of Animal Science and Technology, Guangxi University, Nanning 530004, China; guoruijie210310@163.com (R.G.); yukai@st.gxu.edu.cn (K.Y.); tingliu0123@foxmail.com (T.L.)
2 Department of Omics Technology, Nanning Current Science Biotechnology Co., Ltd., Nanning 530005, China
3 Pearl River Fisheries Research Institute, Chinese Academy of Fishery Sciences, Guangzhou 510380, China
* Correspondence: kaihuangnn1@163.com (K.H.); linq@prfri.ac.cn (Q.L.)

**Abstract:** This study aimed to investigate the immunoprotective effect of *Coptis chinensis* (CC) on *Streptococcus agalactiae* (SA) infection in tilapia. Experimental fish were randomly divided into two groups feeding on a normal diet (ND) and a CC-supplemented diet (CCD) for 2 weeks and then injected with SA. After the inoculation experiment, the ND and CCD groups were named PI_ND and PI_CCD, respectively. CCD increased superoxide dismutase (SOD) activity and decreased malondialdehyde (MDA) activity significantly before and after infection. Immunological assays revealed that the serum interleukin-1β (IL-1β), complement 3 (C3), immunoglobulin M (IgM), Interferon-gamma (IFN-γ), and tumor necrosis factor-α (TNF-α) levels in the CCD group were significantly higher than in the ND group both before and after infection. In addition, proteomics analysis of liver tissue identified 62 differentially expressed proteins (DEPs) in CCD vs. ND, and 36 DEPs in the PI_CCD vs. the PI_ND groups. Furthermore, 80 specific upregulated proteins and 49 specific downregulated proteins were screened in the CCD group. The specific upregulated proteins included important antimicrobial enzymes such as lysozymes and cathepsin D, and antimicrobial peptides such as septins, granulin, and grancalcin, involving multiple KEGG brite categories such as enzymes, exosomes, membrane transport, and proteolipid proteins. Furthermore, specific downregulated proteins were enriched in glycolysis/gluconeogenesis and TCA cycle pathways. In conclusion, CC supplementation effectively enhances the ability of tilapia to resist SA infection by modulating various antioxidant enzymes, immune factors, antimicrobial enzymes, and antimicrobial peptides, and by moderately inhibiting central carbon metabolism. These findings provide a basis for replacing antibiotics with environmentally-friendly functional aquatic feeds to control bacterial diseases.

**Keywords:** tilapia; *Streptococcus agalactiae*; *Coptis chinensis*; proteomic; antimicrobial peptide; immune factor; antioxidant enzyme

**Key Contribution:** *Coptis chinensis* supplementation effectively improved tilapia's ability to resist SA infection. This study provides a basis for replacing antibiotics with environmentally-friendly functional aquafeeds to control bacterial diseases.

**Citation:** Guo, R.; Yu, K.; Huang, K.; Lin, Q.; Liu, T. Immunoprotective Effect of *Coptis chinensis*-Supplemented Diet on *Streptococcus agalactiae* Infection in Tilapia. *Fishes* **2023**, *8*, 370. https://doi.org/10.3390/fishes8070370

Academic Editors: Qiyou Xu, Jianhua Ming, Fei Song, Changle Qi and Chuanpeng Zhou

Received: 11 June 2023
Revised: 11 July 2023
Accepted: 13 July 2023
Published: 19 July 2023



## 1. Introduction

*Streptococcus agalactiae* (SA) (family Streptococcaceae, genus *Streptococcus*), also known as group B *Streptococcus* (GBS), is a facultatively anaerobic, Gram-positive diplococcus and a common pathogenic bacteria in humans, animals, and fish [1]. Streptococcicosis in fish is characterized by wide distribution, high incidence, and severe outcomes. The number of reports of SA infection has increased gradually in recent years, and SA infection affects various marine and freshwater cultured fish worldwide. Among them, tilapia is the most sensitive to SA, and SA infection is the most common and serious infection reported in this species. Since 2009, several serious outbreaks of SA infection in tilapia have been reported in various regions of Asia, including China, which have caused serious damage to the tilapia culture industry owing to the high morbidity and mortality [2,3]. Therefore, research on the pathogenesis and resistance mechanism underlying SA infection in tilapia is essential for prevention and control.

The use of traditional herbal medicines can help reduce indiscriminate overuse of drugs for the suppression of *Streptococcus* resistance to multiple antibiotics and lower the levels of drug residues in aquatic products. Additionally, these medicines help address the issues associated with the arduous developmental process of vaccines. *Coptis chinensis* (CC) is the dried rhizome of the buttercup plants Huanglian, *C. deltoidea*, or *C. teeta* Wall. CC, and can help to reduce heat and purge pathogenic fire (excessive internal heat, which is a concept from traditional Chinese medicine), aid detoxification, and dispel dryness [4]. In addition, CC exerts various biological effects that aid in the treatment of arrhythmia, hyperglycemia, hyperlipidemia, platelet aggregation, and tumors [5]. The experimental results shown by Yu et al. [6] and Wang et al. [7] suggested that CC can improve the resistance of tilapia to various bacteria. Niu et al. [8] conducted a meta-analysis by collecting data on the antibacterial effect of herbs on SA from databases and concluded that CC was more effective in inhibiting SA than *Rheum* and *Rhus chinensi*.

At present, reports on the immune response of CC-treated tilapia to SA and the molecular mechanism of action of herbal medicine on fish immunity are limited. In this study, using iTRAQ-labeled proteomics, we analyzed the differences in protein expression in the liver tissues of SA-infected fish before and after CC treatment to determine the major biological processes and the signaling pathways that differentially expressed proteins (DEPs) are associated with. In addition, we explored the mechanism underlying the immune reaction of tilapia against SA by monitoring the biochemical indexes in blood and related tissues in CC-treated tilapia. Overall, our findings provide the theoretical basis for additional investigations on CC-based treatment for the ecological prevention and control of SA infection.

## 2. Materials and Methods

### 2.1. Experimental Fish and Pathogenic Bacteria

Tilapia and SA strains were provided by the Guangxi Research Institute of Aquatic Sciences and Pearl River Institute of Aquatic Sciences, respectively. The culture experiments were conducted in an indoor recirculating aquaculture system at the aquatic base at Guangxi University. The culture tank was disinfected with 0.5% potassium permanganate solution for 2 h before the SA infection test of tilapia (270 fish, with an average weight of 30 g). The tilapia were temporarily housed in the tank for 14 days and fed with bait twice a day (at 09:00 a.m. and 17:00 p.m.). During cultivation, the water temperature was 30.21 ± 0.75 °C and the dissolved oxygen was 5.37 ± 1.35 mg/L. The water was changed every 2 days (half of the pool at a time, normal aeration). Feeding was stopped at 24 h before sampling.

### 2.2. Experimental Feed

The 1% CC (20:1) added to the feed was provided by Ningxia Guosheng Biotech. The feed preparation process was as follows: (1) CC was ground into a powder, mixed with distilled water in a certain ratio in a round-bottom flask, and extracted in a water bath at 80 °C for 1 h. After centrifugation, the supernatant was extracted, concentrated to 5 mg/mL, and stored at 4 °C for subsequent use; (2) ordinary granular feeds were mechanically ground into a powder, dried, and stored for subsequent use; and (3) the basal feed was mixed with distilled water in equal volumes. In addition, 1 mL of the CC liquid was added per gram of the feed to alter the drug content of the feed to 0.5%. Subsequently, the mixture was dried at 60 °C, cooled, and stored at 4 °C. Feed formulation and proximate composition of experimental diets are shown in Table 1.

**Table 1.** Formulation and proximate analysis of the tilapia diets.

| Material | ND | CCD |
|---|---|---|
| Ingredient (%) | | |
| Fish meal | 8.5 | 8.5 |
| Soybean meal | 40 | 40 |
| Rape seed meal | 20 | 20 |
| Wheat middlings | 13 | 13 |
| Calcium phosphate | 1.5 | 1.5 |
| Choline Chloride | 1.5 | 1.5 |
| Mineral premix [a] | 0.5 | 0.5 |
| Vitamin premix [a] | 0.5 | 0.5 |
| Microcrystalline cellulose | 9 | 9 |
| NaCl | 0.5 | 0.5 |
| Soybean oil | 2 | 2 |
| Carboxymethyl cellulose | 3 | 2 |
| Coptis chinensis powder | 0 | 1 |
| Total | 100 | 100 |
| Proximate composition (%) [b] | | |
| Crude protein | 32.90 | 32.94 |
| Crude lipid | 3.81 | 3.82 |
| Crude fiber | 16.90 | 15.96 |
| Crude ash | 6.40 | 6.42 |

[a] Mineral premix and vitamin premix were purchased from Haibaolu Feeds Co., Ltd., Nanning, China. [b] The proximate composition of diets was calculated using the VF123 software (Beijing Jinmu Times Technology Co., Beijing, China, 2016).

### 2.3. Experimental Stage and Grouping

Prior to the infection experiment, a feeding period of 14 days was carried out. The tilapia were divided into two groups: the normal group (conventional feed + saline, ND group) and the experimental group (conventional feed + 1% CC, CCD group). A total of 3 replicate culture buckets were used for each group (30 fish per bucket, 180 tilapia in all). After the feeding stage, 10 fish were randomly selected from each parallel culture bucket of each group for the 24 h inoculation experiment (named PI_ND and PI_CCD groups, respectively).

### 2.4. Bacterial Culture and Inoculation Experiments

The preserved SA was activated, diluted, and coated on BHI agar plates. Subsequently, monoclones were selected for 16S rRNA gene sequencing (Table S1). The sequencing-cleared monoclones were cultured till the logarithmic growth stage, and the concentration of the bacterial solution was adjusted to $5 \times 10^{11}$ CFUs/mL using saline (0.65%) and diluted in five concentration gradients (1×, 10×, 100×, 1000×, and 10,000×). Subsequently, 120 fish were randomly kept in six experimental pools, and the bacterial solution was injected intraperitoneally (at the base of the ventral fin) in each fish (0.2 mL/fish) in the five groups at the five concentrations (mentioned above). In addition, tilapia in the control group were injected with 0.2 mL of 0.65% saline. After the injection, the fish were placed in the original test tank (water temperature: 30 °C, with normal feeding, aeration, and water replacement). The dead fish were retrieved and recorded for a number every 3 h. After SA inoculation, we counted the number of dead fish and confirmed that the lethal concentration for 50% ($LC_{50}$) of the SA strain was $1.5 \times 10^8$ CFU/mL [7] using the Reed–Muench method [9].

After feeding, the inoculation experiment was conducted. First, ten fish from each parallel group were randomly selected for the inoculation experiment by injecting 0.2 mL of $1.5 \times 10^8$ CFU/mL bacterial solution in the ventral cavity (base of ventral fin). Tilapia in the control group were injected with 0.2 mL of 0.65% saline. After the injection, the fish were placed in the original test bucket. After a duration of 24 h, the tilapia were anesthetized using 200 mg/L MS-222 for sampling.

### 2.5. Sample Collection

After feeding and inoculation, 10 fish were randomly anesthetized with 200 mg/L MS-222 from each of the 3 parallel culture buckets of the 2 feeding groups. Later, fresh blood was collected from the tail vein, placed in a freezing tube for 2 h, and centrifuged at 2000× *g* for 10 min to separate the serum. The extracted serum was then stored in a refrigerator at −20 °C. In addition, liver samples were collected, placed in centrifuge tubes, snap-frozen in liquid nitrogen, and stored in an ultra-low temperature refrigerator (−80 °C) for subsequent biochemical and proteomic assays.

### 2.6. Biochemical and Immunological Index Determination

For examination of the biochemical characteristics, including total protein (TP), alanine transaminase (ALT), aspartate aminotransferase (AST), malondialdehyde (MDA), catalase (CAT), superoxide dismutase (SOD) of the liver, and immunological indexes, such as complement C3 (C3), immunoglobulin M (IgM), tumor necrosis factor-alpha (TNF-$\alpha$), interferon-gamma (IFN-$\gamma$), and Interleukin 1$\beta$ (IL-1$\beta$) of serum samples, the detection kits were used according to the manufacturer's instructions (Nanjing Jiancheng Biology Engineering Institute, Nanjing, China).

### 2.7. iTRAQ Differential Proteomics Assay

The liver tissue samples were ground to a powder form, and the proteins were extracted using UED Lysis Buffer. After determining the protein concentrations using a Bradford kit, samples of 100 µg were reduced, alkylated, and digested into peptides with trypsin. Subsequently, the peptides were labeled by iTRAQ-8plex Amine-Modifying Labeling. The mixed labeled peptides were subjected to HPLC component separation, and the 12 resultant components were analyzed individually using LC–MS/MS. The experiments were performed in three biological replicates.

### 2.8. Database Search and Bioinformatics Analysis

A database search against the tilapia Uniprot proteome sequences was performed using the SEQUEST software of the Proteome Discoverer platform. The database search settings were as follows—trypsin: digestive enzyme, up to two missed cuts allowed; parent ion error: 20 ppm; secondary ion error: 0.05 Da; iTRAQ-8plex (N-term, R, K); Oxidation (M): variable modification; Carbamidomethyl (C): fixed modification; and peptide FDR: 0.01. The identified proteins were annotated using the Uniprot database for GO annotation and the KAAS online tool for KEGG pathway annotation.

### 2.9. Statistical Analysis

Data obtained were recorded and initially stratified using Excel 2010 (Microsoft, Redmond, WA, USA, 2010). One-way analysis of variance (ANOVA) was performed using SPSS version 21 (IBM, Armonk, New York, NY, USA, 2012) to determine the significance of differences ($p < 0.05$). Values were expressed as mean ± standard error (Mean ± SE). DEPs were screened using the criteria of fold change > 1.2 or < 0.83 and $p$ value < 0.05 ($t$-test).

## 3. Results

### 3.1. Effect of CC on Biochemical Indexes in the Liver of Healthy and SA-Infected Tilapia

The results of the liver biochemical indexes suggested that the liver TP level, AST, and SOD activities were significantly increased, whereas the MDA content was significantly decreased in the CCD group compared to the ND group (Figure 1A,C,F). However, the difference in the liver ALT and CAT activities shown was not significant in the CCD group compared to the ND group (Figure 1B,D). Results showed that SA infection decreased the TP level, CAT, and SOD activities, and significantly increased the MDA level and AST activity in the ND group. After SA infection, the PI_CCD group showed a significantly lower ALT activity and MDA content and higher SOD activity than those from the PI_ND group (Figure 1B,E,F). In addition, the liver TP content and CAT activity were higher and the AST activity was lower in the PI_CCD group compared to the PI_ND group; however, the differences were not significant (Figure 1A,C,D).

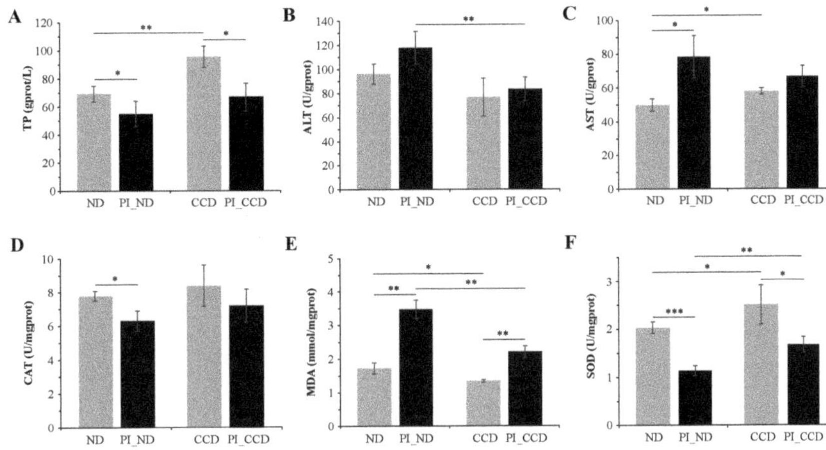

**Figure 1.** Biochemical indexes in tilapia liver in the ND and CCD groups before and after SA infection. (**A**) TP content; (**B**) ALT level; (**C**) AST level; (**D**) CAT level; (**E**) MDA content; (**F**) SOD activity. * indicates significant difference ($p < 0.05$); ** indicates extremely significant difference ($p < 0.01$). *** indicates an extremely significant difference ($p < 0.001$).

### 3.2. *Effect of CC on the Serum Immune Indexes of Healthy and SA-Infected Tilapia*

The serum immune indexes, including C3, IgM, TNF-α, IFN-γ, and IL-1β levels, were higher in the serum of tilapia from the CCD group than in the ND group, especially after SA infection (Figure 2A–E). The increased rates of these immune indexes of the CCD vs. ND comparison were also higher in SA-infected tilapia than in healthy tilapia, especially C3 and IFN-γ (Figure 2F).

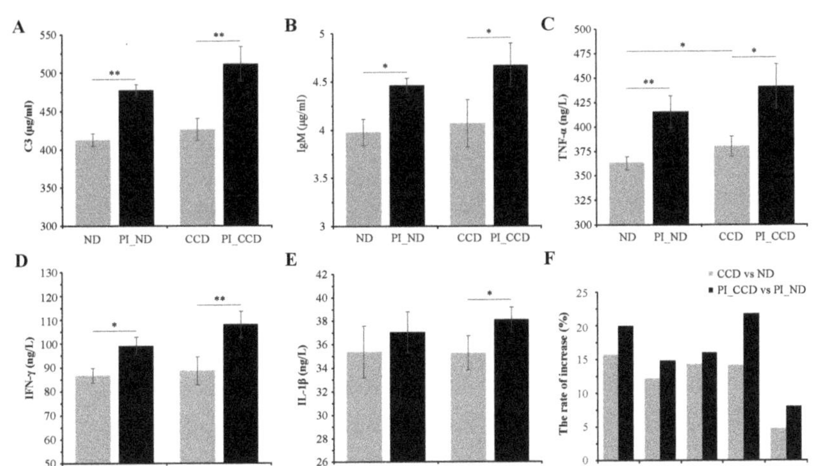

**Figure 2.** Immunological indexes of tilapia liver in the ND and CCD groups before and after SA infection. (**A**) C3 content; (**B**) IgM content; (**C**) TNF-α content; (**D**) IFN-γ content; (**E**) IL-1β content; (**F**) increase rate for each index in the CCD vs. ND comparison before and after SA infection. * indicates significant difference ($p < 0.05$); ** indicates extremely significant difference ($p < 0.01$).

### 3.3. Protein Expression in SA-Infected Tilapia before and after CC Treatment

A high-throughput differential proteomics analysis based on iTRAQ label coupled with LC–MS/MS was performed for multiple groups in this study. Protein identification, quantification, and annotation details were listed in Table S2. The correlation heatmap of the overall protein expression levels in each group indicated a good positive correlation within the groups and low correlation between the groups (especially before and after infection) (Figure 3A). These results indicated that the samples were reproducible and that there were significant differences in the protein expression levels in the liver before and after infection. Expression heatmap clustering analysis of the quantified proteins in each sample suggested that the samples in various groups were clustered together, as were the pre-infection and post-infection samples (Figure 3B). These results further suggested that the samples had good reproducibility and that substantial differences were present in the liver protein expression levels before and after infection. DEP screening (FC > 2 or FC < 0.5, and $p < 0.05$) yielded 15 upregulated DEPs and 47 downregulated DEPs between the CCD and ND groups, 55 upregulated DEPs and 51 downregulated DEPs between the PI_ND and ND groups, 65 upregulated DEPs and 94 downregulated DEPs between the PI_CCD and CCD groups, and 22 upregulated DEPs and 14 downregulated DEPs between the PI_CCD and PI_ND groups (Figure 3C).

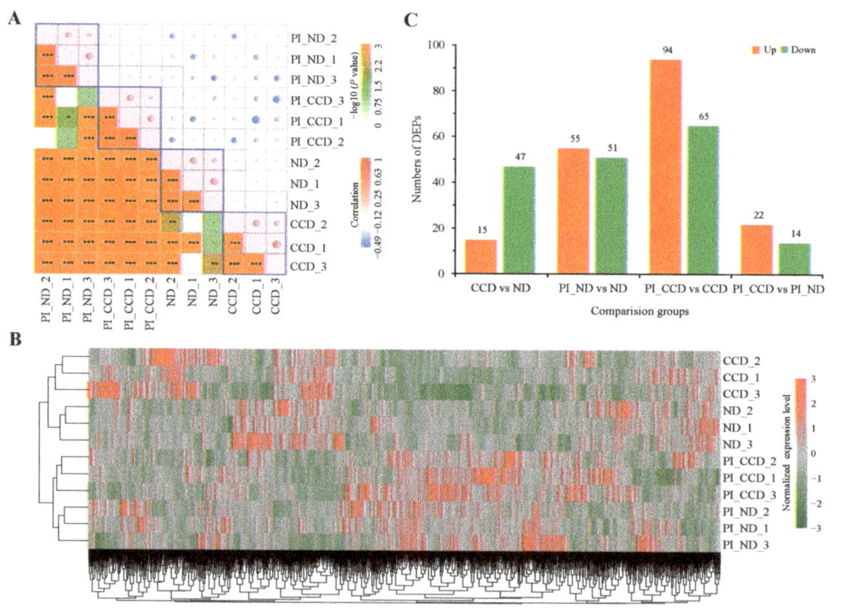


**Figure 3.** Statistical analysis of the results of high-throughput differential proteomics analysis. (**A**) Correlation analysis among samples, asterisks represent significance of differences and dots represent correlation; (**B**) clustering heat map of protein expression for each sample; (**C**) differentially expressed proteins between various groups.

### 3.4. Changes in Protein Expression in CC-Treated Healthy Tilapia

A volcano plot of the differences in proteomic expression data between CCD and ND groups indicated the protein expression patterns in the two groups (Figure 4A). KOG functional classification of 47 downregulated DEPs yielded ten DEPs involved in C (energy production and conversion process) and some proteins involved in J (translation, ribosomal structure, and biogenesis), O (Posttranslational modification, protein turnover, chaperones), and other categories (Figure 4B). The KOG functional classification of

15 upregulated DEPs revealed that three DEPs were involved in C and O functional categories, respectively (Figure 4C).

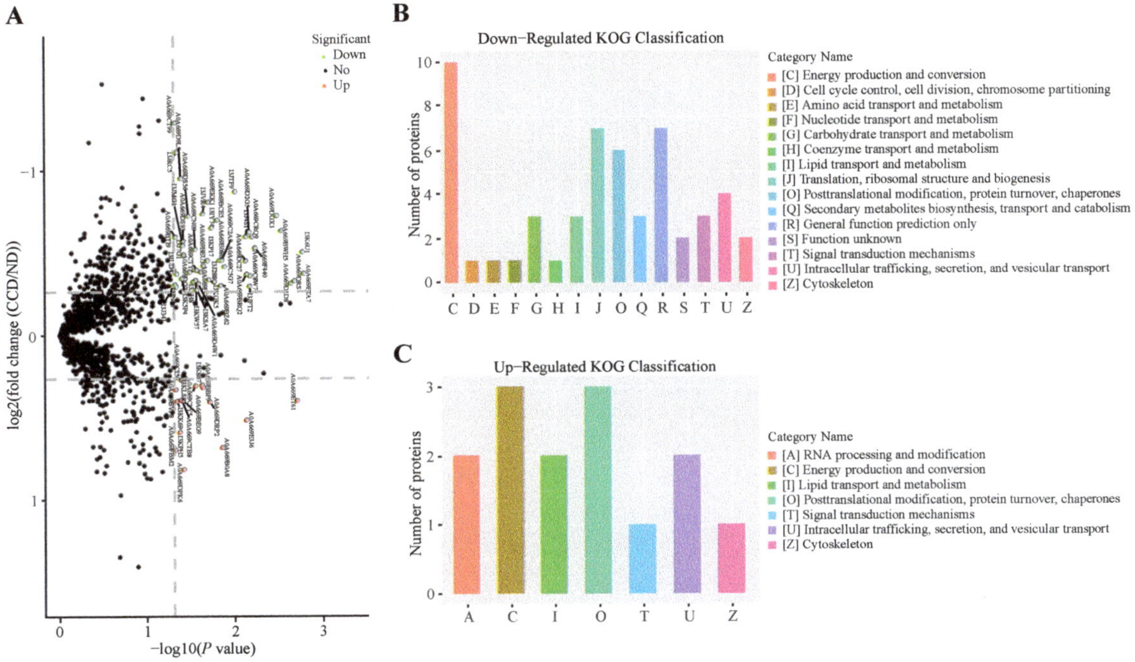


**Figure 4.** Differential protein expression analysis of tilapia liver between the CCD and ND groups. (**A**) Volcano plot of differential protein expression between the CCD and ND groups; (**B**) KOG statistics of downregulated differentially expressed proteins (DEPs) between the CCD and ND groups; (**C**) KOG statistics of upregulated DEPs between the CCD and ND groups.

### *3.5. Changes in Protein Expression in CC-Treated SA-Infected Tilapia*

The volcano plot of the differences in proteomic expression data between the PI_CCD and PI_ND groups indicates the protein expression patterns in the two groups (Figure 5A). KOG functional classification of the 14 downregulated DEPs showed three proteins each in the O (posttranslational modification, protein turnover, and chaperones) and R (general function prediction only) categories (Figure 5B). In addition, KOG functional classification of 22 upregulated DEPs revealed that four DEPs were involved in the O and R functional categories, and that three DEPs were involved in the Q (secondary metabolites biosynthesis, transport, and catabolism) category (Figure 5C).

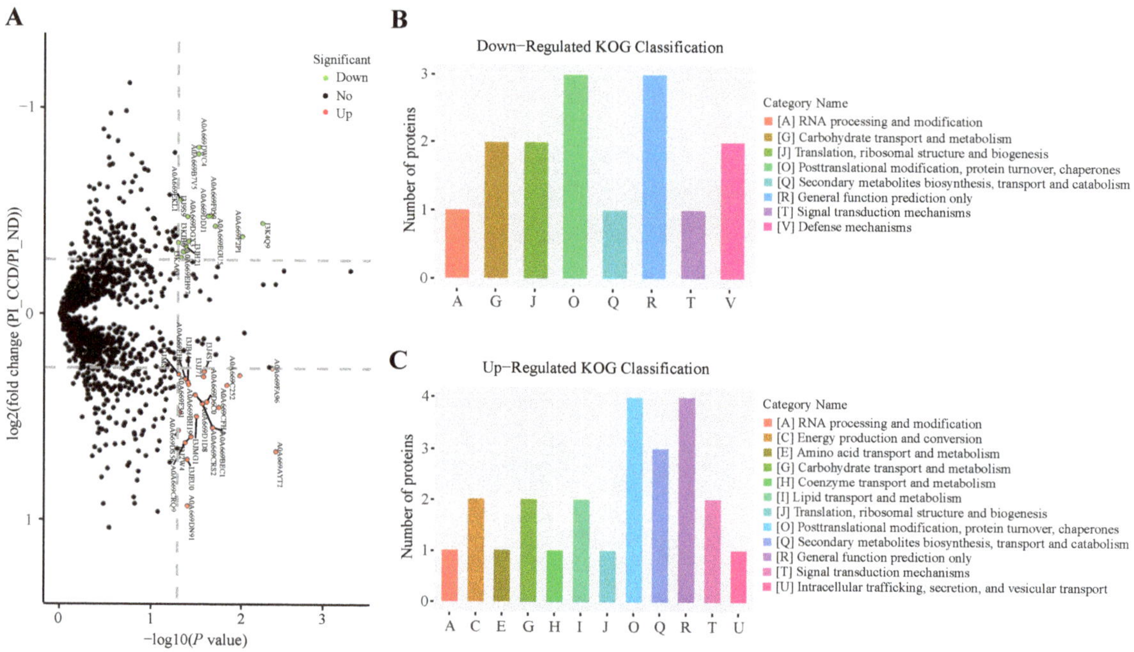


**Figure 5.** Differential analysis of protein expression between PI_CCD group and PI_ND group of tilapia liver infected with *Streptococcus agalactiae*. (**A**) Volcano plot of differential protein expression between the PI_CCD and PI_ND groups; (**B**) KOG statistics of downregulated differentially expressed proteins (DEPs) between the PI_CCD and PI_ND groups; (**C**) KOG statistics of up-regulated DEPs between the PI_CCD and PI_ND groups.

### *3.6. Functional Analysis of CC-Specific Regulatory Proteins in Tilapia before and after SA Infection*

To better understand the protein expression profiles of CC-treated SA-infected tilapia, we defined the DEPs that were upregulated between the PI_CCD and CCD groups but not between the PI_ND and ND groups as CCD-specific upregulated DEPs, and those that were downregulated between the PI_CCD and CCD groups but not between the PI_ND and ND groups as CCD-specific downregulated DEPs. In all, 80 CCD-specific upregulated proteins (Figure 6A) and 49 CCD-specific downregulated DEPs (Figure 6B) were screened from the Venn diagram. According to the results of the KEGG pathway analysis, upregulated DEPs were primarily associated with neurodegeneration, lysosome, lipid metabolism, mitochondrial autophagy, endocytosis, MAPK, and Toll-like receptor, whereas downregulated DEPs were primarily associated with glycolysis/gluconeogenesis, TCA cycle, cGMP-PKG, and pancreatic secretion (Figure 6C). In addition, upregulated DEPs were primarily involved in KEGG BRITE categories, such as exosomes, membrane transport, cytoskeleton, and proteolipid protein. In contrast, the number of downregulated DEPs involved in enzymatic activity-relayed KEGG BRITES was greater than the number of upregulated DEPs (Figure 6D).

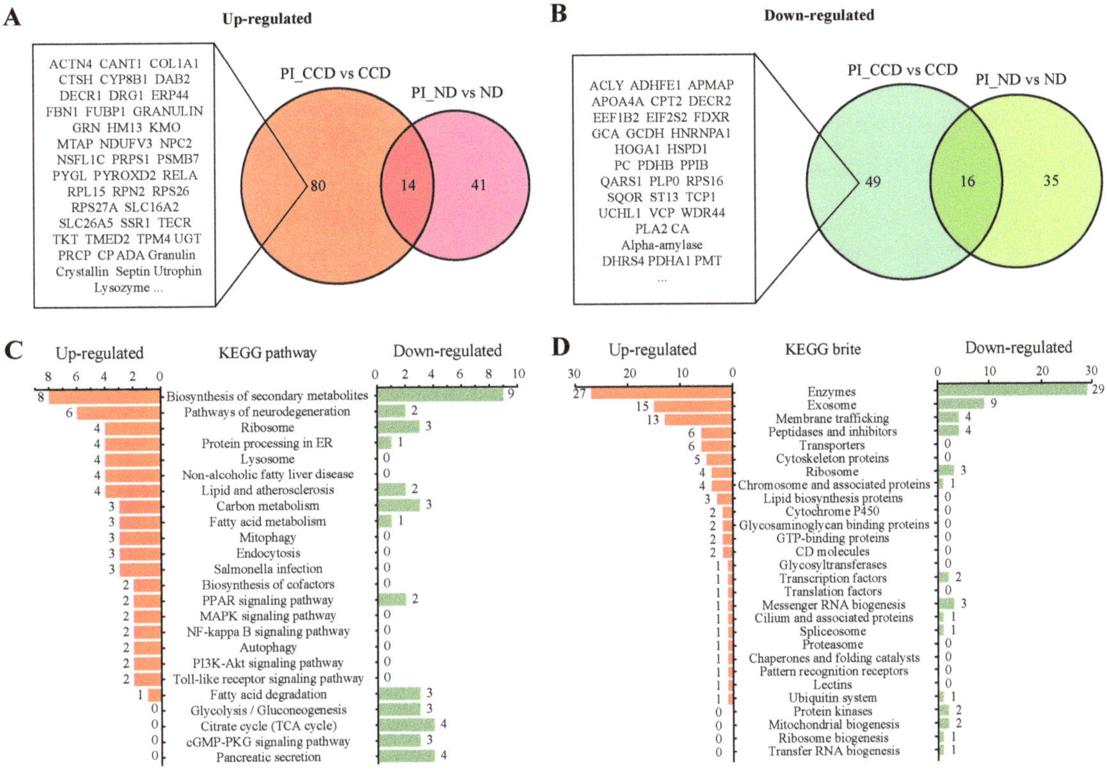


**Figure 6.** Functional analysis of *Streptococcus agalactiae*-infected tilapia before and after *Coptis chinensis* (CC) treatment. (**A**) CCD-specific upregulated differentially expressed proteins (DEPs) in SA-infected tilapia before and after CC treatment and a representative Venn diagram; (**B**) CCD-specific downregulated DEPs in SA-infected tilapia before and after CC treatment and a representative Venn diagram; (**C**) KEGG pathway annotation of CCD-specific regulatory DEPs in SA-infected tilapia before and after CC treatment; (**D**) KEGG BRITE annotation of CCD-specific regulatory DEPs in SA-infected tilapia before and after CC treatment.

## 4. Discussion

CC, a traditional Chinese herb, has been recognized for its immune-enhancing effects in aquaculture. Choi et al. [10] reported that a modified detoxification decoction supplemented with 1% CC significantly enhanced the resistance of grey mullet (*Mugil cephalus*) to *Lactococcus garvieae* infection. However, studies on the mechanism of action of CC in enhancing antioxidant and immune effects in fish have been lacking. In this study, we determined the biochemical and immunological indexes and the differences in the protein expression of SA-infected tilapia before and after CC treatment to identify the potential proteins in CC that can improve the immune function of tilapia.

### *4.1. CC Supplementation in Feed Enhances the Antioxidant Function of Tilapia*

The TP content is closely associated with protein synthesis in the liver, which can be impaired by liver damage [11]. When liver function is severely impaired, protein synthesis is significantly reduced. A study by Xu et al. [12] on *Vibrio harveyi*-infected *Pseudosciaena crocea* revealed that the TP and globulin levels were significantly lower in diseased *P. crocea* than in normal *P. crocea*. In the present study, the TP content in the liver of tilapia from the CCD group was significantly higher than that in those from the control group both before and after SA infection, indicating that CC can promote protein synthesis in the liver to an extent. AST/ALT is an essential aminotransferase involved in amino acid metabolism. When hepatocytes are damaged, hepatocyte membrane permeability increases, resulting in the entry of AST into the bloodstream and a decrease in enzyme activity in the liver [12]. In the present study, liver AST levels were significantly higher after SA infection than before infection. We hypothesized that SA infection affected hepatic amino acid metabolism and disrupted liver function.

SOD, CAT, and MDA are key antioxidant enzymes in living organisms and play a crucial role in the antioxidant system. SOD protects cells from peroxide-mediated damage by scavenging reactive oxygen species through disproportionation. CAT is commonly found in nearly all tissues of the animal body, and it is most abundant in the liver. It provides oxygen to cells while scavenging excess peroxyl radicals to reduce oxidative damage. CAT and SOD are strongly associated and act in conjunction to maintain the dynamic balance in the antioxidant system of organisms. MDA, one of the products of peroxidation by oxygen radicals, severely affects normal functions, such as cellular metabolism, by destroying the structure of the cell membrane by polymerizing with intracellular proteins. The alteration of these factors can be used as a biomarker for oxidative damage in the cells of the organism. Mu et al. [13] revealed that propolis supplementation in a basal diet increased the activity of antioxidant enzymes, such as SOD and CAT, and the expression of several cytokines in turbot (*Scophthalmus maximus*) infected with *Edwardsiella piscicida*. In the present study, the activities of SOD were increased in the CCD group compared to those in the ND group; in addition, the activity of SOD was significantly higher in the CCD group than in the ND group pre- and post-infection, indicating that CC enhanced SOD activity, reduced the oxidation of free radicals in vivo, and protected the body from oxidative damage. In an experiment to study the effects of different strains of SA on blood and related biochemical indexes in tilapia, Ao et al. [14] found that the SOD and CAT levels in the liver of tilapia were lower than normal at 24 h after SA infection. Consistently, results showed that the MDA contents were increased in ND and CCD groups post-SA infection. However, the MDA content was significantly lower in the CCD group than the ND group before and after infection. These results suggest that SA infection enhances lipid oxidation and produces harmful substances, such as MDA. However, CCD alleviates the damage to the antioxidant system by enhancing the activity of antioxidant enzymes, thus establishing dynamic balance.

### 4.2. CC Supplementation in Feed Enhances the Immune Function of Tilapia

Tilapia (a lower, poikilothermic, and bony type of fish) fight against pathogenic bacteria using nonspecific immune mechanisms. Three cytokines, IL-1$\beta$, IFN-$\gamma$, and TNF-$\alpha$, are primarily involved in innate nonspecific immunity. IL-1$\beta$, an important member of the IL-1 family, is a pro-inflammatory factor secreted primarily by activated macrophages and monocytes, and it is involved in the physiological and pathological responses of the body. When the body is stimulated by cytokines, microorganisms, and their products, IL-1$\beta$ bind to cell surface receptors to induce cellular inflammatory responses and modulate other factors to enhance immune regulation [15,16]. IFN-$\gamma$ enhances the phagocytic potential of macrophages and the disruptive power of NK cells and promotes the differentiation of B cells to produce antibodies. In addition, it can influence B and T lymphocytes to regulate the immune response [17]. TNF-$\alpha$, also a pro-inflammatory factor, kills invading pathogens by activating neutrophils, promotes the degradation of extracellular proteins in inflammation, and causes inflammatory cells to release more cytokines and inflammatory mediators [18]. Wang et al. [19] investigated the effect of Chinese yam (*Dioscoreae Rhizoma*) extract on the growth and nonspecific immunity of rainbow trout, and they found that the extract increased the levels of pro-inflammatory factors, such as TNF-$\alpha$, in serum. The results of the present study indicated that CC supplementation in the feed promoted the secretion of cytokines, especially pro-inflammatory factors, to enhance non-specific immunity and improve the immunity of the organism. In addition, the increased ratios of various serum cytokine levels were significantly higher in the comparison of the CCD group vs. ND group after SA infection than before infection. After the SA infection of tilapia, inflammatory cells were stimulated to secrete inflammatory factors that enhanced the pro-inflammatory effect, which promoted the differentiation of B cells, thus boosting the production of antibodies and enhancing phagocytosis. The immunoglobulins of bony fish, including IgM, IgD, and IgT/Z, play a crucial role in humoral immunity [20,21]. IgM, the most abundant and earliest identified immunoglobulin, is primarily found in the serum and mucosa [22]. The level of immunoglobulins in serum can indicate the expression level of immunoglobulins. C3 has functions in bony fish similar to those in mammals and is primarily found in serum. C3 is not biologically active and functions only when cleaved into C3a and C3b by other factors. These factors play an important role in the defense mechanism and homeostasis of the organism and can assist phagocytes and antibodies to eliminate pathogenic microorganisms from the organism [19,23]. Kui and Liu [24] reported the elevation of C3 levels in *Pelteobagrus fulvidraco* injected with an inactivated *E. ictaluri* vaccine, indicating an elevated physiological immune response. In the present study, the C3 levels increased significantly in response to SA injection, indicating an improved immune response in tilapia, which was consistent with the findings of the abovementioned study. In addition, the serum levels of IgM and C3 differed significantly in the pairwise comparisons among the two groups, with the highest levels observed in the PI_CCD group. Therefore, the addition of 1% CC to the feed can significantly improve the humoral immune function of tilapia and enhance their immunity to an extent.

### 4.3. CC Regulates the Expression of Antimicrobial Proteins in the Liver of Tilapia

Lysozymes and cathepsins, two classes of small-molecule lyases, play a crucial role in host innate defense [25]. Lysozymes can provide defense against bacterial infection by cleaving peptidoglycan molecules in cell walls and act as antimicrobial peptides by disrupting the bacterial membrane structure and activating autolytic enzymes in the bacterial cell wall [26]. Extensive studies in the field of aquaculture have shown the essential role of fish lysozymes in their innate immune system. The expression and activities of lysozymes were significantly elevated in catfish (*Ictalurus punctatus*) to provide defense against pathogenic microbial infection [27,28]. The addition of *Glycyrrhiza uralensis* extracts to the feed significantly increased lysozyme activity in the serum and the expression of genes encoding lysozymes in the head kidney of yellow catfish (*P. fulvidraco*), thus enhancing their antimicrobial activity [29]. Moreover, the addition of fermented lemon peel to the diet effectively increased lysozyme activity and enhanced the resistance of orange-spotted grouper to *Photobacterium* [30]. Similarly, in the present study, we found that CC supplementation in the feed was effective in increasing the expression level of hepatic lysozymes, which could enhance the resistance of tilapia to SA to an extent. Lysosomal cathepsins are cysteine peptidases distributed ubiquitously during autophagic apoptosis that are involved in various biological processes, such as extracellular matrix remodeling, inflammation, pathogenic bacterial infection, and cancer [31]. Fish cathepsin B plays a crucial role in the defense of several species against pathogenic microbial infections, including olive flounder (*Paralichthys olivaceus*) against rhabdovirus [32], channel catfish (*I. punctatus*) against *E. ictaluri* and *Flavobacterium* [33], and tilapia (*Oreochromis niloticus*) against *S. agalactiae* [34]. Furthermore, the results of our proteomic analysis also indicated a significantly high expression of cathepsin protein in tilapia infected with SA.

Significant differences were observed in the expression between multiple antimicrobial peptides and antimicrobial proteins during SA infection. Septins, which are highly conserved cytoskeletal proteins, play a vital role in the innate immune and inflammatory response to bacterial infection in fish. According to Maria et al. [35], the depletion of Sept15 and Sept7b significantly increases susceptibility to bacteria in zebrafish. In catfish (*I. punctatus*) and rohu (*Labeo rohita*), the tissue expression levels of several septins are regulated by pathogenic bacterial infection and are closely related to their defense processes [36,37]. We observed that septin proteins were significantly upregulated in the CCD group, which may be attributed to the enhanced resistance to SA infection. In addition, a secreted glycosylated peptide, granulin a, and granulin domain-containing protein was also specifically regulated by CC. Granulin is a cysteine-rich regulatory growth factor that forms multiple pairs of disulfide bonds and regulates various processes, such as cell proliferation, tumor invasion, and immune response [38]. Potent peptide GRN-41 induces the expression of multiple immune factors, such as TNF-$\alpha$, TNF-$\beta$, IL-8, IL-1$\beta$, IL-6, IL-26, IL-21, and IL-10 in transgenic zebrafish, thus effectively preventing *V. vulnificus* infection [39]. The above research team further identified the good antimicrobial activity of GRN-41 against *Vibrio* species in Mozambique tilapia [40]. Saleh et al. [41], in their proteomic study of *Ichthyophthirius multifiliis*-infected carp, found that granulin was significantly upregulated in the skin mucus, which was also consistent with the results of the present study. Moreover, a member of the penta-EF-hand protein family, Grancalcin, was reported to be translocated in macrophages in response to stimulation with bacterial LPS [42]. We also found that this protein was upregulated in the livers of tilapia fed with CC.

## 5. Conclusions

In conclusion, we applied a comprehensive iTRAQ-based liver proteomic analysis to evaluate the anti-SA effects on the molecular level of CCD. The results showed that CC supplementation can effectively improve the ability of tilapia to resist SA infection by modulating various antioxidant enzymes, immune factors, antimicrobial enzymes, and antimicrobial peptides, and can moderately inhibit central carbon metabolism. The results of the study provide a basis for replacing antibiotics with environmentally-friendly functional feed additives for the control of bacterial diseases.

**Supplementary Materials:** The following supporting information can be downloaded at: https://www.mdpi.com/article/10.3390/fishes8070370/s1, Table S1: The sequence of 16S rDNA of Streptococcus agalactiae strain. Table S2: Liver proteomic details in ND and CCD groups before and after SA infection of tilapia by LC-MS/MS.

**Author Contributions:** Conceptualization, K.H. and Q.L.; methodology, K.Y., R.G. and T.L.; funding acquisition, K.H.; proteomic analysis and bioinformatic analysis, K.Y.; writing—original draft preparation, R.G. and K.Y.; writing—review and editing, K.H. and K.Y. All authors have read and agreed to the published version of the manuscript.

**Funding:** This work was supported by the Development Project of Key Laboratory on the Creation of Fishery Drugs of the Ministry of Chinese Agriculture and Rural Affairs (201807) and the National Natural Science Foundation of China (No. 32160865).

**Institutional Review Board Statement:** The animal experiments were approved and monitored by the Animal Experiments Ethical Review Committee of Guangxi University (approval code: GXU-2022-255).

**Data Availability Statement:** The original proteomic data (PXD00000, the accession No. will provide during review) can by fully accessed from the ProteomeXchange consortium via the PRIDE partner repository.

**Acknowledgments:** We acknowledge the State Key Laboratory for Conservation and Utilization of Subtropical Agro-bioresources for the support of LC-MS/MS platform. We thank Nanning Current Science Biotechnology Co., Ltd. for optimizing the figures and editing this manuscript.

**Conflicts of Interest:** The authors declare no conflict of interest.

## References

1. Zhang, Z. Research Advances on Tilapia Streptococcosis. *Pathogens* **2021**, *10*, 558. [CrossRef] [PubMed]
2. Liu, G.; Zhu, J.; Chen, K.; Gao, T.; Yao, H.; Liu, Y.; Zhang, W.; Lu, C. Development of Streptococcus agalactiae vaccines for tilapia. *Dis. Aquat. Org.* **2016**, *122*, 163–170. [CrossRef]

3. Barkham, T.; Zadoks, R.N.; Azmai, M.N.A.; Baker, S.; Bich, V.T.N.; Chalker, V.; Chau, M.L.; Dance, D.; Deepak, R.N.; van Doorn, H.R.; et al. One hypervirulent clone, sequence type 283, accounts for a large proportion of invasive Streptococcus agalactiae isolated from humans and diseased tilapia in Southeast Asia. *PLOS Negl. Trop. Dis.* **2019**, *13*, e0007421. [CrossRef] [PubMed]
4. Wang, J.; Wang, L.; Lou, G.; Zeng, H.-R.; Hu, J.; Huang, Q.; Peng, W.; Yang, X.-B. Coptidis Rhizoma: A comprehensive review of its traditional uses, botany, phytochemistry, pharmacology and toxicology. *Pharm. Biol.* **2019**, *57*, 193–225. [CrossRef] [PubMed]
5. Ma, G.Q.; Liu, D.L. Progress in pharmacological research of Coptis chinensis. *Gansu Agric.* **2019**, *10*, 97–99.
6. Yu, F.X.; Cao, M.X.; Liang, W.W.; Wang, C.H.; Wang, S.; Wei, Y.Y.; Hu, T.J. Clinical application of Sanhuanglian powder in prevention of Streptococcicosis in tilapia. *Anim. Husb. Feed Sci.* **2020**, *41*, 86–91.
7. Wang, C.H.; Yu, F.X.; Zhao, Y.; Liang, W.W.; Wei, Y.Y.; Hu, T.J. Clinical application effect of Sanhuanglian mixture in preventing tilapia against Streptococcosis. *Guangxi J. Anim. Husb. Vet. Med.* **2020**, *36*, 147–150.
8. Niu, Z.W.; Fan, H.L.; Long, S.; Huang, J.; Zhu, H.F.; Huang, G.C.; Liang, J.Z. A Meta-analysis: The effect of three kinds of Chinese herbalmedicines for tilapia Streptococcus agalactiae. *J. Fish. Res.* **2016**, *38*, 409–414.
9. Matumoto, M. A note on some points of calculation method of LD50 by Reed and Muench. *Jpn. J. Exp. Med.* **1949**, *20*, 175–179.
10. Choi, W.; Lam, C.; Mo, W.; Cheng, Z.; Mak, N.; Bian, Z.; Wong, M. Effects of the modified Huanglian Jiedu decoction on the disease resistance in grey mullet (*Mugil cephalus*) to *Lactococcus garvieae*. *Mar. Pollut. Bull.* **2014**, *85*, 816–823. [CrossRef]
11. Ji, D.W.; Li, M.Y.; Wang, T.Z.; Zhang, C.N.; Xu, Z.; Xu, W.T. Effects of low temperature stress periods on serum biochemical indexes in large yellow croaker Pseudosciaena crocea. *Fish. Sci.* **2009**, *28*, 1–4.
12. Xu, X.-J.; Xu, B.; Wang, J.; Su, Y.-Q.; Zhang, Z.-W.; Chen, X. Studies on blood chemistry indices and histopathology of Pseudosciaena crocea artificially challenged with Vibrio harveyi. *J. Fish. China* **2010**, *34*, 618–625. [CrossRef]
13. Mu, D.; Jiang, Y.; He, J.; Zhang, Y.; Yang, D.; Liu, Q.; Wang, Z. Dietary supplementation of propolis enhanced the innate immune response against Edwardsiella piscicida challenge in turbot (*Scophthalmus maximus*). *Fish Shellfish Immunol.* **2022**, *124*, 273–279. [CrossRef]
14. Ao, Q.W.; Lu, Y.J.; Lu, M.; Zhu, J.J. Effects of Streptococcus agalactiae infection on blood and hepatopancreatic tissue biochemical indices in different species of tilapia. *Prog. Fish. Sci.* **2020**, *41*, 167–173.
15. Bird, S.; Zou, J.; Wang, T.; Munday, B.; Cunningham, C.; Secombes, C.J. Evolution of interleukin-1β. *Cytokine Growth Factor Rev.* **2002**, *13*, 483–502. [CrossRef] [PubMed]
16. Danilko, K.V.; Korytina, G.F.; Akhmidishina, L.Z.; Ianbaeva, D.G.; Zagidullin, S.Z.; Victorova, T.V. Association of cytokines genes (IL1B, IL1RN, TNFA, LTA, IL6, IL8, and IL10) polymorphic markers with chronic obstructive pulmonary disease. *Mol. Biol.* **2007**, *41*, 26–36. [CrossRef]
17. Roberts-Thomson, I.C.; Fon, J.; Uylaki, W.; Cummins, A.G.; Barry, S. Cells, cytokines and inflammatory bowel disease: A clinical perspective. *Expert Rev. Gastroenterol. Hepatol.* **2011**, *5*, 703–716. [CrossRef] [PubMed]
18. Salomon, B.L. Insights into the biology and therapeutic implications of TNF and regulatory T cells. *Nat. Rev. Rheumatol.* **2021**, *17*, 487–504. [CrossRef] [PubMed]
19. Wang, F.; Li, C.G.; Ma, Y.W.; Liu, F.; Meng, X.J. Effects of yam extract on growth performance and non-specific immune parameters of rainbow trout (*Oncorhynchus mykiss*). *J. Henan Agric. Sci.* **2020**, *49*, 167–172. [CrossRef]
20. Castro, R.; Bromage, E.; Abós, B.; Pignatelli, J.; Granja, A.G.; Luque, A.; Tafalla, C. CCR7 Is Mainly Expressed in Teleost Gills, Where It Defines an IgD+IgM− B Lymphocyte Subset. *J. Immunol.* **2014**, *192*, 1257–1266. [CrossRef]
21. Zhang, Y.-A.; Salinas, I.; Li, J.; Parra, D.; Bjork, S.; Xu, Z.; La Patra, S.E.; Bartholomew, J.; Sunyer, J.O. IgT, a primitive immunoglobulin class specialized in mucosal immunity. *Nat. Immunol.* **2010**, *11*, 827–835. [CrossRef] [PubMed]
22. Ehrenstein, M.R.; Notley, C.A. The importance of natural IgM: Scavenger, protector and regulator. *Nat. Rev. Immunol.* **2010**, *10*, 778–786. [CrossRef]
23. Li, X.T.; Ke, X.L.; Lu, M.X.; Liu, Z.G.; Wang, M.; Pang, J.C. Cloning and tissues expression analysis of complement C3 of tilapia. *Biotechnology* **2016**, *26*, 205–212. [CrossRef]
24. Kui, L.L.; Liu, Y. Effects of Edwardsiella ictaluri inactivated vaccine on immune gene wxpression and activities of CAT, SOD and complement C3 in yellow catfish Pelteobagrus fulvidraco. *Fish. Sci.* **2015**, *34*, 150–154. [CrossRef]
25. Magnadóttir, B. Innate immunity of fish (overview). *Fish Shellfish Immunol.* **2006**, *20*, 137–151. [CrossRef]
26. Farnsworth, R.H.; Achen, M.G.; Stacker, S.A. Lymphatic endothelium: An important interactive surface for malignant cells. *Pulm. Pharmacol. Ther.* **2006**, *19*, 51–60. [CrossRef]
27. Bilodeau-Bourgeois, L.; Bosworth, B.G.; Peterson, B.C. Differences in mortality, growth, lysozyme, and Toll-like receptor gene expression among genetic groups of catfish exposed to virulent *Edwardsiella ictaluri*. *Fish Shellfish Immunol.* **2008**, *24*, 82–89. [CrossRef]
28. Small, B.C.; Bilodeau, A.L. Effects of cortisol and stress on channel catfish (*Ictalurus punctatus*) pathogen susceptibility and lysozyme activity following exposure to Edwardsiella ictaluri. *Gen. Comp. Endocrinol.* **2005**, *142*, 256–262. [CrossRef]
29. Wang, Q.; Shen, J.; Yan, Z.; Xiang, X.; Mu, R.; Zhu, P.; Yao, Y.; Zhu, F.; Chen, K.; Chi, S.; et al. Dietary Glycyrrhiza uralensis extracts supplementation elevated growth performance, immune responses and disease resistance against Flavobacterium columnare in yellow catfish (*Pelteobagrus fulvidraco*). *Fish Shellfish Immunol.* **2020**, *97*, 153–164. [CrossRef]
30. Zhuo, L.-C.; Chen, C.-F.; Lin, Y.-H. Dietary supplementation of fermented lemon peel enhances lysozyme activity and susceptibility to Photobacterium damselae for orange-spotted grouper, *Epinephelus coioides*. *Fish Shellfish Immunol.* **2021**, *117*, 248–252. [CrossRef]

31. Gobec, S. Inhibitors of Cathepsin B. *Curr. Med. Chem.* **2006**, *13*, 2309–2327. [CrossRef]
32. Zhang, F.; Zhang, Y.; Chen, Y.; Zhu, R.; Dong, C.; Li, Y.; Zhang, Q.; Gui, J. Expressional induction of Paralichthys olivaceus cathepsin B gene in response to virus, poly I:C and lipopolysaccharide. *Fish Shellfish Immunol.* **2008**, *25*, 542–549. [CrossRef]
33. Li, C.; Song, L.; Tan, F.; Su, B.; Zhang, D.; Zhao, H.; Peatman, E. Identification and mucosal expression analysis of cathepsin B in channel catfish (*Ictalurus punctatus*) following bacterial challenge. *Fish Shellfish Immunol.* **2015**, *47*, 751–757. [CrossRef]
34. Li, L.; Shi, Y.; Wang, R.; Huang, T.; Liang, W.; Luo, H.; Gan, X.; Huang, W.; Li, J.; Lei, A.; et al. Proteomic analysis of tilapia Oreochromis niloticus *Streptococcus agalactiae* strains with different genotypes and serotypes. *J. Fish Biol.* **2015**, *86*, 615–636. [CrossRef]
35. Mazon-Moya, M.J.; Willis, A.R.; Torraca, V.; Boucontet, L.; Shenoy, A.R.; Colucci-Guyon, E.; Mostowy, S. Septins restrict inflammation and protect zebrafish larvae from Shigella infection. *PLOS Pathog.* **2017**, *13*, e1006467. [CrossRef] [PubMed]
36. Fu, Q.; Li, Y.; Yang, Y.; Li, C.; Yao, J.; Zeng, Q.; Qin, Z.; Liu, S.; Li, D.; Liu, Z. Septin genes in channel catfish (*Ictalurus punctatus*) and their involvement in disease defense responses. *Fish Shellfish Immunol.* **2016**, *49*, 110–121. [CrossRef] [PubMed]
37. Das, S.; Chhottaray, C.; Das Mahapatra, K.; Saha, J.N.; Baranski, M.; Robinson, N.; Sahoo, P.K. Analysis of immune-related ESTs and differential expression analysis of few important genes in lines of rohu (*Labeo rohita*) selected for resistance and susceptibility to Aeromonas hydrophila infection. *Mol. Biol. Rep.* **2014**, *41*, 7361–7371. [CrossRef] [PubMed]
38. Klupp, F.; Kahlert, C.; Franz, C.; Halama, N.; Schleussner, N.; Wirsik, N.M.; Warth, A.; Schmidt, T.; Ulrich, A.B. Granulin: An Invasive and Survival-Determining Marker in Colorectal Cancer Patients. *Int. J. Mol. Sci.* **2021**, *22*, 6436. [CrossRef] [PubMed]
39. Wu, S.-H.; Lin, H.-J.; Lin, W.-F.; Wu, J.-L.; Gong, H.-Y. A potent tilapia secreted granulin peptide enhances the survival of transgenic zebrafish infected by Vibrio vulnificus via modulation of innate immunity. *Fish Shellfish Immunol.* **2018**, *75*, 74–90. [CrossRef]
40. Wu, S.-H.; Chou, H.-Y.; Liu, P.-C.; Wu, J.-L.; Gong, H.-Y. Granulin peptide GRN-41 of Mozambique tilapia is a novel antimicrobial peptide against Vibrio species. *Biochem. Biophys. Res. Commun.* **2019**, *515*, 706–711. [CrossRef]
41. Saleh, M.; Kumar, G.; Abdel-Baki, A.-A.S.; Dkhil, M.A.; El-Matbouli, M.; Al-Quraishy, S. Quantitative proteomic profiling of immune responses to Ichthyophthirius multifiliis in common carp skin mucus. *Fish Shellfish Immunol.* **2019**, *84*, 834–842. [CrossRef] [PubMed]
42. Liu, F.; Shinomiya, H.; Kirikae, T.; Hirata, H.; Asano, Y. Characterization of Murine Grancalcin Specifically Expressed in Leukocytes and Its Possible Role in Host Defense against Bacterial Infection. *Biosci. Biotechnol. Biochem.* **2004**, *68*, 894–902. [CrossRef] [PubMed]

fishes

Article

# Effects of Dietary *Lentinus edodes* Fermentation Supplementation on Digestive Enzyme Activity, Antioxidant Capacity and Morphology of the Liver and Intestine in Largemouth Bass (*Micropterus salmoides*) Fed High Plant Protein Diets

**Jiaming Xu [1], Zhoulin Yu [1], Guangye Liu [1], Sijie Li [1], Guoyong Zhou [2], Hanhua Wang [2], Yewei Dong [1], Cuihong You [1], Weidong Bai [3,4], Meng Zhou [1,*], Yanhua Huang [1,*] and Xiaohong Tan [1,*]**

1 Innovative Institute of Animal Healthy Breeding, College of Animal Sciences and Technology, Zhongkai University of Agriculture and Engineering, Guangzhou 510225, China; jadexu777@outlook.com (J.X.); yuzhoulin99@outlook.com (Z.Y.); liuguangye01@outlook.com (G.L.); lsijieqqq@outlook.com (S.L.); sunshinewonder@163.com (Y.D.); youcuihong@zhku.edu.cn (C.Y.)

2 YueHao Biotechnology (Guangzhou) Co., Ltd., Guangzhou 510663, China; zgy@yuehaobio.om (G.Z.); whh@yuehaobio.com (H.W.)

3 Guangdong Provincial Key Laboratory of Lingnan Specialty Food Science and Technology, Guangzhou 510225, China; whitebai2001@163.com

4 Key Laboratory of Green Processing and Intelligent Manufacturing of Lingnan Specialty Food, Ministry of Agriculture, Guangzhou 510225, China

* Correspondence: zhoumeng@zhku.edu.cn (M.Z.); huangyh111@126.com (Y.H.); tanxiaohong@zhku.edu.cn (X.T.)

**Citation:** Xu, J.; Yu, Z.; Liu, G.; Li, S.; Zhou, G.; Wang, H.; Dong, Y.; You, C.; Bai, W.; Zhou, M.; et al. Effects of Dietary *Lentinus edodes* Fermentation Supplementation on Digestive Enzyme Activity, Antioxidant Capacity and Morphology of the Liver and Intestine in Largemouth Bass (*Micropterus salmoides*) Fed High Plant Protein Diets. *Fishes* **2023**, *8*, 329. https://doi.org/10.3390/fishes8060329

Academic Editors: Qiyou Xu, Jianhua Ming, Fei Song, Changle Qi and Chuanpeng Zhou

Received: 5 June 2023
Revised: 16 June 2023
Accepted: 16 June 2023
Published: 20 June 2023



**Abstract:** This study evaluated the effect of *Lentinus edodes* fermentation (LEF) on digestive enzyme activity, antioxidant capacity and morphology of the liver and intestine in largemouth bass (*Micropterus salmoides*) fed high plant protein diets (HPPD). LEF was supplemented in HPPD with 0 g kg$^{-1}$ (LEF0), 10 g kg$^{-1}$ (LEF1), 20 g kg$^{-1}$ (LEF2), 30 g kg$^{-1}$ (LEF3), 40 g kg$^{-1}$ (LEF4), 50 g kg$^{-1}$ (LEF5), respectively, and then the six diets were fed to largemouth bass with a body weight of 28.8 ± 0.05 g for eight weeks. Juvenile fish were randomized into 6 groups and each group had 4 replicates with 40 fish. Dietary LEF supplementation alleviated the liver inflammatory reaction of largemouth bass caused by HPPD and improved liver morphology. Goblet cells multiplied and the gut muscle layer thickened after LEF supplementation. The LEF significantly increased amylase activity in the liver and intestine of largemouth bass in individual experimental groups. The LEF could increase the activity of catalase in the liver and intestine of largemouth bass ($p < 0.05$). The content of malondialdehyde was significantly lower than that in the control group ($p < 0.05$). Dietary LEF supplementation had no significant effect on the intestinal flora of largemouth bass. These findings imply that LEF supplementation can reduce liver inflammation, enhance intestinal tissue morphology, and eventually benefit largemouth bass health.

**Keywords:** largemouth bass; *Lentinus edodes* fermentation; high plant protein diets; antioxidant capacity; gut microbiota

**Key Contribution:** A. Low dose of LEF supplementation in high plant protein diet can reduce liver inflammatory infiltration. B. LEF supplementation has a potential future in liver protection as a new functional feed addition.

## 1. Introduction

The largemouth bass is an important freshwater economic fish in China, with good meat quality, rich nutrition and strong adaptability [1]. The normal growth of carnivorous fish requires feed containing 30~55% crude protein. As a result, fishmeal is often chosen as the main protein source for carnivorous economic fish feeds such as largemouth bass. As one of the main raw materials for terrestrial and aquatic animal feed, fishmeal has the advantages of good palatability, balanced amino acid composition, and ease of animal digestion and absorption [2,3]. While the demand for fishmeal has increased due to the global aquaculture industry's quick development, other factors such as declining fishmeal production and rising prices have forced people to look for high-yield, low-cost protein alternatives to fishmeal [3–6], and plant protein has received widespread attention as an ideal alternative to fishmeal.

Certain plant proteins are more advantageous in terms of amino acid composition and nutrient content, so they are widely used in aquaculture [7–9]. However, numerous studies have also noted that using plant protein as the main protein might have a detrimental effect on fish development, liver function, and gastrointestinal health. Studies on Japanese sea bass (*Lateolabrax japonicus*), Florida pompano (*Trachinotus carolinus*), and carp (*Carassius auratus gibelio*) have inflammation occurs in both the liver and intestines of fish found that after ingesting a high proportion of soy protein feed [10–12]. These negative effects are mainly caused by anti-nutritional factors (ANFs) in plant proteins [1]. The study demonstrates that high plant protein diets are difficult to meet the nutritional needs of largemouth bass [13,14], and have a negative impact on the growth and development of largemouth bass, liver and intestinal health [15–19].

*Lentinus edodes* (*L. edodes*) belongs to the basidiomycetes order, which grows naturally in warm and humid environments, and is one of the most popular large edible fungi [20]. Mycelium and fruiting bodies of *L. edodes* are sources of a variety of nutrients and bioactive compounds with numerous positive health effects. [21]. As a highly nutritious functional food, various studies have confirmed that *Lentinus edodes* have anti-inflammatory and liver protection, anti-tumor, hypoglycemic and hypolipidemic effects, and antibacterial and other pharmacological effects [22–24]. At present, few studies have been related to the effects of *L. edodes* on the growth and health of aquatic animals, and the research on *L. edodes* first fermented and treated as an economic fish feed additive has rarely been reported. Based on the above results, LEF can be used as a functional additive in commercial fish culture to reduce the liver damage induced by HPPD.

## 2. Materials and Methods

### 2.1. Preparation of Experimental Diet

Six experimental diets were formulated to contain graded levels of LEF (YueHao Biotechnology (Guangzhou) Co., Ltd., Guangzhou, China). Fish meal, soybean meal, corn gluten meal and peanut bran are the protein sources of the experimental base diet (LEF0). The other five diets (LEF1, LEF2, LEF3, LEF4, LEF5) were supplemented with 10 g $kg^{-1}$ (LEF1), 20 g $kg^{-1}$ (LEF2), 30 g $kg^{-1}$ (LEF3), 40 g $kg^{-1}$ (LEF4), 50 g $kg^{-1}$ (LEF5) LEF, respectively. The formulation and composition of experimental diets are presented in Table 1. All the ingredients in the experimental diet were ground into a powder and mixed, then separately mixed with fish oil and water to make a dough, which was then extruded into a 3-mm granulator (F-26, South China University of Technology, Guangzhou, China). After drying, all diets were stored at −20 °C until use.

According to AOAC (Association of Official Analytical Chemists, 2005), the composition of feed nutrients was analyzed. Moisture was determined using the 105 °C drying constant weight method (for 24 h); crude protein was determined using the Kjeldahl method; crude fat was determined using the Soxhlet extraction method and the ash determination used the chamber resistance furnace burning method.

**Table 1.** Composition and nutrient levels of experimental diets (g $kg^{-1}$).

| Material | LEF0 | LEF1 | LEF2 | LEF3 | LEF4 | LEF5 |
|---|---|---|---|---|---|---|
| fish meal | 280.0 | 280.0 | 280.0 | 280.0 | 280.0 | 280.0 |
| Soybean protein concentrate | 100.0 | 100.0 | 100.0 | 100.0 | 100.0 | 100.0 |
| Corn gluten meal | 140.0 | 140.0 | 140.0 | 140.0 | 140.0 | 140.0 |
| Soybean meal | 78.0 | 78.0 | 78.0 | 78.0 | 78.0 | 78.0 |
| Peanut bran | 60.0 | 60.0 | 60.0 | 60.0 | 60.0 | 60.0 |
| flour | 205.9 | 205.9 | 205.9 | 205.9 | 205.9 | 205.9 |
| Beer yeast powder | 50.0 | 50.0 | 50.0 | 50.0 | 50.0 | 50.0 |
| fish oil | 51.0 | 51.0 | 51.0 | 51.0 | 51.0 | 51.0 |
| lecithin | 10.0 | 10.0 | 10.0 | 10.0 | 10.0 | 10.0 |
| Vitamin and mineral premix [1] | 10.0 | 10.0 | 10.0 | 10.0 | 10.0 | 10.0 |
| Choline chloride (50%) | 5.0 | 5.0 | 5.0 | 5.0 | 5.0 | 5.0 |
| antioxidant | 0.1 | 0.1 | 0.1 | 0.1 | 0.1 | 0.1 |
| Calcium dihydrogen phosphate | 10.0 | 10.0 | 10.0 | 10.0 | 10.0 | 10.0 |
| Mushroom leftover fermentation product [2] | 0 | 10 | 20 | 30 | 40 | 50 |
| Proximate composition (% dry matter) | | | | | | |
| crude protein | 46.05 | 46.15 | 46.26 | 46.36 | 46.46 | 46.57 |
| crude lipid | 10.08 | 10.09 | 10.10 | 10.11 | 10.13 | 10.14 |
| Ash | 7.07 | 7.12 | 7.17 | 7.22 | 7.27 | 7.32 |
| Gross energy (MJ $kg^{-1}$) [3] | 18.40 | 18.43 | 18.46 | 18.48 | 18.51 | 18.54 |

[1] vitamin and mineral premix Provided by Shenzhen Jingji Zhinong Times Co., Ltd. (mg $kg^{-1}$ diet): Vitamin A ≥ 450,000 IU/kg, Vitamin B1 ≥ 1000 mg/kg, Vitamin B2 ≥ 1000 mg/kg, Vitamin B6 ≥ 1500 mg/kg, Vitamin B12 ≥ 5 mg/kg, Vitamin K3 ≥ 800 mg/kg, inositol ≥ 12,000 mg/kg, D-Pantothenic acid ≥ 3500 mg/kg, nicotinic acid ≥ 2000 mg/kg, folic acid ≥ 500 mg/kg, D-Biotin ≥ 5 mg/kg, Vitamin D3 300,000~400,000 IU/kg, Vitamin E ≥ 8000 IU/kg, $Na_2SeO_3$ 20 mg, $CuSO_4{\cdot}5H_2O$ 24 mg, $FeSO_4{\cdot}H_2O$ 266.65 mg, $ZnSO_4{\cdot}H_2O$ 100 mg, $MnSO_4{\cdot}H_2O$ 120 mg, $Ca\ (IO_3)_2$ 50 mg, $CoSO_4{\cdot}7H_2O$ 10 mg, Mg 20g, zeolite 4380.55 mg. [2] YueHao Biotechnology (Guangzhou) Co., Ltd. [3] Energy equivalents of 23.64, 39.54, and 17.15 kJ g1 for protein, fat, and digestible carbohydrates, respectively, were used to compute gross energy. LEF0 (0 g $kg^{-1}$), LEF1 (10 g $kg^{-1}$), LEF2 (20 g $kg^{-1}$), LEF3 (30 g $kg^{-1}$), LEF4 (40 g $kg^{-1}$), LEF5 (50 g $kg^{-1}$).

### 2.2. Experimental Fish and Samples Collection

This experiment was conducted at Baiyun Base of Guangdong Academy of Agricultural Sciences (Guangzhou, China). Experimental fish were provided by the base and domesticated for two weeks using basic feed under experimental conditions. After domestication, fish were fasted for 24 h, then weighed and grouped. Experimental fish were assigned to six treatments with four replicates per treatment and released 40 fish with an average weight of 28.8 g ± 0.2 g into cages (1 m × 1 m × 1.5 m). Experimental fish were fed twice a day at 8:00 and 16:00 until significant satiety was observed. Experimental fish were fed for 8 weeks under a natural light cycle. During the experiment, dissolved oxygen was higher than 6.0 mg $L^{-1}$, ammonia nitrogen was lower than 0.05 mg $L^{-1}$, and the water temperature was between 28 °C and 33 °C.

After the 8-week feeding experiment, experimental fish were fasted for 24 h first, then weighed according to experimental groups. In each replicate, 8 fish were randomly selected for anesthesia with eugenol in dosage 150 mL/L, and 5 fish were selected to be weighed the body weight, visceral weight and liver weight respectively. The liver and part of the intestine were removed and then stored at −80 °C for subsequent analysis; the Caudal vein of the fish was punctured, blood samples were obtained, centrifuged (3000 rpm, 15 min, 4 °C), separated and stored at −80 °C.

### 2.3. Growth Performance

At the end (after 8 weeks) of the feeding trial, fish of each cage were weighed and growth parameters including weight gain rate (WGR, %), specific growth rate (SGR, % $day^{-1}$), feed conversion ratio (FCR), hepatosomatic index (HSI, %), viscerosomatic index (VSI, %), condition factor (CF, g $cm^{-3}$) and survival rate (SR, %) were calculated.

WGR = (final body weight (g) − initial body weight (g))/initial body weight (g) × 100%;

SGR = (ln final body weight (g) − ln initial body weight (g))/experimental period (d) × 100%;

FCR = feed consumption (g)/(final body weight (g) − initial body weight (g));

HSI = liver weight (g)/whole body weight (g) × 100%;

VSI = viscera weight (g)/whole body weight (g) × 100%;

CF = body weight (g)/(body length (cm))$^3$ × 100%;

SR = final number of the fish/initial number of the fish × 100%.

*2.4. Morphology of Liver and Intestine*

Hematoxylin eosin staining: the liver and intestinal tissue samples of three experimental fish in each group were cut, washed in pre-cooled PBS solution, fixed in 4% paraformaldehyde solution for 24~48 h, dehydrated with gradient alcohol, transparent in xylene, embedded in paraffin, sliced with a microtome (5 μm), stained with hematoxylin and eosin, and sealed with neutral gum. The liver and intestinal tissues were observed and photographed with a microscope. The height, width and thickness of the muscularis of the intestinal were measured with a micrometer.

Ultrastection: the intestinal samples of fish in each replicate were cut into small pieces less than 1 mm$^3$, fixed with 2.5% glutaraldehyde and 1% osmium tetroxide, dehydrated with gradient alcohol, embedded with resin, sliced with an ultrathin microtome, stained with uranyl acetate and lead citrate, observed and photographed with a transmission electron microscope.

*2.5. Activities of Digestive Enzymes and Antioxidant Enzymes*

The tissues of the liver and intestinal were cut and the blood was washed with pre-cooled normal saline, the surface water was absorbed with filter paper, weighed, cut with scissors, and transferred into the homogenizing tube. Nine times the volume of normal saline was added to the tube for homogenization and then centrifuged at 2000 rpm for 15 min, and the supernatant was collected for standby.

The activities of amylase (AMS), lipase (LPS) and trypsin (TRY) were measured by enzyme labeling instrument, and the detection was carried out through the assay kit with strict instructions (Nanjing Jiancheng Bioengineering Institute, Wuhan, China).

Total antioxidant capacity (T-AOC), the activities of peroxidase (CAT) and superoxide dismutase (SOD), and the amounts of malondialdehyde (MDA) of liver and intestinal homogenates were measured in strict accordance with the operating instructions of the kit (Nanjing Jiancheng Bioengineering Institute, Wuhan, China).

*2.6. Intestinal Microbiota Analysis*

Gut contents were collected from 5 fish in each cage, these samples were sent for sequencing (OmicShare Co., Ltd., Guangzhou, China). According to the commercial DNA kit instructions, the total DNA of bacterial communities was extracted from the gut contents of experimental fish using the TIANamp Stool DNA Kit (Omega Biotek, Norcross, GA, USA). After DNA extraction, DNA concentration was measured using a NanoDrop 8000 spectrophotometer and integrity was detected by electrophoresis of a 1% agarose gel. Then, a pair of universal primes F341 (5-CCTAYGGRBGCASCAG-3) and R806 (5-GGACTACNNGGGTATCTAAT-3), designed based on the V3 + V4 region of 16s rDNA, were used for DNA amplification of gut bacteria. Qualification and sequencing of amplicons using illumine Hiseq2500, then raw sequences were analyzed according to Dong [25].

*2.7. Statistical Analysis*

The experimental data were statistically analyzed using Excel 2016 and SPSS 26.0 software. All data results were expressed as "mean ± SD". The normality assumption of the data was tested using Kolmogorov–Smirnov test. One-way analysis of variance

(ANOVA) was performed on the test data. Duncan's method was used for multiple comparisons to test the difference between groups. $p < 0.05$ meant a significant difference.

## 3. Results

### 3.1. Growth Properties

The SR, WGR, FCR, VSI and HSI of largemouth bass are shown in Table 2. There were no significant differences in WGR and SGR between any of the groups ($p > 0.05$). Fish in Group LEF2 had the highest WGR and SGR. Nonetheless, the FCR of the five treatments was lower than that of the control group, even though there was no statistically significant difference between the groups ($p > 0.05$). In addition, the VSI, HSI, and SR of the six treatments were not significantly different from each other ($p > 0.05$).

**Table 2.** Effects of LEF on growth performance, feed utilization and morphological index of largemouth bass.

| Indexes | LEF0 | LEF1 | LEF2 | LEF3 | LEF4 | LEF5 |
|---|---|---|---|---|---|---|
| IBW (g) | 28.8 ± 0.04 | 28.8 ± 0.04 | 28.8 ± 0.04 | 28.8 ± 0.04 | 28.8 ± 0.04 | 28.8 ± 0.04 |
| FBW (g) | 64.58 ± 3.09 $^{ab}$ | 62.99 ± 2.21 $^{ab}$ | 68.24 ± 2.44 $^{a}$ | 64.51 ± 1.45 $^{ab}$ | 59.82 ± 3.35 $^{b}$ | 64.04 ± 1.82 $^{ab}$ |
| WGR (%) | 123.99 ± 10.70 $^{ab}$ | 118.49 ± 7.65 $^{ab}$ | 136.74 ± 8.43 $^{a}$ | 123.77 ± 5.01 $^{ab}$ | 107.50 ± 11.63 $^{b}$ | 122.13 ± 6.32 $^{ab}$ |
| SGR (%day$^{-1}$) | 1.34 ± 0.08 $^{ab}$ | 1.30 ± 0.06 $^{ab}$ | 1.43 ± 0.06 $^{a}$ | 1.34 ± 0.04 $^{ab}$ | 1.21 ± 0.09 $^{b}$ | 1.33 ± 0.05 $^{ab}$ |
| FCR | 1.45 ± 0.14 | 1.37 ± 0.06 | 1.42 ± 0.11 | 1.29 ± 0.02 | 1.40 ± 0.05 | 1.40 ± 0.06 |
| CF (%) | 1.85 ± 0.03 $^{ab}$ | 1.84 ± 0.02 $^{ab}$ | 1.88 ± 0.03 $^{ab}$ | 1.93 ± 0.03 $^{a}$ | 1.84 ± 0.05 $^{ab}$ | 1.82 ± 0.03 $^{b}$ |
| VSI (%) | 5.57 ± 0.09 | 5.18 ± 0.17 | 5.37 ± 0.13 | 5.36 ± 0.17 | 5.31 ± 0.14 | 5.53 ± 0.11 |
| HSI (%) | 1.48 ± 0.07 | 1.38 ± 0.07 | 1.52 ± 0.07 | 1.52 ± 0.09 | 1.52 ± 0.09 | 1.54 ± 0.06 |
| SR (%) | 89.38 ± 2.77 | 90.63 ± 0.63 | 90.00 ± 5.20 | 91.88 ± 1.57 | 91.25 ± 2.39 | 88.75 ± 2.39 |

IBW: initial body weight; FBW: final body weight; WGR: weight gain rate; SGR: specific growth rate; FCR: feed conversion coefficient; CF: conditional factor; VSI: viscerosomatic index; HSI: hepatosomatic index; SR: survival rate. Means in the same raw with different superscripts are significantly different (mean ± SD; ANOVA, $p < 0.05$; $n = 4$). LEF0 (0 g kg$^{-1}$), LEF1 (10 g kg$^{-1}$), LEF2 (20 g kg$^{-1}$), LEF3 (30 g kg$^{-1}$), LEF4 (40 g kg$^{-1}$), LEF5 (50 g kg$^{-1}$).

### 3.2. Antioxidant Capacity

As shown in Figure 1, largemouth bass had the lowest liver T-AOC in the LEF3 group, with no discernible difference in liver T-AOC amongst the other groups ($p > 0.05$). The fish of LEF2 treatment had considerably higher liver CAT activity compared with others ($p < 0.05$). The MDA levels of the LEF0 treatment were significantly higher than other treatments ($p < 0.05$).

As is shown in Figure 2, no significant difference in intestinal T-AOC and SOD activity between all diets was found ($p > 0.05$). Compared with the LEF1 treatment, the activity of CAT was notably higher in the fish of the LEF2 and LEF5 groups ($p < 0.05$). However, gut MDA levels did not differ significantly across groups ($p > 0.05$).

### 3.3. Digestive Enzyme Activity

The liver AMS activity of the LEF1 group was considerably higher with the exception of the LEF3 group ($p < 0.05$). No discernible effect was found in the LPS activity of the liver among all treatments ($p > 0.05$). The LEF4 group had the highest trypsin activity, but there was no statistically significant difference in TRY activity between groups LEF0, LEF1, LEF2, LEF4 and LEF5 ($p > 0.05$, Figure 3).

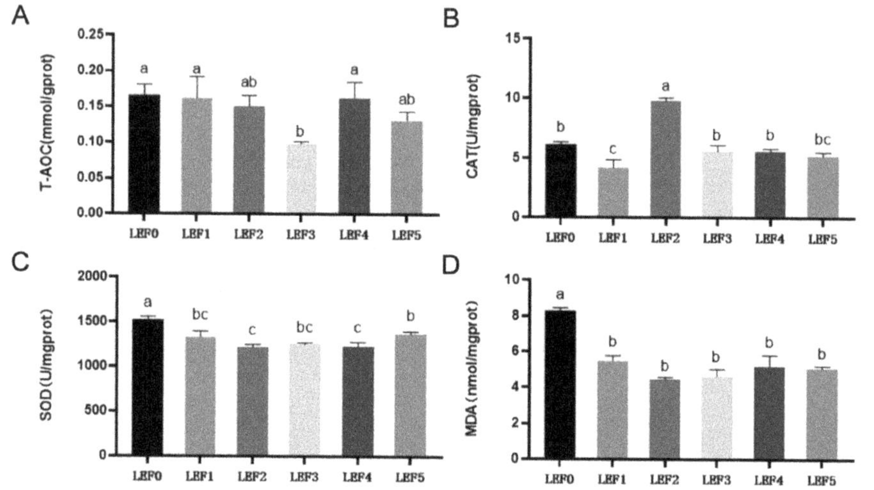


**Figure 1.** Effect of LEF on hepatic antioxidative activities of largemouth bass. Total antioxidant capacity (T-AOC, **A**), catalase (CAT, **B**), superoxide dismutase (SOD, **C**), malondialdehyde (MDA, **D**). Bars with different letters are significantly different (mean ± SD; ANOVA, $p < 0.05$; $n = 4$). LEF0 (0 g kg$^{-1}$), LEF1 (10 g kg$^{-1}$), LEF2 (20 g kg$^{-1}$), LEF3 (30 g kg$^{-1}$), LEF4 (40 g kg$^{-1}$), LEF5 (50 g kg$^{-1}$).

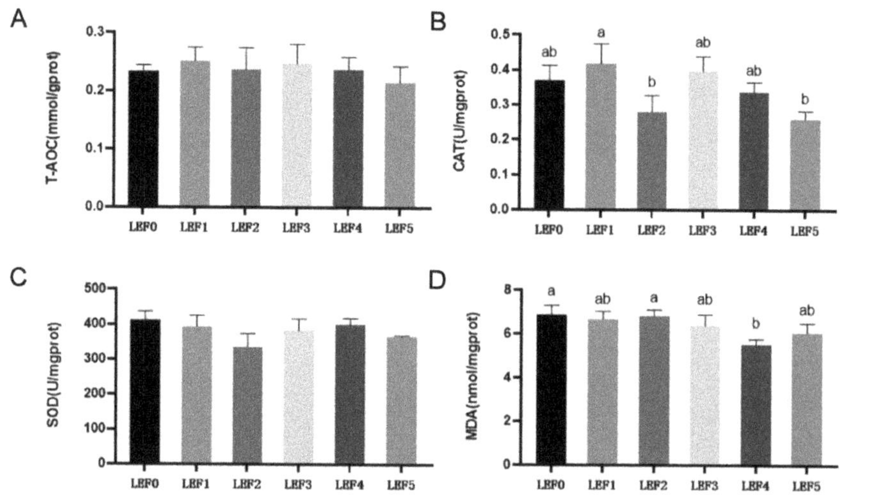


**Figure 2.** Effect of LEF on intestinal antioxidative activities of largemouth bass. Total antioxidant capacity (T-AOC, **A**), catalase (CAT, **B**), superoxide dismutase (SOD, **C**), malondialdehyde (MDA, **D**). Bars with different letters are significantly different (mean ± SD; ANOVA, $p < 0.05$; $n = 4$). LEF0 (0 g kg$^{-1}$), LEF1 (10 g kg$^{-1}$), LEF2 (20 g kg$^{-1}$), LEF3 (30 g kg$^{-1}$), LEF4 (40 g kg$^{-1}$), LEF5 (50 g kg$^{-1}$).

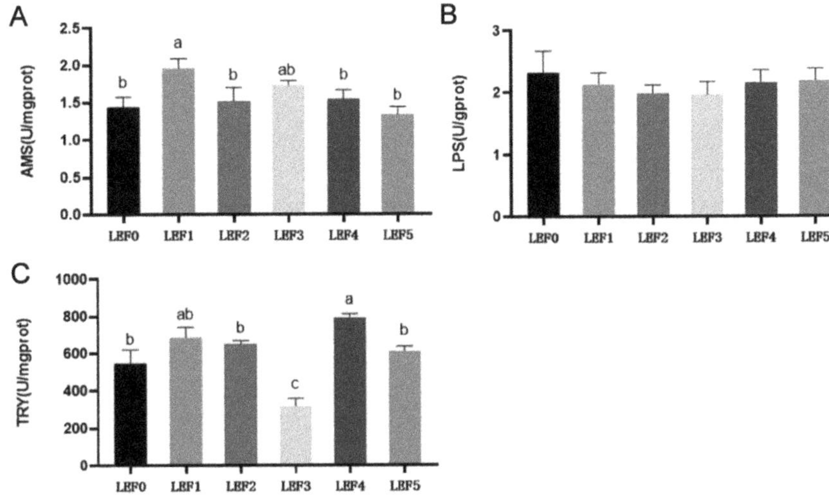


**Figure 3.** Effect of LEF on hepatic digestive activities of largemouth bass. Amylase (AMS, **A**), lipase (LPS, **B**), trypsin (TRY, **C**). Bars with different letters are significantly different (mean ± SD; ANOVA, $p < 0.05$; $n = 4$). LEF0 (0 g kg$^{-1}$), LEF1 (10 g kg$^{-1}$), LEF2 (20 g kg$^{-1}$), LEF3 (30 g kg$^{-1}$), LEF4 (40 g kg$^{-1}$), LEF5 (50 g kg$^{-1}$).

The intestinal AMS activity of groups LEF1, LEF3 and LEF4 increased significantly ($p > 0.05$). The intestinal LPS activity of group LEF0 was significantly higher than that of the other groups ($p < 0.05$); the activity of TRY in the intestine of largemouth bass in groups LEF0 and LEF1 had no statistically significant difference, but it was significantly higher than that of the other groups ($p < 0.05$, Figure 4).

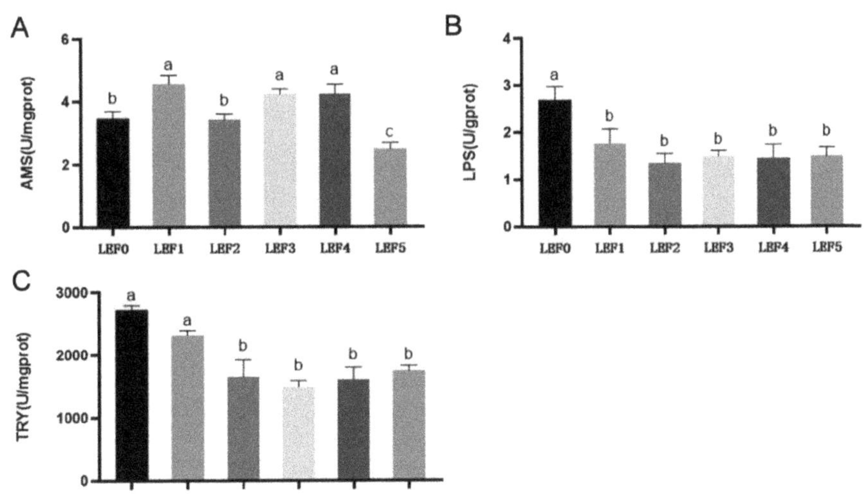


**Figure 4.** Effect of LEF on intestinal digestive activities of largemouth bass. Amylase (AMS, **A**), lipase (LPS, **B**), trypsin (TRY, **C**). Bars with different letters are significantly different (mean ± SD; ANOVA, $p < 0.05$; $n = 4$). LEF0 (0 g kg$^{-1}$), LEF1 (10 g kg$^{-1}$), LEF2 (20 g kg$^{-1}$), LEF3 (30 g kg$^{-1}$), LEF4 (40 g kg$^{-1}$), LEF5 (50 g kg$^{-1}$).

*3.4. Morphological Structure of Liver and Intestine*

The effects of LEF on the liver and intestinal morphology of largemouth bass were analyzed by H&E staining. The liver morphology is shown in Figure 5. In LEF0 treatment, widened hepatic sinus space, and leukocyte infiltration was observed, while in the LEF1, LEF2 and LEF3 groups, the situation improved. In the LEF5 group, however, leukocyte infiltration reoccurred.

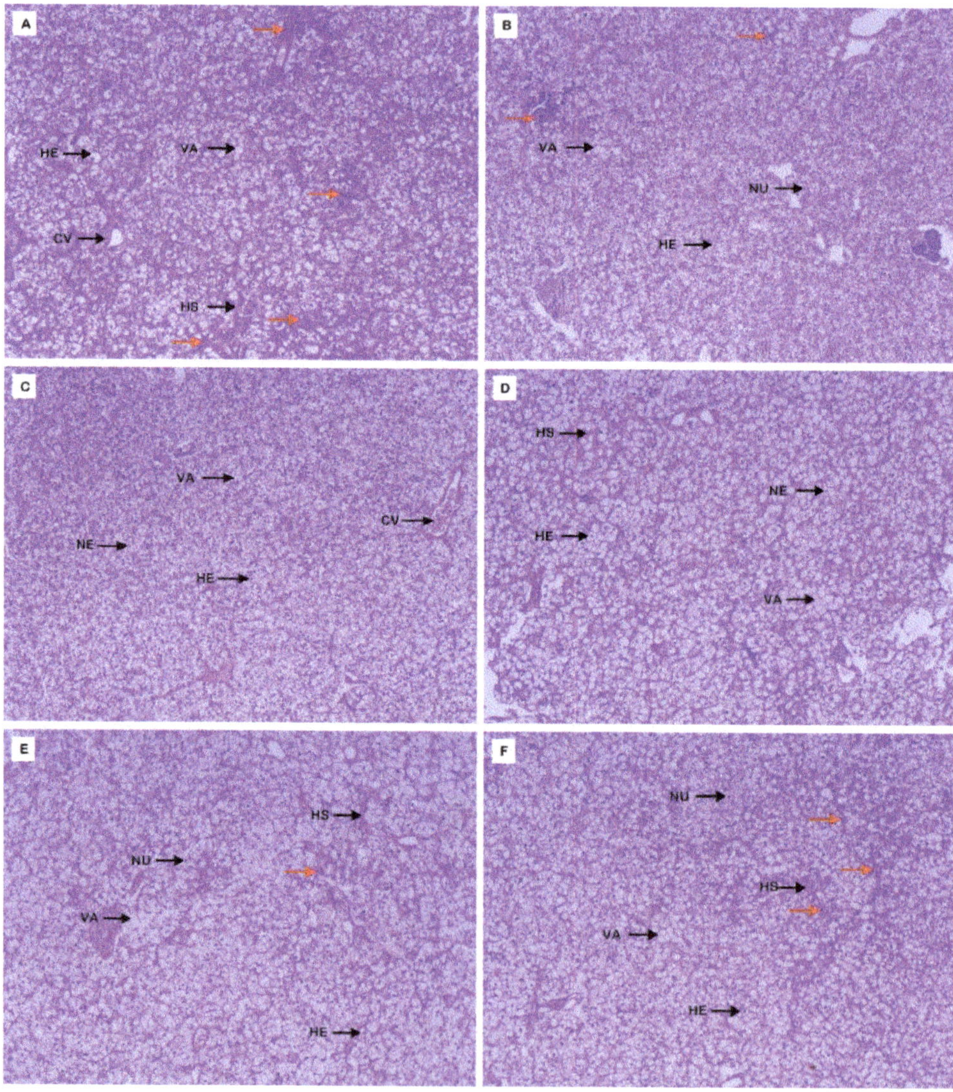


**Figure 5.** Effect of LEF on liver morphology of largemouth bass. (**A**): LEF0; (**B**): LEF1; (**C**): LEF2; (**D**): LEF3; (**E**): LEF4; (**F**): LEF5. (HE: hepatocytes; NU: nucleus; HS: hepatic sinusoid; CV: center vain; VA: vacuoles; Red Arrow: inflammatory infiltration). LEF0 (0 g $kg^{-1}$, **A**), LEF1 (10 g $kg^{-1}$, **B**), LEF2 (20 g $kg^{-1}$, **C**), LEF3 (30 g $kg^{-1}$, **D**), LEF4 (40 g $kg^{-1}$, **E**), LEF5 (50 g $kg^{-1}$, **F**).

Figure 6 depicts the morphological structure of the intestine. No significant pathological changes were observed in the midgut tissue in any of the dietary treatments, and no significant pathological changes were observed in the midgut group. The ultrastructure of the intestine is shown in Figure 7. There were microvilli closely arranged on the top of absorption cells in each group. There were abundant mitochondria and tubular systems near the top of the cells. The structure of mitochondria was normal, the ultrastructure of cells was normal, and the boundary was clear. It can be seen from Table 3 that there was no significant difference in the statistical results of the number of villi in each group ($p > 0.05$). Compared with the LEF5 group, the muscularis of fish in the LEF0, LEF1 and LEF3 groups decreased significantly ($p < 0.05$). In addition, the number of goblet cells and the height of the villus in the intestine of largemouth bass in the LEF5 group were also higher than those in other groups.

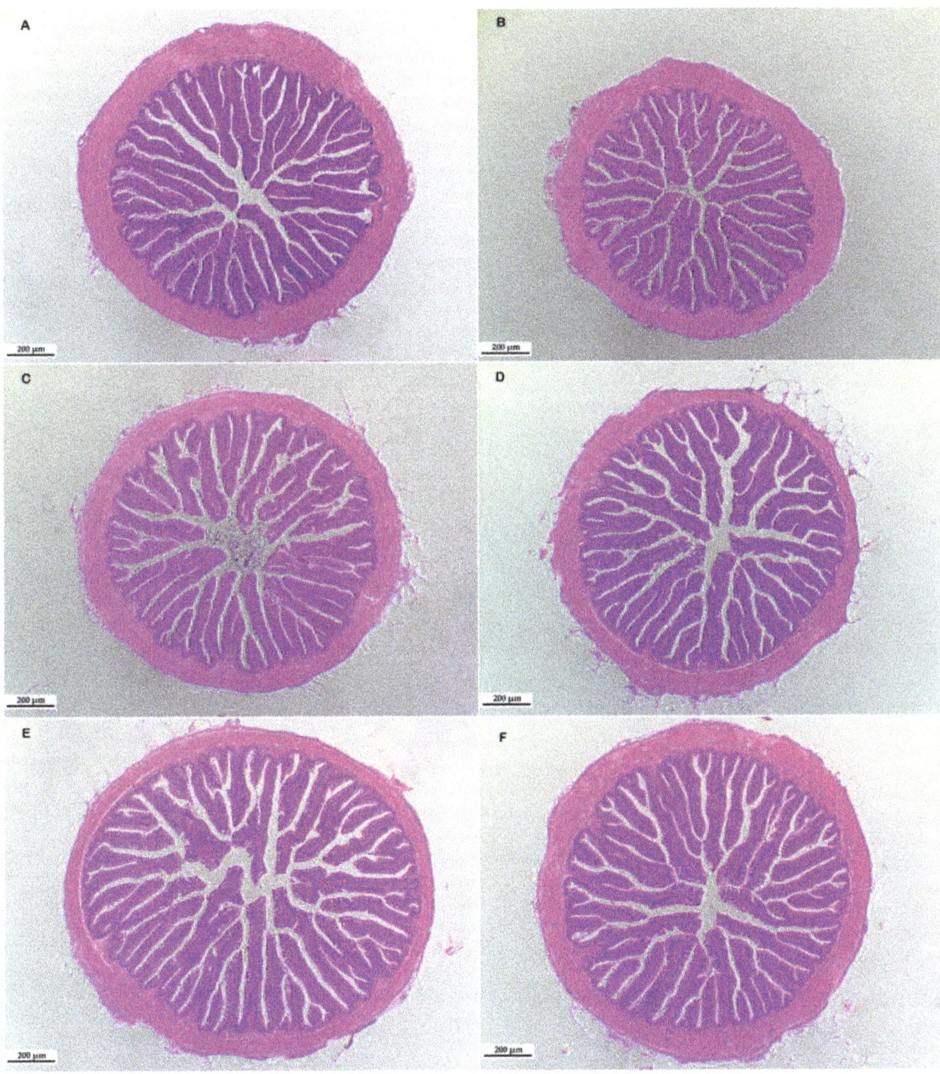


**Figure 6.** Effect of LEF on gut morphology of largemouth bass. LEF0 (0 g $kg^{-1}$, **A**), LEF1 (10 g $kg^{-1}$, **B**), LEF2 (20 g $kg^{-1}$, **C**), LEF3 (30 g $kg^{-1}$, **D**), LEF4 (40 g $kg^{-1}$, **E**), LEF5 (50 g $kg^{-1}$, **F**).

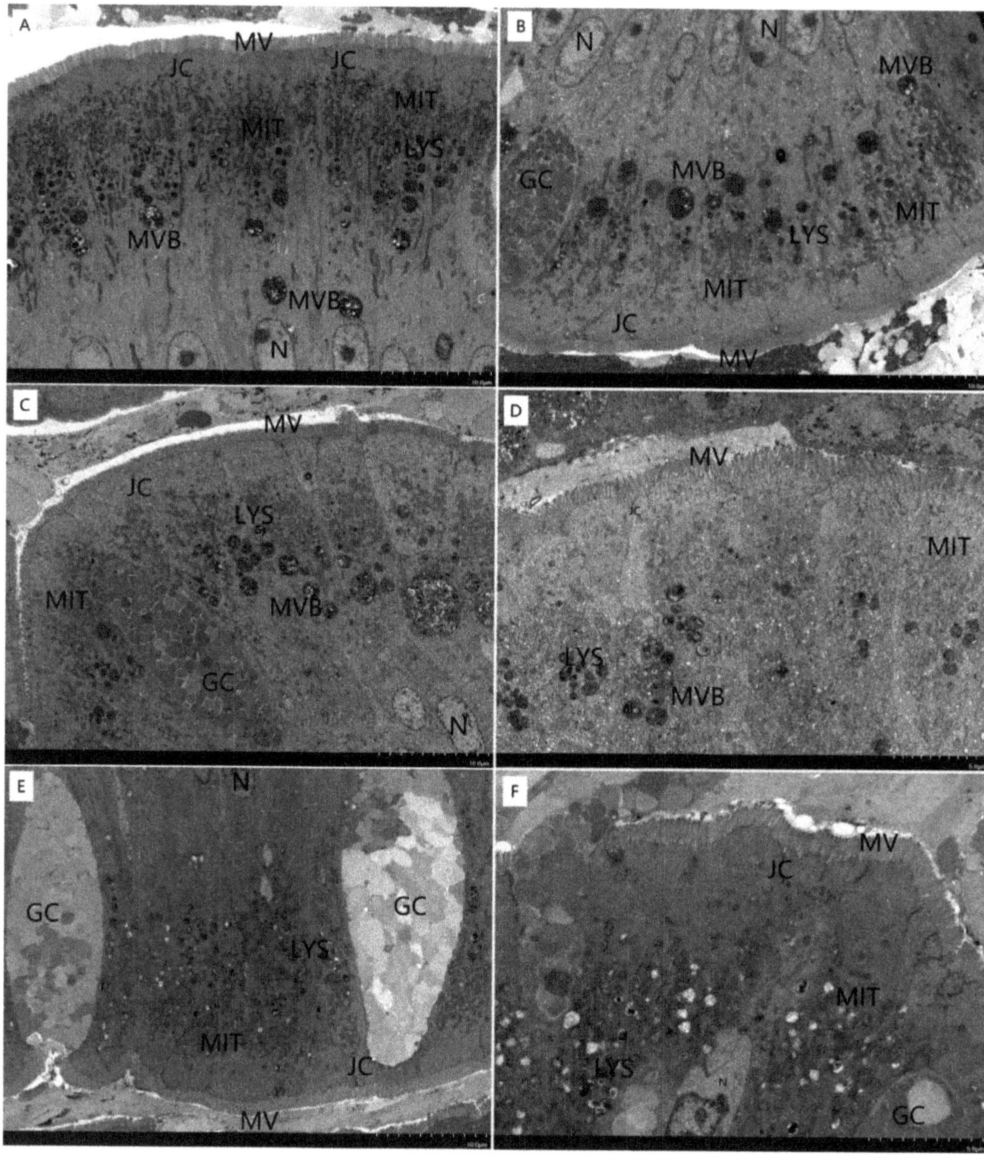


**Figure 7.** Effect of LEF on intestinal ultrastructure of largemouth bass. MV: micro villi; MVB: Multi-vesicle; MIT: Mitochondrial; GC: goblet cells; LYS: Lysosomal; N: Nucleus. LEF0 (0 g $kg^{-1}$, **A**), LEF1 (10 g $kg^{-1}$, **B**), LEF2 (20 g $kg^{-1}$, **C**), LEF3 (30 g $kg^{-1}$, **D**), LEF4 (40 g $kg^{-1}$, **E**), LEF5 (50 g $kg^{-1}$, **F**).

**Table 3.** Effect of LEF on intestinal morphology of largemouth bass.

| Indexes | LEF0 | LEF1 | LEF2 | LEF3 | LEF4 | LEF5 |
|---|---|---|---|---|---|---|
| Villus number | 27.75 ± 0.85 | 28.25 ± 1.97 | 29.00 ± 2.27 | 28.25 ± 1.71 | 29.75 ± 1.49 | 30.25 ± 0.62 |
| Muscularis (um) | 106.08 ± 4.12 $^{b}$ | 103.51 ± 8.93 $^{b}$ | 114.79 ± 7.27 $^{ab}$ | 104.18 ± 5.91 $^{b}$ | 116.48 ± 7.68 $^{ab}$ | 128.64 ± 5.15 $^{a}$ |
| Goblet cell number | 11.78 ± 1.42 $^{b}$ | 15.11 ± 1.23 $^{ab}$ | 13.89 ± 2.87 $^{b}$ | 14.72 ± 1.68 $^{ab}$ | 15.28 ± 2.252 $^{ab}$ | 20.25 ± 1.42 $^{a}$ |
| Villus length (um) | 613.75 ± 17.67 $^{ab}$ | 634.52 ± 30.79 $^{a}$ | 597.31 ± 24.78 $^{ab}$ | 583.57 ± 15.44 $^{ab}$ | 536.88 ± 32.68 $^{b}$ | 636.95 ± 23.53 $^{a}$ |

Means in the same raw with different superscripts are significantly different (mean ± SD; ANOVA, $p < 0.05$; $n = 4$). LEF0 (0 g $kg^{-1}$), LEF1 (10 g $kg^{-1}$), LEF2 (20 g $kg^{-1}$), LEF3 (30 g $kg^{-1}$), LEF4 (40 g $kg^{-1}$), LEF5 (50 g $kg^{-1}$).

### 3.5. Intestinal Microbiota

The OTU VENN diagram visually showed the number of unique and common OTUs in each group. The number of OTUs in each group was 185, 304, 143, 162, 342 and 131, respectively. The total number of OTUs was highest in group LEF4. The rarefaction curve indicated that each sample had reached a sufficiently deep sampling depth (Figure 8). The α-diversity indexes were investigated to explore the bacterial community diversity of all groups (Table 4). The result showed that the microbial richness of group LEF4 significantly increased ($p < 0.05$), but there were no significant differences in Chao1, ACE, Shannon and Simpsons indices among all groups ($p > 0.05$).

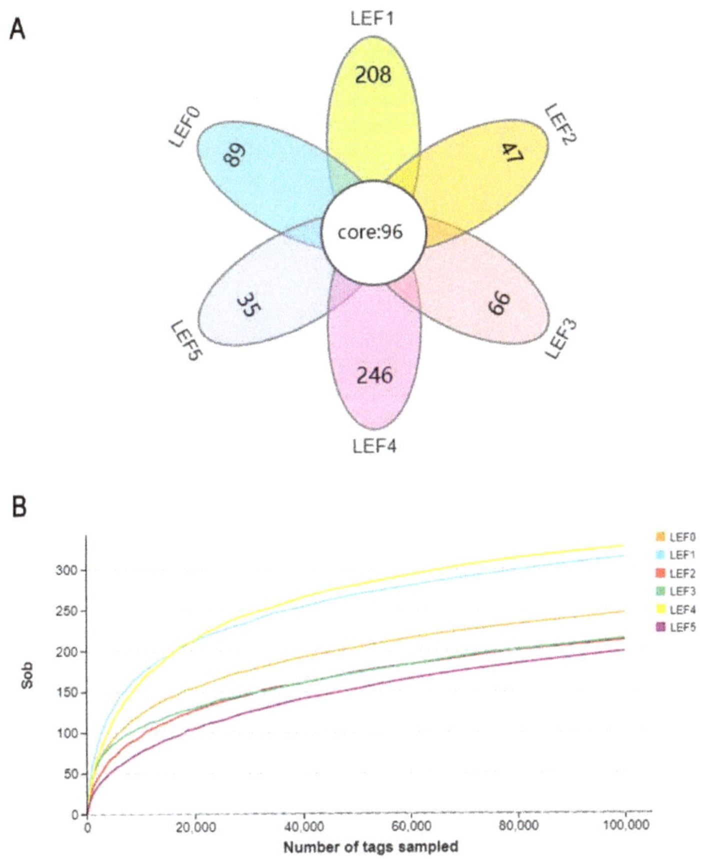


**Figure 8.** OTU VENN analysis and α Diversity dilution curve of gut microbiota in largemouth bass fed with different levels of LEF. (**A**): OTU Venn diagram; (**B**): rarefaction curve. LEF0 (0 g $kg^{-1}$), LEF1 (10 g $kg^{-1}$), LEF2 (20 g $kg^{-1}$), LEF3 (30 g $kg^{-1}$), LEF4 (40 g $kg^{-1}$), LEF5 (50 g $kg^{-1}$).

Principal coordinates analysis (PCoA) plots were generated from weighted-UniFrac distances to assess community composition (Figure 9). The contribution rates for the first and second principal components are 51.61% and 21.78%, respectively. The figure showed that there was no clear dissimilarity among groups LEF0, LEF1, LEF2, LEF4, LEF5, which meant that the composition of bacterial communities in these groups was similar. However, the samples of LEF3 were mainly centered in the first quadrants, while samples of other groups were near the second quadrant.

**Table 4.** Effect of LEF on intestinal microbiota α-diversity of largemouth bass.

| Diets | Sob | Chao1 | ACE | Shannon | Simpson | Goods-Coverage |
|---|---|---|---|---|---|---|
| LEF0 | 255.00 $^{abc}$ | 341.48 | 321.34 | 2.30 | 0.59 | 99.94 |
| LEF1 | 321.00 $^{ab}$ | 376.04 | 389.52 | 2.96 | 0.74 | 99.94 |
| LEF2 | 217.67 $^{bc}$ | 285.29 | 279.94 | 1.91 | 0.55 | 99.94 |
| LEF3 | 226.00 $^{abc}$ | 297.42 | 315.24 | 2.13 | 0.52 | 99.94 |
| LEF4 | 334.67 $^{a}$ | 396.91 | 381.50 | 2.14 | 0.53 | 99.94 |
| LEF5 | 207.00 $^{c}$ | 283.75 | 281.52 | 1.81 | 0.59 | 99.94 |

Means in the same raw with different superscripts are significantly different (mean ± SD; ANOVA, $p < 0.05$; $n = 4$). Sob: Number of species observed in sample, Chao1: The number of OTU in the community was estimated by Chao1 algorithm, and the value of Chao1 was positively correlated with the total number of species], ACE: An index used to estimate the number of OTUs in a community and the ACE values positively correlated with the richness of the microbial community, Shannon: The index takes into account the abundance and evenness of the community, and its value is positively correlated with the richness and evenness of the community, Simpson: The index was used to estimate microbial diversity in samples, and the values were negatively correlated with community diversity. LEF0 (0 g kg$^{-1}$), LEF1 (10 g kg$^{-1}$), LEF2 (20 g kg$^{-1}$), LEF3 (30 g kg$^{-1}$), LEF4 (40 g kg$^{-1}$), LEF5 (50 g kg$^{-1}$).

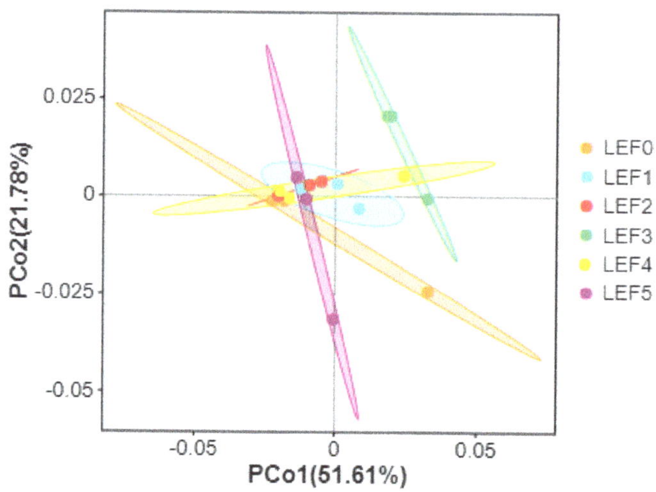


**Figure 9.** Principal coordinates analysis (PCoA) analysis based on weighted-unifrac method.

The OTUs were identified into 12 phyla. At the phylum level, *Fusobacteria* was the predominant microflora in group LEF0 (46.13%), LEF1 (36.43%), LEF2 (55.43%), LEF4 (53.31%), LEF5 (39.14%), followed by *Tenericutes*, *Firmicutes*, *Proteobacteria*, *Bacteroidetes*, while the predominant microflora of group LEF3 (64.25%) was *Tenericutes* (Table 5; Figure 10A). At the genus level, the percentage of *Clostridium* increased in groups LEF1, LEF2, LEF5, and the percentage of *Cetobacterium* also increased in groups LEF2, LEF4 (Figure 10B).

**Table 5.** Relative abundance of intestinal microbiota (Phylum) of largemouth bass.

| Phylum | LEF0 | LEF1 | LEF2 | LEF3 | LEF4 | LEF5 |
|---|---|---|---|---|---|---|
| Fusobacteria | 46.13 | 36.43 | 55.43 | 4.41 | 53.31 | 39.14 |
| Tenericutes | 6.21 | 21.34 | 16.53 | 64.25 | 22.95 | 11.27 |
| Firmicutes | 23.52 | 23.10 | 19.94 | 3.33 | 6.65 | 25.02 |
| Proteobacteria | 13.87 | 15.70 | 2.57 | 22.46 | 12.64 | 23.76 |
| Bacteroidetes | 7.69 | 2.28 | 4.65 | 3.58 | 3.11 | 0.34 |

LEF0 (0 g kg$^{-1}$), LEF1 (10 g kg$^{-1}$), LEF2 (20 g kg$^{-1}$), LEF3 (30 g kg$^{-1}$), LEF4 (40 g kg$^{-1}$), LEF5 (50 g kg$^{-1}$).

Tax4Fun functional prediction analysis showed that KEGG related to membrane transport, carbohydrate metabolism, amino acid metabolism, cofactor and vitamin metabolism, energy metabolism, nucleotide metabolism, signal transduction, transcription, replication and repair, and lipid metabolism were enriched in all treatments, and there was no significant difference between all groups ($p > 0.05$; Figure 11). Welch's $t$-test, on the other hand, revealed that certain predicted pathways between LEF0, LEF1, LEF3 and LEF4 were significantly different ($p < 0.05$). Several predicted pathways were significantly enriched in the LEF3 group versus the LEF0 and LEF1 groups ($p < 0.05$). In KEGG level 3 analysis, lysine degradation, fatty acid metabolism, geraniol degradation, tuberculosis and toluene degradation were significantly more abundant in group LEF3 compared with LEF0 ($p < 0.05$; Figure 11C); pyruvate metabolism, starch and sucrose metabolism and pentose phosphate pathway were more abundant in group LEF3 ($p < 0.05$; Figure 11D), while butanoate metabolism, thiamine metabolism enriched in group LEF1; pyruvate metabolism, starch and sucrose metabolism, lipopolysaccharide biosynthesis, thiamine metabolism enriched in group LEF2, while histidine metabolism, pyruvate metabolism, starch and sucrose metabolism were more abundant in group LEF3 ($p < 0.05$; Figure 11E); ribosome, lipopolysaccharide biosynthesis, histidine metabolism are more abundant in group LEF5 ($p < 0.05$; Figure 11F).

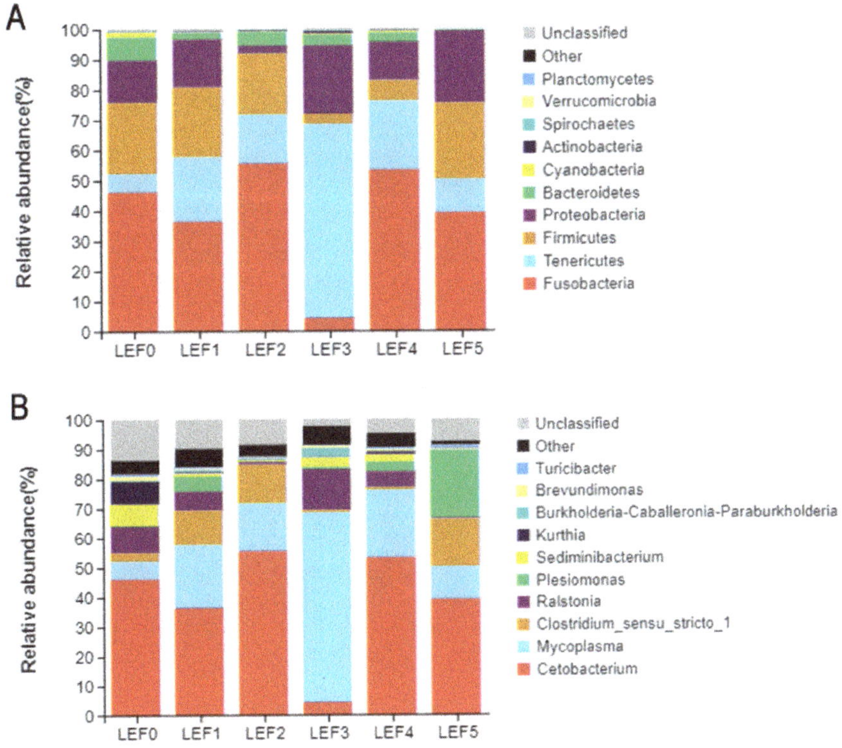


**Figure 10.** The relative abundance of bacteria in the intestine of largemouth bass. (**A**): phylum level; (**B**): genus level. LEF0 (0 g $kg^{-1}$), LEF1 (10 g $kg^{-1}$), LEF2 (20 g $kg^{-1}$), LEF3 (30 g $kg^{-1}$), LEF4 (40 g $kg^{-1}$), LEF5 (50 g $kg^{-1}$).

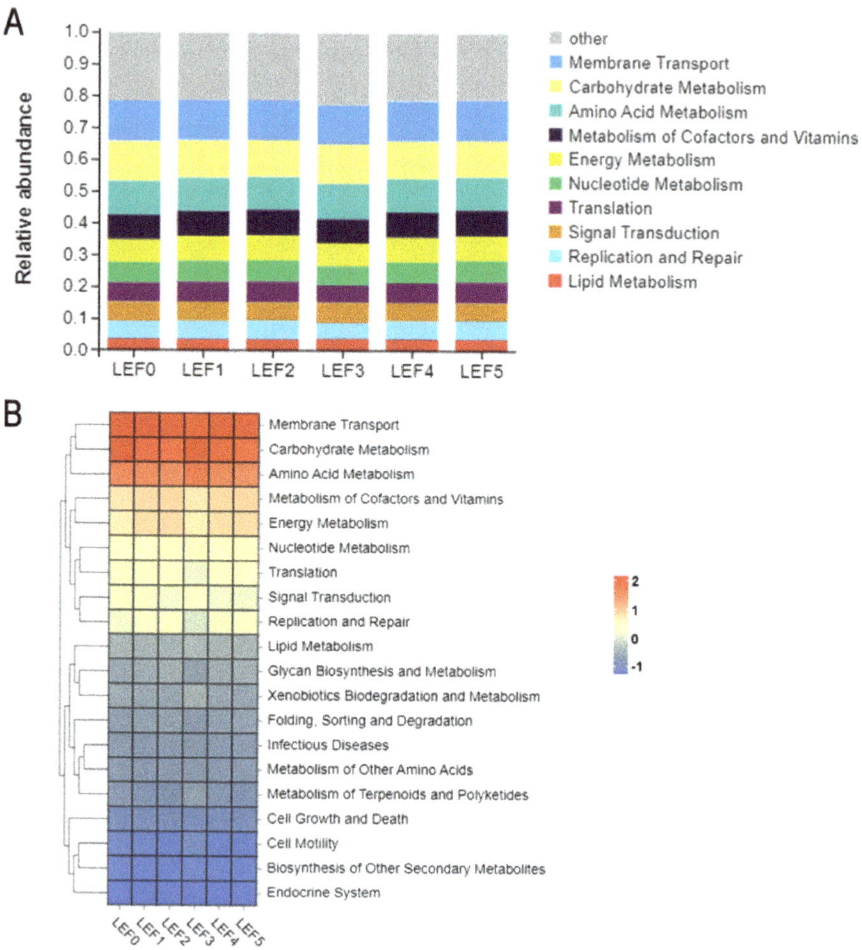


**Figure 11.** *Cont.*

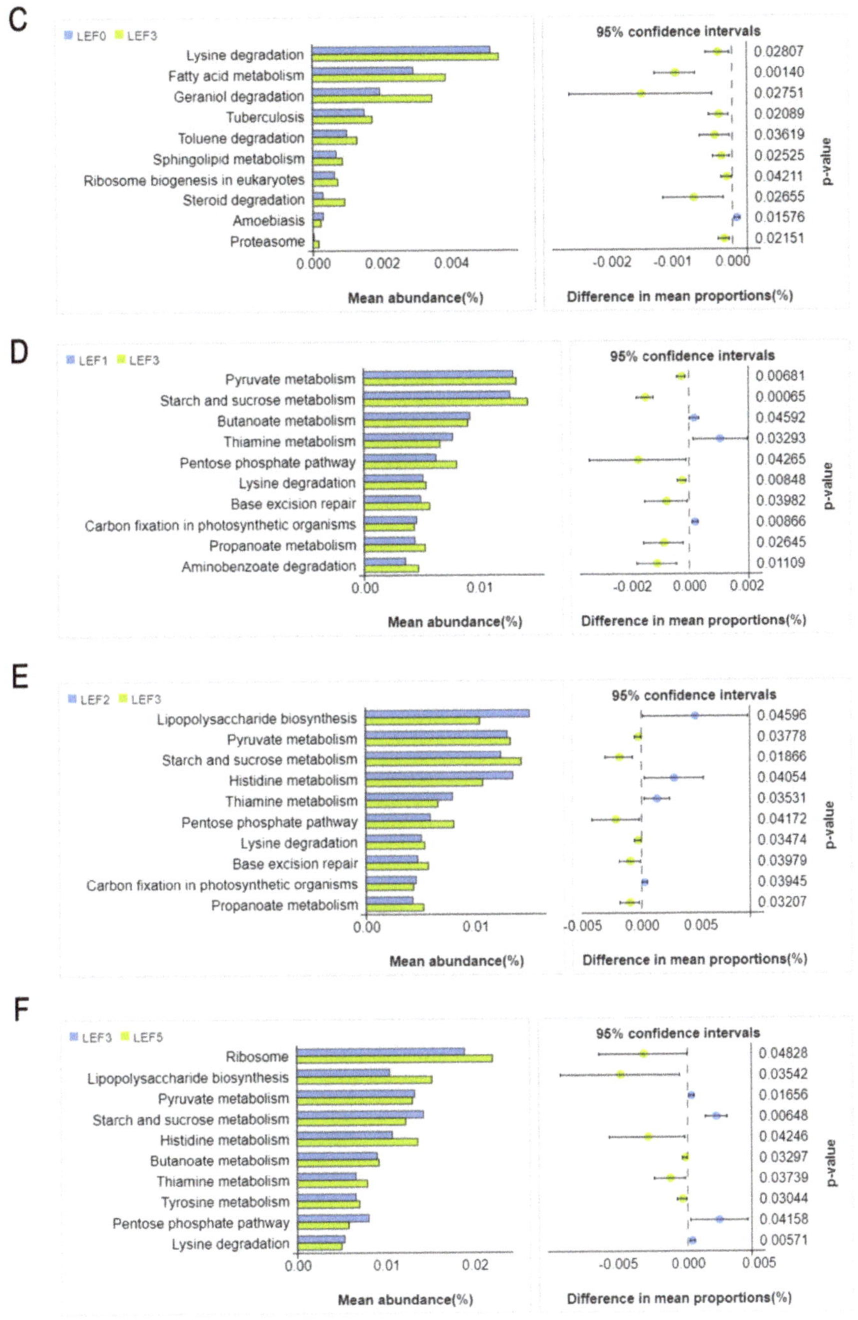


**Figure 11.** Tax4fun function prediction of gut microbiota in largemouth bass. (**A**): Relative enrichment of KEGG level 1; (**B**): Corresponding heatmaps for KEGG Level 1 pathways; (**C**–**F**): The top 10 pathways of predicted functions of intestinal microbiota in KEGG level 3. The Welch's *t*-test was used in groups LEF0, LEF1, LEF2, LEF3, LEF5 ($p < 0.05$). LEF0 (0 g $kg^{-1}$), LEF1 (10 g $kg^{-1}$), LEF2 (20 g $kg^{-1}$), LEF3 (30 g $kg^{-1}$), LEF4 (40 g $kg^{-1}$), LEF5 (50 g $kg^{-1}$).

## 4. Discussion

Adopting plant protein as a substitute for fishmeal to save cost has been used in aquaculture commonly [3–6], but this is detrimental to the growth performance of carnivorous fish such as largemouth bass. The inclusion of LEF did not significantly improve the growth performance of largemouth bass fed HPPD diets, according to the current findings. Furthermore, there was a significant decrease in the activity of TRY and LPS in the intestine. The three main digestive enzymes in the intestine are AMS, LPS, and TRY, and measuring the activity of these enzymes can provide an indirect view of nutrient digestion and absorption in the intestine [26,27]. In this trial, HPPD was used as the basal diet for largemouth bass, therefore, the above findings may be related to factors such as poor palatability and anti-nutritional factors in HPPD. Amiri et al. [28] found no improvement in Rainbow Trout (Oncorhynchus mykiss) growth parameters compared to fish that fed non-supplemental feed after 8 weeks feeding of White Button Mushroom (*Agaricus bisporus*) Powder (WBMP). Similarly, a study on holothurian (*Apostichopus japonicus*) also showed that there was no significant effect on weight gain and specific growth rate after dietary administration of *Pleurotus ostreatus* polysaccharides (POPS) [29]. In contrast, Mohan et al. [30] found that dietary Ganoderma lucidum polysaccharides (GLPs) at 2.5 g $kg^{-1}$ had a positive effect on the growth of freshwater shrimp. Additionally, Doan et al. [31] reported that SGR, WG, FW and FCR were remarkably improved in spent mushroom (*cordyceps militaris*) substrate treatment groups. Dietary mushrooms also have a positive effect on the growth performance of piglets. Research on the effect of dietary spent mushroom (*cordyceps militaris*) on the growth performance of weaning pigs showed that spent mushrooms at 1.5 g $kg^{-1}$ improved the body weight, average daily weight gain and average daily feed [32]. Another, research on the impacts of fermented spent mushroom substrates on piglets also received similar results [33]. These results indicated that dietary supplementation of mushrooms had different effects on the growth performance of different animals, which may be related to the species, processing method and dosage of mushrooms.

The overproduction of free radicals can produce adverse reactions in different tissues [34–36]. The liver is one of the susceptible tissues to oxidative stress [37]. Supplementation of LEF in the present study significantly reduced the hepatic MDA level. Previous research discovered that supplementing *Agaricus bisporus* mushrooms increased GSH-Px, GR and GST activity while decreasing MDA production in turkey poult liver [38]. Liu et al. [39] also reported that the application of the deep root mushroom reduced the lipid peroxidation of the liver in mice. The CAT is an important antioxidant enzyme that removes the reactive oxygen species (ROS) intermediate $H_2O_2$ and protects the body from ROS [40–42]. In the present study, significantly increased CAT activity was observed in the LEF2 group. This result agreed with Liu et al. [39] and Mahfuz et al. [43], who found that the CAT activity of animal liver increased after the administration of mushrooms. Furthermore, a significant improvement in liver morphology was also observed in the LEF2 treatment. This study showed that the hepatic sinusoidal space decreased and the infiltration of macrophages and neutrophils reduced, which indicated an improvement in the liver morphology of group LEF2. Drori et al. [44] also revealed a corresponding improvement of the immune-mediated liver injury of mice after feeding vitamin D-rich mushroom extract. In addition, Kuang et al. [45] reported that the injury degree of the liver in mice with gut-origin sepsis was improved with the increase in lentinan concentration. In addition, it was found that lentinan sulfate polysaccharide can improve the multiple organ dysfunction syndrome induced by yeast polysaccharides in mice and reduce the vacuolization of mouse liver 46]. The improvement in liver morphology can be attributed to the bioactive compounds in mushrooms such as B vitamins, minerals, polysaccharides, lentinan and so on, which have the effect of liver protection and enhancing the antioxidant capacity of the liver [43,46–54]. The present study showed that LEF can reduce hepatic MDA levels and enhance the ability of hepatocytes to remove hydrogen peroxide.

The composition of gut microbial communities is closely related to the health of the host. Parasitic microbial communities in the host are involved in many physiological processes [55]. Many reports suggested that mushroom supplementation in diets changed the intestinal morphology and the composition of the microbiota. Nevel et al. [23] found that supplementation with *Lentinus edodes* affects the composition of hindgut microorganisms and change the intestinal morphology of piglet. Giannenas et al. [56] also found that *Lactobacilli* spp. were higher in the ileum of broiler chickens fed a diet containing *Agaricus bisporus*. Another study by Giannenas showed that mucosal architecture was influenced by mushroom consumption in terms of villus height [57]. Similar to the above findings, the present study found that the relative abundance of *Clostridium*, a member of *Firmicutes*, significantly increased in groups LEF1, LEF2 and LEF5. Besides, the muscularis thickness and goblet cell number increased significantly in group LEF5 compared to the control. Similar findings had also been reported for Rainbow Trout (*Oncorhynchus mykiss*), crude lentinan significantly promoted the growth of short-chain fatty-acid-producing bacteria. In addition, Yang et al. [51] claimed that the application of lentinan supplementation can reduce *Proteobacteria* and increase the abundance of Actinomycetes to improve intestinal microflora imbalance in mice. In the gut, *Clostridium* digests polysaccharides, producing many short-chain fatty acids (SCFA) that are beneficial to animal health [58]. SCFA, and specifically butyrate, play an important role in mediating the effects of the gut microbiome on local and systemic immunity [59]. Alves jesus et al. [60] reported that adding sodium butyrate to the diet significantly reduced liver lymphocyte infiltration and improved intestinal integrity in tilapia. Similarly, the villus height (VH) to crypt depth (CD) ratio (VH:CD ratio) of the ileum were higher in piglet fed with butyrate [61]. Therefore, this study suggested that LEF may have a potential prebiotic effect and that the bioactive compounds in LEF may be fermented by gut microbes to produce short-chain fatty acids by substrate fermentation, thus affecting the composition of gut microbial communities, and this change ultimately affects the morphology of the intestine [62–65].

## 5. Conclusions

In conclusion, this study provided the experimental basis for HPPD to develop mushroom feed additives. Dietary LEF supplementation can improve antioxidant activity in the liver and gut of largemouth bass, relieve liver inflammation, improve intestinal tissue morphology, and ultimately have a positive effect on the health of largemouth bass. Dietary LEF is recommended at required at 20 g $kg^{-1}$.

**Author Contributions:** Conceptualization, J.X. and G.L.; methodology, J.X., Z.Y., G.L., W.B. and S.L.; validation, G.Z., H.W. and Z.Y.; investigation, J.X., X.T., G.L., S.L., Y.D. and C.Y.; data curation, J.X.; writing—original draft preparation, J.X. and Z.Y.; writing—review and editing, J.X., X.T. and W.B.; visualization, J.X.; supervision and project administration, Y.H., X.T and M.Z.; funding acquisition, M.Z., X.T. and Y.H. All authors have read and agreed to the published version of the manuscript.

**Funding:** The research was supported by Research, development and application of quality assurance technology for Lingnan aquatic products prefabricated dishes (2022ZDJS022), Key Realm R&D Program of GuangDong Province (2021B0202050002), Innovative Team Projects of Ordinary Colleges and Universities in Guangdong Province (2020KCXTD019), Provincial Agricultural Science and Technology Development and Resources Environmental Protection Management Projects (2022KJ115) and Guangdong Maoming Binhai New Area Marine Fishery Industrial Park Project (0835-220FA8102621).

**Institutional Review Board Statement:** The experimental design was approved by the Animal Care Committee of Zhongkai University of Agriculture and Engineering (Ethical approval number: 20210726).

**Informed Consent Statement:** Not applicable, as there is no human involved in this study.

**Data Availability Statement:** The datasets of the current study are available from the corresponding author upon reasonable request.

**Conflicts of Interest:** The authors declare no conflict of interest.

## References


1. He, M.; Yu, Y.; Li, X.; Poolsawat, L.; Yang, P.; Bian, Y.; Guo, Z.; Leng, X. An evaluation of replacing fish meal with fermented soybean meal in the diets of largemouth bass (*Micropterus salmoides*): Growth, nutrition utilization and intestinal histology. *Aquac. Res.* **2020**, *51*, 4302–4314. [CrossRef]
2. Ma, Z.; Hassan, M.; Allais, L.; He, T.; Leterme, S.; Ellis, A.V.; McGraw, B.; Qin, J.G. Replacement of fishmeal with commercial soybean meal and EnzoMeal in juvenile barramundi *Lates calcarifer*. *Aquac. Res.* **2018**, *49*, 3258–3269. [CrossRef]
3. Jannathulla, R.; Rajaram, V.; Kalanjiam, R.; Ambasankar, K.; Muralidhar, M.; Dayal, J.S. Fishmeal availability in the scenarios of climate change: Inevitability of fishmeal replacement in aquafeeds and approaches for the utilization of plant protein sources. *Aquac. Res.* **2019**, *50*, 3493–3506. [CrossRef]
4. Kobayashi, M.; Msangi, S.; Batka, M.; Vannuccini, S.; Dey, M.M.; Anderson, J.L. Fish to 2030: The role and opportunity for aquaculture. *Aquac. Econ. Manag.* **2015**, *19*, 282–300. [CrossRef]
5. Dawood, M.A.O.; Koshio, S. Application of fermentation strategy in aquafeed for sustainable aquaculture. *Rev. Aquac.* **2020**, *12*, 987–1002. [CrossRef]
6. Jha, R.; Fouhse, J.M.; Tiwari, U.P.; Li, L.; Willing, B.P. Dietary Fiber and Intestinal Health of Monogastric Animals. *Front. Vet. Sci.* **2019**, *6*, 48. [CrossRef]
7. Parma, L.; Candela, M.; Soverini, M.; Turroni, S.; Consolandi, C.; Brigidi, P.; Mandrioli, L.; Sirri, R.; Fontanillas, R.; Gatta, P.P.; et al. Next-generation sequencing characterization of the gut bacterial community of gilthead sea bream (*Sparus aurata*, L.) fed low fishmeal based diets with increasing soybean meal levels. *Anim. Feed. Sci. Technol.* **2016**, *222*, 204–216. [CrossRef]
8. Liang, X.; Han, J.; Xue, M.; Yu, H.; Huang, H.; Wu, X.; Zheng, Y.; Qin, Y.; Liang, X. Growth and feed intake regulation responses to anorexia, adaptation and fasting in Japanese seabss, Lateolabrax japonicas when fishmeal is totally replaced by plant protein. *Aquaculture* **2019**, *498*, 528–538. [CrossRef]
9. Zhu, R.; Duan, J.; Li, L.; Li, M.; Yu, Z.; Wang, H.; Quan, Y.; Wu, L. Effects of Substituting Fish Meal with Two Types of Soybean Meal on Growth, Enzyme Activities, and Intestinal Morphology in Juvenile *Rhynchocypris lagowskii*. *N. Am. J. Aquac.* **2021**, *83*, 267–282. [CrossRef]
10. Wang, J.; Mai, K.; Ai, Q. Conventional soybean meal as fishmeal alternative in diets of Japanese Seabass (*Lateolabrax japonicus*): Effects of functional additives on growth, immunity, antioxidant capacity and disease resistance. *Antioxidants* **2022**, *11*, 951. [CrossRef] [PubMed]
11. Novriadi, R.; Rhodes, M.; Powell, M.; Hanson, T.; Davis, D.A. Effects of soybean meal replacement with fermented soybean meal on growth, serum biochemistry and morphological condition of liver and distal intestine of Florida pompano Trachinotus carolinus. *Aquac. Nutr.* **2018**, *24*, 1066–1075. [CrossRef]
12. Liu, H.; Jin, J.; Zhu, X.; Han, D.; Yang, Y.; Xie, S. Effect of substitution of dietary fish meal by soybean meal on different sizes of gibel carp (*Carassius auratus gibelio*): Digestive enzyme gene expressions and activities, and intestinal and hepatic histology. *Aquac. Nutr.* **2017**, *23*, 129–147. [CrossRef]
13. Huang, D.; Wu, Y.; Lin, Y.; Chen, J.; Karrow, N.; Ren, X.; Wang, Y. Dietary Protein and Lipid Requirements for Juvenile Largemouth Bass, *Micropterus salmoides*. *J. World Aquac. Soc.* **2017**, *48*, 782–790. [CrossRef]
14. Sampaio-Oliveira, A.; Cyrino, J. Digestibility of plant protein-based diets by largemouth bass *Micropterus salmoides*. *Aquac. Nutr.* **2008**, *14*, 318–323. [CrossRef]
15. Ma, D.; Fan, J.; Tian, Y.; Jiang, P.; Wang, J.; Zhu, H.; Bai, J. Selection of reference genes for quantitative real-time PCR normalisation in largemouth bass *Micropterus salmoides* fed on alternative diets. *J. Fish Biol.* **2019**, *95*, 393–400. [CrossRef] [PubMed]
16. Amoah, A.; Coyle, S.D.; Webster, C.D.; Durborow, R.M.; Bright, L.A.; Tidwell, J.H. Effects of Graded Levels of Carbohydrate on Growth and Survival of Largemouth Bass, Micropterus salmoides. *J. World Aquac. Soc.* **2008**, *39*, 397–405. [CrossRef]
17. Li, X.; Zheng, S.; Ma, X.; Cheng, K.; Wu, G. Use of alternative protein sources for fishmeal replacement in the diet of largemouth bass (*Micropterus salmoides*). Part I: Effects of poultry by-product meal and soybean meal on growth, feed utilization, and health. *Amino Acids* **2021**, *53*, 33–47. [CrossRef]
18. Li, X.; Zheng, S.; Cheng, K.; Ma, X.; Wu, G. Use of alternative protein sources for fishmeal replacement in the diet of largemouth bass (*Micropterus salmoides*). Part II: Effects of supplementation with methionine or taurine on growth, feed utilization, and health. *Amino Acids* **2021**, *53*, 49–62. [CrossRef]
19. Rossi, W., Jr.; Allen, K.M.; Habte-Tsion, H.-M.; Meesala, K.-M. Supplementation of glycine, prebiotic, and nucleotides in soybean meal-based diets for largemouth bass (*Micropterus salmoides*): Effects on production performance, whole-body nutrient composition and retention, and intestinal histopathology. *Aquaculture* **2021**, *532*, 736031. [CrossRef]
20. Gaitán-Hernández, R.; López-Peña, D.; Esqueda, M.; Gutiérrez, A. Review of Bioactive Molecules Production, Biomass, and Basidiomata of Shiitake Culinary-Medicinal Mushrooms, *Lentinus edodes* (*Agaricomycetes*). *Int. J. Med. Mushrooms* **2019**, *21*, 841–850. [CrossRef]
21. Bisen, P.; Baghel, R.; Sanodiya, B.; Thakur, G.; Prasad, G. *Lentinus edodes*: A Macrofungus with Pharmacological Activities. *Curr. Med. Chem.* **2010**, *17*, 2419–2430. [CrossRef] [PubMed]
22. Conway, E.; Sweeney, T.; Dowley, A.; Maher, S.; Rajauria, G.; Yadav, S.; Wilson, J.; Gabrielli, W.; O'doherty, J.V. The effects of mushroom powder and vitamin $D_2$-enriched mushroom powder supplementation on the growth performance and health of newly weaned pigs. *J. Anim. Physiol. Anim. Nutr.* **2022**, *106*, 517–527. [CrossRef]

23. Van Nevel, C.J.; Decuypere, J.A.; Dierick, N.; Molly, K. The influence of *Lentinus edodes* (*Shiitake mushroom*) preparations on bacteriological and morphological aspects of the small intestine in piglets. *Arch. Anim. Nutr.* **2003**, *57*, 399–412. [CrossRef] [PubMed]
24. Chen, H.; Li, D.; Chang, B.; Gong, L.; Piao, X.; Yi, G.; Zhang, J. Effects of lentinan on broiler splenocyte proliferation, interleukin-2 production, and signal transduction. *Poult. Sci.* **2003**, *82*, 760–766. [CrossRef]
25. Dong, L.; Zhang, H.; Gao, Y.; Diao, Q. Dynamic profiles of fermentation characteristics and bacterial community composition of Broussonetia papyrifera ensiled with perennial ryegrass. *Bioresour. Technol.* **2020**, *310*, 123396. [CrossRef]
26. Spim, S.R.V.; de Oliveira, B.G.C.C.; Leite, F.G.; Gerenutti, M.; Grotto, D. Effects of Lentinula edodes consumption on biochemical, hematologic and oxidative stress parameters in rats receiving high-fat diet. *Eur. J. Nutr.* **2017**, *56*, 2255–2264. [CrossRef] [PubMed]
27. Cahu, C.; Rønnestad, I.; Grangier, V.; Infante, J.Z. Expression and activities of pancreatic enzymes in developing sea bass larvae (*Dicentrarchus labrax*) in relation to intact and hydrolyzed dietary protein; involvement of cholecystokinin. *Aquaculture* **2004**, *238*, 295–308. [CrossRef]
28. Amiri, O.; Miandare, H.K.; Hoseinifar, S.H.; Shabni, A.; Safari, R. Skin mucus protein profile, immune parameters, immune-related gene expression, and growth performance of rainbow rrout (*Oncorhynchus mykiss*) fed white button mushroom (*Agaricus bisporus*) powder. *Int. J. Med. Mushrooms* **2018**, *20*, 337–347. [CrossRef]
29. Song, X.; Feng, Z.; Tan, J.; Wang, Z.; Zhu, W. Dietary administration of Pleurotus ostreatus polysaccharides (POPS) modulates the non-specific immune response and gut microbiota diversity of Apostichopus japonicus. *Aquac. Rep.* **2021**, *19*, 100578. [CrossRef]
30. Chang, S.-C.; Lin, M.-J.; Chao, Y.-P.; Chiang, C.-J.; Jea, Y.-S.; Lee, T.-T. Effects of spent mushroom compost meal on growth performance and meat characteristics of grower geese. *Rev. Bras. de Zootec.* **2016**, *45*, 281–287. [CrossRef]
31. Mohan, K.; Padmanaban, A.; Uthayakumar, V.; Chandirasekar, R.; Muralisankar, T.; Santhanam, P. Effect of dietary Ganoderma lucidum polysaccharides on biological and physiological responses of the giant freshwater prawn Macrobrachium rosenbergii. *Aquaculture* **2016**, *464*, 42–49. [CrossRef]
32. Van Doan, H.; Hoseinifar, S.H.; Tapingkae, W.; Chitmanat, C.; Mekchay, S. Effects of Cordyceps militaris spent mushroom substrate on mucosal and serum immune parameters, disease resistance and growth performance of Nile tilapia, (*Oreochromis niloticus*). *Fish Shellfish. Immunol.* **2017**, *67*, 78–85. [CrossRef]
33. Boontiam, W.; Wachirapakorn, C.; Phaengphairee, P.; Wattanachai, S. Effect of spent mushroom (*Cordyceps militaris*) on growth performance, immunity, and intestinal microflora in weaning pigs. *Animals* **2020**, *10*, 2360. [CrossRef] [PubMed]
34. Qi, Q.; Peng, Q.; Tang, M.; Chen, D.; Zhang, H. Microbiome Analysis Investigating the Impacts of Fermented Spent Mushroom Substrates on the Composition of Microbiota in Weaned Piglets Hindgut. *Front. Veter-Sci.* **2020**, *7*, 584243. [CrossRef] [PubMed]
35. Klein, R.D.; Borges, V.D.; Rosa, C.E.; Colares, E.P.; Robaldo, R.B.; Martinez, P.E.; Bianchini, A. Effects of increasing temperature on antioxidant defense system and oxidative stress parameters in the Antarctic fish Notothenia coriiceps and Notothenia rossii. *J. Therm. Biol.* **2017**, *68*, 110–118. [CrossRef] [PubMed]
36. Furukawa, S.; Fujita, T.; Shimabukuro, M.; Iwaki, M.; Yamada, Y.; Nakajima, Y.; Nakayama, O.; Makishima, M.; Matsuda, M.; Shimomura, I. Increased oxidative stress in obesity and its impact on metabolic syndrome. *J. Clin. Investig.* **2017**, *114*, 1752–1761. [CrossRef]
37. Bosoi, C.R.; Rose, C.F. Oxidative stress: A systemic factor implicated in the pathogenesis of hepatic encephalopathy. *Metab. Brain Dis.* **2013**, *28*, 175–178. [CrossRef]
38. Cichoż-Lach, H.; Michalak, A. Oxidative stress as a crucial factor in liver diseases. *World J. Gastroenterol. WJG* **2014**, *20*, 8082. [CrossRef]
39. Liu, H.; Yang, Q.; Gao, Z.; Zhu, Y.; Zhang, J.; Jia, L. The Antioxidative, Anti-inflammatory, and Liver-Protective Effects of Mycelia Selenium Polysaccharides from the Deep Root Mushroom, *Oudemansiella radicata* (Agaricomycetes). *Int. J. Med. Mushrooms* **2019**, *21*, 1019–1031. [CrossRef]
40. Miller, A.-F. Superoxide dismutases: Active sites that save, but a protein that kills. *Curr. Opin. Chem. Biol.* **2004**, *8*, 162–168. [CrossRef]
41. Ezraty, B.; Gennaris, A.; Barras, F.; Collet, J.F. Oxidative stress, protein damage and repair in bacteria. *Nat. Rev. Microbiol.* **2017**, *15*, 385–396. [CrossRef] [PubMed]
42. Kirkman, H.N.; Gaetani, G.F. Mammalian catalase: A venerable enzyme with new mysteries. *Trends Biochem. Sci.* **2007**, *32*, 44–50. [CrossRef] [PubMed]
43. Mahfuz, S.; He, T.; Ma, J.; Liu, H.; Long, S.; Shang, Q.; Zhang, L.; Yin, J.; Piao, X. Mushroom (*Flammulina velutipes*) stem residue on growth performance, meat quality, antioxidant status and lipid metabolism of broilers. *Ital. J. Anim. Sci.* **2020**, *19*, 803–812. [CrossRef]
44. Drori, A.; Shabat, Y.; Ben Ya'Acov, A.; Danay, O.; Levanon, D.; Zolotarov, L.; Ilan, Y. Extracts from *Lentinula edodes* (Shiitake) Edible Mushrooms Enriched with Vitamin D Exert an Anti-Inflammatory Hepatoprotective Effect. *J. Med. Food* **2016**, *19*, 383–389. [CrossRef]
45. Kuang, Z.; Jin, T.; Wu, C.; Zong, Y.; Yin, P.; Dong, W.; Lin, X.; You, W.; Zhang, C.; Wang, L.; et al. Lentinan Attenuates Damage of the Small Intestinal Mucosa, Liver, and Lung in Mice with Gut-Origin Sepsis. *J. Immunol. Res.* **2021**, *2021*, 2052757. [CrossRef]
46. Sun, W.; Feng, Y.; Zhang, M.; Song, X.; Jia, L. Protective effects of sulfated polysaccharides from *Lentinula edodes* on the lung and liver of MODS mice. *Food Funct.* **2021**, *12*, 6389–6402. [CrossRef] [PubMed]

47. Hamza, R.G.; El Shahat, A.; Al-Seeni, M. Studying the antioxidant role of mushroom against hazards of gamma radiation exposure in male rats. *Indian J. Anim. Res.* **2018**, *52*, 226–231. [CrossRef]
48. Dogan, A.; Dalar, A.; Sadullahoglu, C.; Battal, A.; Uzun, Y.; Celik, I.; Demirel, K. Investigation of the protective effects of horse mushroom (*Agaricus arvensis Schaeff.*) against carbon tetrachloride-induced oxidative stress in rats. *Mol. Biol. Rep.* **2018**, *45*, 787–797. [CrossRef] [PubMed]
49. Chuang, W.Y.; Lin, L.J.; Der Shih, H.; Shy, Y.M.; Chang, S.C.; Lee, T.T. Intestinal Microbiota, Anti-Inflammatory, and Anti-Oxidative Status of Broiler Chickens Fed Diets Containing Mushroom Waste Compost By-Products. *Animals* **2021**, *11*, 2550. [CrossRef]
50. Zhao, H.; Lan, Y.; Liu, H.; Zhu, Y.; Liu, W.; Zhang, J.; Jia, L. Antioxidant and Hepatoprotective Activities of Polysaccharides from Spent Mushroom Substrates (*Laetiporus sulphureus*) in Acute Alcohol-Induced Mice. *Oxidative Med. Cell. Longev.* **2017**, *2017*, 5863523. [CrossRef]
51. Yang, X.; Zheng, M.; Zhou, M.; Zhou, L.; Ge, X.; Pang, N.; Li, H.; Li, X.; Li, M.; Zhang, J.; et al. Lentinan Supplementation Protects the Gut–Liver Axis and Prevents Steatohepatitis: The Role of Gut Microbiota Involved. *Front. Nutr.* **2022**, *8*, 803691. [CrossRef]
52. Du, T.; Fang, Q.; Zhang, Z.; Zhu, C.; Xu, R.; Chen, G.; Wang, Y. Lentinan Protects against Nonalcoholic Fatty Liver Disease by Reducing Oxidative Stress and Apoptosis via the PPARα Pathway. *Metabolites* **2022**, *12*, 55. [CrossRef] [PubMed]
53. Ulukõy, G.; Baba, E.; Birincioğlu, S.S. Effects of dietary lentinula edodes extract on liver and gut histology of rainbow trout (Oncorhynchus mykiss, Walbaum 1792). *Fish Shellfish. Immunol.* **2016**, *53*, 106. [CrossRef]
54. Song, X.; Cai, W.; Ren, Z.; Jia, L.; Zhang, J. Antioxidant and Hepatoprotective Effects of Acidic-Hydrolysis Residue Polysaccharides from Shiitake Culinary-Medicinal Mushroom *Lentinus edodes* (*Agaricomycetes*) in Mice. *Int. J. Med. Mushrooms* **2021**, *23*, 85–96. [CrossRef]
55. Hertli, S.; Zimmermann, P. Molecular interactions between the intestinal microbiota and the host. *Mol. Microbiol.* **2022**, *117*, 1297–1307. [CrossRef] [PubMed]
56. Giannenas, I.; Tontis, D.; Tsalie, E.; Chronis, E.; Doukas, D.; Kyriazakis, I. Influence of dietary mushroom Agaricus bisporus on intestinal morphology and microflora composition in broiler chickens. *Res. Veter-Sci.* **2010**, *89*, 78–84. [CrossRef]
57. Giannenas, I.; Tsalie, E.; Chronis, E.; Mavridis, S.; Tontis, D.; Kyriazakis, I. Consumption of Agaricus bisporus mushroom affects the performance, intestinal microbiota composition and morphology, and antioxidant status of turkey poults. *Anim. Feed. Sci. Technol.* **2011**, *165*, 218–229. [CrossRef]
58. Guo, P.; Zhang, K.; Ma, X.; He, P. Clostridium species as probiotics: Potentials and challenges. *J. Anim. Sci. Biotechnol.* **2020**, *11*, 1–10. [CrossRef]
59. Siddiqui, M.T.; Cresci, G.A.M. The Immunomodulatory Functions of Butyrate. *J. Inflamm. Res.* **2021**, *14*, 6025–6041. [CrossRef]
60. Jesus, G.F.A.; Owatari, M.S.; Pereira, S.A.; Silva, B.C.; Syracuse, N.M.; Lopes, G.R.; Addam, K.; Cardoso, L.; Mouriño, J.L.P.; Martins, M.L. Effects of sodium butyrate and Lippia origanoides essential oil blend on growth, intestinal microbiota, histology, and haemato-immunological response of Nile tilapia. *Fish Shellfish. Immunol.* **2021**, *117*, 62–69. [CrossRef]
61. Zeng, X.; Yang, Y.; Wang, J.; Wang, Z.; Li, J.; Yin, Y.; Yang, H. Dietary butyrate, lauric acid and stearic acid improve gut morphology and epithelial cell turnover in weaned piglets. *Anim. Nutr.* **2022**, *11*, 276–282. [CrossRef] [PubMed]
62. Gibson, G.R.; Roberfroid, M.B. Dietary Modulation of the Human Colonic Microbiota: Introducing the Concept of Prebiotics. *J. Nutr.* **1995**, *125*, 1401–1412. [CrossRef]
63. Moreau, M.C.; Coste, M. Immune responses to dietary protein antigens. *World Rev. Nutr. Diet.* **1993**, *74*, 22–57. [PubMed]
64. Sharma, R.; Schumacher, U.; Ronaasen, V.; Coates, M. Rat intestinal mucosal responses to a microbial flora and different diets. *Gut* **1995**, *36*, 209–214. [CrossRef] [PubMed]
65. Hamann, L.; El-Samalouti, V.; Ulmer, A.J.; Flad, H.-D.; Rietschel, E.T. Components of gut bacteria as immunomodulators. *Int. J. Food Microbiol.* **1998**, *41*, 141–154. [CrossRef]

*fishes*

MDPI

*Article*

# Weight–Length Relationship and Condition Factor of Gibel Carp (*Carassius auratus gibelio* var. CAS V) at Different Growth Stages and Feed Formulations

**Han Zhang [1,2,3], Shouqi Xie [3] and Sheng Wang [1,2,*]**

[1] Key Laboratory of Microecological Resources and Utilization in Breeding Industry, Ministry of Agriculture and Rural Affairs, Guangzhou 511400, China; zhangh_scholar@126.com
[2] Guangdong Haid Group Co., Ltd., Guangzhou 511400, China
[3] Institute of Hydrobiology, Chinese Academy of Sciences, Wuhan 430072, China; sqxie@ihb.ac.cn
* Correspondence: wangsheng@haid.com.cn

**Abstract:** Accurate recording of growth indicators for aquaculture species at different stages is essential for evaluating aquaculture production effectiveness and the rationality of feed formulations. Due to their convenience and non-invasiveness, weight, length, and condition factor are commonly used to assess the growth of aquaculture species. However, fish growth indicators, can vary significantly with age structure and feed components (nutrition), and relying on a single indicator may lead to misjudgments. In this study, we investigated the growth indicators of Gibel carp (*Carassius auratus gibelio* var. CAS V) at different growth stages (juvenile and adult) and different feed formulations. Meanwhile, the fish weight–length relationship ($W = bL^a$) was used to assess the growth indicators. The results showed that the weight–length relationship of Gibel carp varied significantly with age and feed formulation. Additionally, the condition factor calculated depended on the weight–length relationship being more similar with weight and length change trend than the condition factor was measured. Weight analysis indicated that weight, length, depth, body width, and carcass ratio had higher weights when analyzing fish growth. Therefore, during aquaculture production, fish growth evaluating requires considering diverse indicators such as weight, length, body depth, body width, and carcass ratio, as well as the condition factor, to avoid misjudging the actual growth situation. Meanwhile, the use of the condition factor should consider the sufficient number of data and whether the assumptions (such as being in an isometric growth period) are met.

**Keywords:** Gibel carp; growth indicators; weight–length relationship; condition factor; growth stage; feed formulation

**Key Contribution:** This study suggested that fish growth evaluation requires considering diverse indicators. Meanwhile, the use of the condition factor should consider the sufficient number of data and whether the assumptions (such as being in a isometric growth period) are met.

**Citation:** Zhang, H.; Xie, S.; Wang, S. Weight–Length Relationship and Condition Factor of Gibel Carp (*Carassius auratus gibelio var.* CAS V) at Different Growth Stages and Feed Formulations. *Fishes* **2023**, *8*, 439. https://doi.org/10.3390/fishes8090439

Academic Editors: Qiyou Xu, Jianhua Ming, Fei Song, Changle Qi and Chuanpeng Zhou

Received: 7 August 2023
Revised: 24 August 2023
Accepted: 25 August 2023
Published: 29 August 2023



## 1. Introduction

In aquaculture, morphological and physiological indices of cultured species are used to assess their physiological or nutritional status. Weight, length, and condition factor are commonly used evaluation indicators due to their convenience and non-invasiveness. The weight (W) of fishes is exponentially related to their length (L) according to the weight–length relationship (WLR) equation $W = bL^a$ [1], which was established by Keys, 1928 [2] and has been widely used since then for determining fish condition and comparing fish growth ($\boldsymbol{a} = 3$, isometric growth; $\boldsymbol{a} < 3$, negative allometric growth; $\boldsymbol{a} > 3$, positive allometric growth) [3]. Another indicator, the condition factor(also known as Fulton's condition factor $K$), was intertwined with WLR because Fulton (1904) noted this factor as $K = 100\frac{W}{L^3}$ [4], which was somehow restricted to the condition of isometric growth. The condition factor is an empirical indicator based on the positive correlation between the physiological or nutritional status of animals and the energy storage in their bodies, which, in turn, is positively correlated with body weight [5]. Among individuals of the same body size (length), those with higher weight are generally considered to have better physiological and nutritional status [6]. However, in practical production, the growth stages of fish, such as the juvenile and adult stages, and different feed formulations can lead to deviations from this ideal growth status [7], and relying solely on the condition factor to assess the growth of cultured species may cause misjudgments. Therefore, research suggested using a range of diverse indicators to evaluate cultured species [8].

Thus, the aim of this research is to provide a more precise assessment of WLR and applicable conditions of the condition factor, as well as various indicators, such as body weight, length, depth, body width, and carcass ratio for evaluating Gibel carp (*Carassius auratus gibelio* var. CAS V) growth characteristics at different growth stages (juvenile and adult) and under different feed formulations (fish meal replacement, FMR; plant protein replacement, PPR). This research has the potential to enhance the production management and feed formulation design in aquaculture.

## 2. Materials and Methods

### 2.1. Ethical Statement

This research was approved by the Animal Ethics Committee of the Key Laboratory of microecological resources and utilization in the breeding industry, the Ministry of Agriculture and Rural Affairs, and all experiments were conducted according to the protocols and procedures of the Laboratory Animal Management Ordinance of China.

### 2.2. Feeding Management

#### 2.2.1. Juvenile Stage

Larva Gibel carp (yolk sac) were purchased from Bairong Company (Huanggang, China), transported to Xinghua nursery factory (Xinghua, China), and divided into self-sufficient farm (Xinghua nursery factory) and commercial farm (Yancheng aquaculture farm, Yanchen, China) parts, with a stocking density of 450 fish/$m^2$. After temporary culture for 10 days, commercial feed purchased from Taizhou Biological Feed Co., Ltd. (Xinhua, China) was used with a feeding rate of 6% for another 120 days. To ensure the stability of feeding, the feeding frequency and amount were consistent between the self-sufficient farm and the commercial farm.

#### 2.2.2. Adult Stage

After being raised to the adult stage (63.04 $\pm$ 6.70 g), the Gibel carp were transferred to net cages (2.5 m $\times$ 2.5 m $\times$ 3 m) for cultivation, with 45 fish per cage and a total of 76 cages. These cages were randomly divided into 19 groups, with 4 replicates per group. A total of 19 different formulation feeds were fed at 4% body weight/day, 3 times/day for 60 days.

#### 2.2.3. Feed Formulation

Fish meal replacement is one of the important research directions for coping with the current shortage of fish meal. At the same time, different plant proteins are widely used in feed formulation design to avoid nutritional and cost imbalances caused by single-plant protein source [9]. Therefore, this experiment used two approaches, fish meal replacement (FMR) and plant protein replacement (PPR), to create different feed formulations. The feed formulation design approach for the adult stage is shown in Table 1. The details of feed formulations for the adult fish were as shown in Tables S1 and S2 and the proximate composition of commercial feed for juvenile fish was as shown in Table S3.

**Table 1.** Feed formulation design approach for the adult stage.

| Main Approach | Control Group | Groups | Feed Formulation Design Approach |
|---|---|---|---|
| Fish meal replacement | 1 | 2 | 50% fish meal replaced by expanded soybean |
| | | 3 | Fish meal replaced by a combination of expanded soybean and corn protein powder |
| | | 4 | Fish meal replaced by cottonseed meal |
| | | 5 | Fish meal replaced by corn gluten meal |
| Plant protein replacement | 4 | 5 | Cottonseed meal replaced by corn gluten meal |
| | | 6, 7, 8 | Soybean meal replaced gradually by peanut meal |
| | | 9, 10 | Rapeseed meal (Canada) replaced gradually by sunflower seed meal |
| | | 11, 12, 13 | Soybean meal replaced gradually by sunflower seed meal |
| | | 14, 15 | Soybean oil replaced by wheat with equal energy |
| | | 16, 17, 18, 19 | Rapeseed meal (Canada) replaced by fermented rapeseed meal |

### 2.3. *Data Collection and Processing*

#### 2.3.1. Juvenile Stage

Data Collection and Analysis

Fish were taken after 120 days and were anesthetized with MS-222 (25 mg/L). Then fish were measured for weight and length and were classified by cultivation area. The recorded and measured data were statistically analyzed using SPSS (R26.0.0.0 version X.X ) and EXCEL (version 16.76(23081101) version XX) software. The parameters (***a*** and ***b***) of the Weight–Length Relationship (WLR) equation in aquaculture (Equation (1)) was calculated by SPSS regression analysis [10].

Weight–length relationship

$$W = bL^a \tag{1}$$

W: weight in grams; L: length in centimeters; ***a*** and ***b*** are parameters

In order to evaluate whether the WLR equation could fit the fish during the juvenile stage, the fitted weight data were inversely calculated from the measured length collected by a commercial farm using the WLR equation. Independent sample *t*-test analysis was performed with the measured weight and fitted weight to evaluate the differences.

#### 2.3.2. Adult Stage

Data Collection and Analysis

The net cage experiment lasted for 60 days. Ten fish were randomly selected from each cage and were anesthetized with MS-222 (25 mg/L) and then killed by a lethal blow to the head. Then the fish were measured for weight, length, depth, body width, and carcass ratio, and the Measured Condition Factor (MCF) was calculated. The length (AB) was determined by rule with an accuracy of 0.1 mm; depth (CD) and body width (EF) were determined by electronic digital calipers with an accuracy of 0.01 mm (Figure 1). The equations for carcass ratio and MCF are as follows:

$$\text{Carcass ratio} = \frac{\text{body weight} - \text{viscera weight}}{\text{body weight}} \times 100\% \tag{2}$$

$$\text{Condition factor} = \frac{\text{body weight}}{\text{length}^3} \times 100\% \tag{3}$$

The recorded and measured data were statistically analyzed using SPSS and EXCEL software. Growth indicator data from each group and the entire population were used individually to calculate the parameters of WLR (Equation (1)) by regression analysis. Afterwards, weight was fitted by the WLR equation for the entire population and each group, respectively. Along with the measured length, the entire population fitted condition factor (ECF) and group fitted condition factor (GCF) were calculated using fitted weight and Equations (3). One-way ANOVA was used to test the differences in various indicators of Gibel carp under different feed formulations. Principal component analysis (PCA) was used to analyze the weight (normalized) of weight, length, depth, body width, MCF, ECF and GCF. All the data sources can be seen in supplements, as described in Table 2.

**Figure 1.** Measurement of morphological indicators (Gibel Carp is shown as an example). AB, body length; CD, body depth; EF, body width (F is the point E on the other side of the fish).

**Table 2.** Data source of Figure/Table/Equations.

| Figure/Table/Equation | Data Source | Sheet |
|---|---|---|
| Equation (4)/Figure S1 | Data source for Juvenile fish.xlsx | sheet 1: Data from self-sufficient farm |
| Table 3–Juvenile fish | Data source for Juvenile fish.xlsx | sheet 2: Data from commercial farm |
| Table 3–Adult fish | Data source for Adult fish.xlsx | Sheet 1: Weight analysis |
| Tables 4–9 | Data source for Adult fish.xlsx | sheet 2: All data |
| Table 6 | Data source for Adult fish.xlsx | sheet 2: All data/sheet3: WLR |

## 3. Results

### 3.1. *Juvenile Stage*

Weight–Length Relationship in Juvenile Stage

A regression analysis was performed on the data sampled from the self-sufficient farm, and the values of ***a*** and ***b*** were calculated by regression analysis, resulting in an exponential relationship ($r = 0.978$) (Equation (4)). This exponential equation fits well with Equation (1) (Figure S1). When the data collected by commercial farm were used, the fitted weight based on the WLR was not significantly different from the measured weight ($p = 0.87$, Table 3), indicating that Equation (4) is consistent with the actual situation.

$$W = 3.023 \times 10^{-5} L^{3.023} (r = 0.978) \quad (4)$$

**Table 3.** Statistical significance between measured and fitted weight of juvenile and adult fish.

| Data Source | Statistical Significance between Measured and Fitted Weight | |
|---|---|---|
| Juvenile fish from commercial farm | - | $p = 0.93$ |
| Adult fish | ** | $p = 0.000008$ |

-: $p > 0.05$; **: $p < 0.01$.

## 3.2. Adult Stage

### 3.2.1. Length–Weight Relationship of Juvenile Fish Did Not Fit the Adult Fish

The weight data of the adult fish were fitted using the WLR of Gibel carp at the juvenile stage (Equation (4)). The difference between the fitted weight and the measured weight was analyzed and found to be significantly different ($p < 0.01$, Table 3), indicating that the WLR established during the juvenile stage was not applicable to the adult stage of Gibel carp.

### 3.2.2. Growth Indicators under Different Feed Formulations Had Varying Degrees of Impact

The results showed that different feed formulations had varying degrees of impact on several commonly used fish growth indicators. However, the trends in the MCF were inconsistent with other indicators (weight, length, depth, carcass ratio, and body width). In the FMR group, groups 2 and 3 showed significant differences in weight, length, depth, carcass ratio, and body width, while the MCF showed no significant difference (Table 4). In the PPR group, all groups except group 9 and group 11 showed significant differences in the MCF compared to the control group (Table 5), while most groups showed no difference in weight, length, depth, carcass ratio, and body width compared to the control group. This indicates that using the condition factor alone to assess the growth characteristics of Gibel carp under different feed formulations may cause misjudgments.

**Table 4.** The effect of fish meal replacement on Gibel carp growth.

| Control | Group | Statistical Significance | | | | | |
|---|---|---|---|---|---|---|---|
| | | Weight | Length | Body Depth | Carcass Ratio | Body Width | MCF |
| 1 | 2 | ** | ** | ** | ** | ** | - |
| | 3 | ** | ** | ** | ** | ** | - |
| | 4 | - | * | ** | ** | ** | ** |
| | 5 | - | - | - | * | - | - |
| total of sd/n-sd group | | 2/2 | 3/1 | 3/1 | 4/0 | 3/1 | 1/3 |

-: $p > 0.05$; *: $0.01 < p < 0.05$; **: $p < 0.01$; sd: Significant differences; n-sd: Non-Significant differences.

**Table 5.** The effect of plant protein replacement on Gibel carp growth.

| Control | Group | Statistical Significance | | | | | |
|---|---|---|---|---|---|---|---|
| | | Weight | Length | Body Depth | Carcass Ratio | Body Width | MCF |
| 4 | 5 | - | - | * | - | ** | * |
| | 6 | - | - | - | - | - | * |
| | 7 | - | - | - | - | * | ** |
| | 8 | - | - | ** | - | ** | ** |
| | 9 | * | * | - | * | - | - |
| | 10 | - | - | ** | - | ** | ** |
| | 11 | - | - | - | - | - | - |
| | 12 | - | - | - | - | ** | ** |
| | 13 | - | - | - | - | * | ** |
| | 14 | - | - | - | - | - | ** |
| | 15 | - | ** | - | - | - | ** |
| | 16 | - | - | * | - | - | * |
| | 17 | - | - | ** | - | ** | ** |
| | 18 | - | - | - | - | - | ** |
| | 19 | - | ** | ** | * | ** | ** |
| total of sd/n-sd group | | 1/14 | 3/12 | 6/9 | 2/13 | 8/7 | 13/2 |

-: $p > 0.05$; *: $0.01 < p < 0.05$; **: $p < 0.01$; MCF: Measured Condition Factor; sd: Significant differences; n-sd: Non-Significant differences.

#### 3.2.3. Weight–Length Relationship Fitted Condition Factor Showed Similar Trends to Growth Indicators

The WLR of the entire population and each group of adult fish were established (Table 6). In the WLR, the dimensional of constant ***b*** is consistent with the condition factor, and when ***a*** = 3, ***b***'s definition premise and biological significance are consistent with the condition factor [11]. Weight was fitted by the WLR equation of the entire population (Entire population WLR, Table 6) and of each group (group 1–19 WLR, Table 6) by measured length data. Then, ECF and GCF was calculated using fitted weight, measured length and Equation (3). The results showed that the GCF had similar trends to those of weight and length. In contrast, the trend for the MCF and ECF differed significantly from these indicators (Tables 7 and 8).

**Table 6.** Weight–length relationships of entire population and each group.

| Groups | Weight–Length Relationships | *r*-Value |
|---|---|---|
| Entire population | $W = 7.178 \times 10^{-5}L^{2.826}$ | 0.872 |
| Group1 | $W = 8.164 \times 10^{-5}L^{2.800}$ | 0.794 |
| Group2 | $W = 4.251 \times 10^{-5}L^{2.930}$ | 0.879 |
| Group3 | $W = 1.397 \times 10^{-4}L^{2.693}$ | 0.760 |
| Group4 | $W = 1.416 \times 10^{-4}L^{2.694}$ | 0.839 |
| Group5 | $W = 1.019 \times 10^{-4}L^{2.756}$ | 0.815 |
| Group6 | $W = 1.846 \times 10^{-5}L^{3.102}$ | 0.948 |
| Group7 | $W = 7.178 \times 10^{-5}L^{2.826}$ | 0.908 |
| Group8 | $W = 1.319 \times 10^{-4}L^{2.699}$ | 0.854 |
| Group9 | $W = 8.384 \times 10^{-5}L^{3.267}$ | 0.903 |
| Group10 | $W = 1.095 \times 10^{-4}L^{2.738}$ | 0.938 |
| Group11 | $W = 8.620 \times 10^{-4}L^{2.326}$ | 0.815 |
| Group12 | $W = 5.712 \times 10^{-5}L^{2.872}$ | 0.929 |
| Group13 | $W = 3.142 \times 10^{-5}L^{2.988}$ | 0.909 |
| Group14 | $W = 2.038 \times 10^{-5}L^{3.078}$ | 0.837 |
| Group15 | $W = 5.021 \times 10^{-5}L^{2.913}$ | 0.892 |
| Group16 | $W = 5.063 \times 10^{-5}L^{2.896}$ | 0.930 |
| Group17 | $W = 3.826 \times 10^{-5}L^{2.950}$ | 0.926 |
| Group18 | $W = 7.178 \times 10^{-5}L^{2.826}$ | 0.902 |
| Group19 | $W = 1.124 \times 10^{-4}L^{2.734}$ | 0.875 |

**Table 7.** The effect of fish meal replacement on weight, length and condition factors.

| Control | Group | Statistical Significance | | | | |
|---|---|---|---|---|---|---|
| | | Weight | Length | MCF | GCF | ECF |
| 1 | 2 | ** | ** | - | - | ** |
| | 3 | ** | ** | - | ** | ** |
| | 4 | - | * | ** | ** | * |
| | 5 | - | - | - | ** | - |
| total of sd/n-sd group | | 2/2 | 3/1 | 1/3 | 3/1 | 3/1 |

-: $p > 0.05$; *: $0.01 < p < 0.05$; **: $p < 0.01$; MCF: Measured Condition Factor; GCF: Group Fitted Condition Factor; ECF: Entire Population Fitted Condition Factor; sd: Significant differences; n-sd: Non-Significant differences.

#### 3.2.4. Weight Analysis of Growth Indicators

Weight Analysis of Growth Indicators indicated that body weight, length, depth, body width, and carcass ratio had higher weights when analyzing fish growth characteristics (Table 9).

**Table 8.** The effect of plant protein replacement on weight, length and condition factors.

| Control | Group | Statistical Significance | | | | |
|---|---|---|---|---|---|---|
| | | Weight | Length | MCF | GCF | ECF |
| 4 | 5 | - | - | * | ** | - |
| | 6 | - | - | * | ** | - |
| | 7 | - | - | ** | ** | - |
| | 8 | - | - | ** | ** | - |
| | 9 | * | * | - | ** | * |
| | 10 | - | - | ** | ** | - |
| | 11 | - | - | - | * | - |
| | 12 | - | - | ** | ** | - |
| | 13 | - | - | ** | ** | - |
| | 14 | - | - | ** | ** | - |
| | 15 | - | ** | ** | ** | ** |
| | 16 | - | - | * | ** | - |
| | 17 | - | - | ** | ** | - |
| | 18 | - | - | ** | ** | - |
| | 19 | - | ** | ** | ** | ** |
| total of sd/n-sd group | | 1/14 | 3/12 | 13/2 | 15/0 | 3/12 |

-: $p > 0.05$; *: $0.01 < p < 0.05$; **: $p < 0.01$; MCF: Measured Condition Factor; GCF: Group Fitted Condition Factor; ECF: Entire Population Fitted Condition Factor; sd: Significant differences; n-sd: Non-Significant differences.

**Table 9.** Weight analysis of Gibel carp growth indicators.

| Factors | Weight (Normalization) |
|---|---|
| Body weight | 0.211 |
| Carcass ratio | 0.211 |
| Depth | 0.179 |
| Length | 0.175 |
| Body width | 0.124 |
| Measured Condition Factor | 0.99 |

## 4. Discussion

### *4.1. Weight–Length Relationship of Gibel Carp Juvenile and Adult Stages*

This study collected growth data of Gibel carp from the Xinghua nursery factory, calculated the WLR of the Gibel carp, and verified it using the growth data collected from a commercial farm. The results showed that the WLR (Equation (1)) can fit well, with a value of $a = 3.023$. From the definition of the equation, the coefficient $a$ represents the ratio of weight gain to length growth in fish. From a mathematical perspective, when $a$ undergoes a small change, $b$ will change significantly, and even when a is relatively stable, $b$ can still undergo frequent changes. Therefore, the value of $a$ reflects the growth characteristics of fish in different stages and environments [1,3]. Another physical interpretation of Equation (1) is that the relationship between fish weight (W), fish density (ρ), and volume (V) is $W = \rho V$, and the volume is an exponential function of length. Generally, the power exponent is close to 3, indicating that fish growth is isometric [12]. In this study, the value of $a$ is 3.023, indicating that Gibel carp is close to isometric growth during the juvenile stage, which may explain why the weight–length relationship fits well with Equation (4) for both farms.

The results also showed significant differences in the WLR between the juvenile and adult stages. This is because factors such as fish species, age, and food can influence the value of $a$ and $b$ of WLR [8,13–16]. The results also showed that the $a$ values of individual groups and the entire population data of the adult stage were smaller than 3 (Table 6), possibly because the Gibel carp was in a hypoallometric growth period.

### 4.2. Limitations of the Condition Factor in Evaluating Fish Growth Characteristics under Different Feed Formulations

The condition factor is often used to analyze the growth status or reproductive capacity of fish under different conditions [17]. If assuming that the expected weight of animals increases approximately uniformly with length growth, for example, with the increase of the long axis of the body, and the body radial size increases proportionally, then the expected weight will be positively correlated with the cube of the body, which is the original concept of the condition factor (also known as the Fulton index) [11]. Combining Equations (1) and (3), when fish are in an isometric growth period ($\boldsymbol{a} = 3$), the biological significance of $\boldsymbol{b}$ and the assumption premise of the condition factor are consistent and numerically identical. However, in practical production, fish growth can deviate from this ideal growth state due to changes in environmental conditions such as temperature and feed [7,18,19]. As shown in this study, the WLR of various groups and the entire data in the adult stage showed $\boldsymbol{a} < 3$, indicating that Gibel carp may be in a hypoallometric growth period under different feed formulations. The MCF of various groups did not show significant differences, and its trend was inconsistent with other indicators (weight, length, depth, body width, and carcass ratio). This further indicates that the use of the condition factor should consider whether its premise assumption (e.g., isometric growth stage or $\boldsymbol{a} = 3$) is met to avoid misjudgments.

In this study, the regression equation of the entire population of adult fish WLR was used to calculate the fitted weight based on the measured length and then calculate the fitted condition factor. The trend of this ECF description was shown to be more similar to other growth indicators (weight, length,) than GCF and MCF. This result indicates that a sufficient number of data is an important condition for the biological significance of the condition factor. Therefore, when using the condition factor for evaluation, a sufficient number of data need to be considered to ensure the accuracy of this indicator's evaluation.

The indicators of fish weight, length, depth, body width, and carcass ratio were also analyzed for evaluating fish growth. The results showed that these indicators had higher weights. Therefore, in aquaculture production, the evaluation of fish growth under different feed formulations should comprehensively consider diverse indicators such as weight, length, depth, body width, carcass ratio, and condition factor to avoid misjudgments caused by a single indicator.

## 5. Conclusions

This study discussed the application limitations of WLR and CF under different growth periods and feed formulations, as well as the application of other indicators, and demonstrated that, to support an accurate assessment of fish growth in aquaculture production, a comprehensive growth indicator system was needed, such as weight, length, depth, body width, carcass ratio, and condition factor. However, the application of the condition factor in practice needs to consider the sufficient number of data and whether its premise assumption (e.g., isometric growth stage) is met. In other words, with a thorough understanding of the relationship between CF and WLR, as well as their biological significance, CF could be more meaningful and credible during the evaluation of fish growth and their economic value.

**Supplementary Materials:** The following supporting information can be downloaded at: https://www.mdpi.com/article/10.3390/fishes8090439/s1, Table S1: Feed formulations for fish meal replacement group; Table S2: Feed formulations for plant protein replacement group; Table S3: Proximate composition of commercial feed for juvenile fish; Figure S1: Weight-length regression analysis of Juvenile fish.

**Author Contributions:** Conceptualization, S.X. and S.W.; methodology, S.W; funding acquisition, S.W.; data analysis, H.Z.; writing—original draft preparation, H.Z.; writing—review and editing, S.X. and S.W. All authors have read and agreed to the published version of the manuscript.

**Funding:** This work was supported by the 2023 Rural Revitalization Strategy Special Project—Agricultural Science and Technology Development and Resource Environmental Protection Management Project (2023KJ115).

**Institutional Review Board Statement:** This research was approved by the Animal Ethical and Welfare Committee of the Institute of Hydrobiology, Chinese Academy of Sciences, China (approval number: IHBCAS-20210601 approved on 1 June 2021).

**Data Availability Statement:** All the research data have been upload in supplementary materials. The original data could be find as Table 2 described.

**Acknowledgments:** We acknowledge the Guangdong Haid Group Co., Ltd. for the experiment platform. We thank the Key Laboratory of microecological resources and utilization in breeding industry, Ministry of Agriculture and Rural Affairs and Institute of Hydrobiology, Chinese Academy of Sciences for funding and technical support.

**Conflicts of Interest:** The authors declare no conflict of interest.

## References

1. Froese, R.; Tsikliras, A.C.; Stergiou, K.I. Editorial note on weight—Length relations of fishes. *Acta Ichthyol. Piscat.* **2011**, *41*, 261–263. [CrossRef]
2. Keys, A.B. The weight-length relation in fishes. *Proc. Natl. Acad. Sci. USA* **1928**, *14*, 922–925. [CrossRef] [PubMed]
3. Froese, R. Cube law, condition factor and weight–length relationships: History, meta-analysis and recommendations. *J. Appl. Ichthyol.* **2006**, *22*, 241–253. [CrossRef]
4. Fulton, T. The rate of growth of fishes. In *Twenty-Second Annual Report of the Fishery Board of Scotland*; Fishery Board for Scotland: Edinburgh, UK, 1904; pp. 141–241.
5. Schulte-Hostedde, A.I.; Zinner, B.; Millar, J.S.; Hickling, G.J. Restitution of mass–size residuals: Validating body condition indices. *Ecology* **2005**, *86*, 155–163. [CrossRef]
6. Jones, R.; Petrell, R.; Pauly, D. Using modified length–weight relationships to assess the condition of fish. *Aquac. Eng.* **1999**, *20*, 261–276. [CrossRef]
7. Cone, R.S. The need to reconsider the use of condition indices in fishery science. *Trans. Am. Fish. Soc.* **1989**, *118*, 510–514. [CrossRef]
8. Du, Z.; Turchini, G.M. Are we actually measuring growth?—An appeal to use a more comprehensive growth index system for advancing aquaculture research. *Rev. Aquac.* **2021**, *14*, 525–527. [CrossRef]
9. Jannathulla, R.; Rajaram, V.; Kalanjiam, R.; Ambasankar, K.; Muralidhar, M.; Dayal, J.S. Fishmeal availability in the scenarios of climate change: Inevitability of fishmeal replacement in aquafeeds and approaches for the utilization of plant protein sources. *Aquac. Res.* **2019**, *50*, 3493–3506. [CrossRef]
10. Le Cren, E.D. The length-weight relationship and seasonal cycle in gonad weight and condition in the perch (*Perca fluviatilis*). *J. Anim. Ecol.* **1951**, *20*, 201. [CrossRef]
11. Birnie. The Sovereignty of the Sea. *Int. Law Forum Du Droit Int.* **2002**, *4*, 45–47. [CrossRef]
12. Timothy, B. *Methods for Assessment of Fish Production in Fresh Waters*; Blackwell Science Inc.: New York, NY, USA, 1978.
13. Koutrakis, E.T.; Tsikliras, A.C. Length-weight relationships of fishes from three northern Aegean estuarine systems (Greece). *J. Appl. Ichthyol.* **2003**, *19*, 258–260. [CrossRef]
14. Otieno, O.N.; Kitaka, N.; Njiru, J. Length-weight relationship, condition factor, length at first maturity and sex ratio of *Nile tilapia, Oreochromis niloticus* in Lake Naivasha, Kenya. *Int. J. Fish. Aquat. Stud.* **2014**, *2*, 67–72.
15. Samat, A.; Shukor, M.; Mazlan, A.; Arshad, A.; Fatimah, M. Length-weight relationship and condition factor of *Pterygoplichthys pardalis* (Pisces: Loricariidae) in Malaysia Peninsula. *Res. J. Fish. Hydrobiol.* **2008**, *3*, 48–53.
16. Anibeze, C. Length-weight relationship and relative condition of *Heterobranchus longifilis* (Valenciennes) from Idodo River, Nigeria. *Naga ICLARM Q.* **2000**, *23*, 34–35. Available online: https://aquadocs.org/handle/1834/25662 (accessed on 10 August 2023).
17. Badiani, A.; Stipa, S.; Nanni, N.; Gatta, P.P.; Manfredini, M. Physical Indices, Processing Yields, Compositional Parameters and Fatty Acid Profile of Three Species of Cultured Sturgeon (Genus Acipenser). *J. Sci. Food Agric.* **1997**, *74*, 257–264. [CrossRef]
18. Mazumder, S.K.; Das, S.K.; Bakar, Y.; Ghaffar, M.A. Effects of temperature and diet on length-weight relationship and condition factor of the juvenile Malabar blood snapper (Lutjanus malabaricus Bloch & Schneider, 1801). *J. Zhejiang Univ. B* **2016**, *17*, 580–590. [CrossRef]
19. De Giosa, M.; Czerniejewski, P.; Rybczyk, A. Seasonal changes in condition factor and weight-length relationship of invasive Carassius gibelio (Bloch, 1782) from Leszczynskie Lakeland, Poland. *Adv. Zool.* **2014**, *2014*, 678763. [CrossRef]

*fishes*

MDPI

*Article*

# Effects of High Dietary Carbohydrate Levels on Growth Performance, Enzyme Activities, Expression of Genes Related to Liver Glucose Metabolism, and the Intestinal Microbiota of *Lateolabrax maculatus* Juveniles

**Luzhe Zheng [1,2], Zhanzhan Wang [3,4], Bo Zhang [2], Lulu Yan [2], Pengfei Wang [2], Chao Zhao [2], Heizhao Lin [4,5], Lihua Qiu [2,*] and Chuanpeng Zhou [3,4,*]**

1 College of Fisheries and Life Sciences, Dalian Ocean University, Dalian 116023, China
2 Key Laboratory of Fishery Resources Development and Utilization in South China Sea, South China Sea Fisheries Research Institute, Ministry of Agriculture and Rural Affairs, Guangzhou 510300, China
3 Key Laboratory of Aquatic Product Processing, Ministry of Agriculture and Rural Affairs, South China Sea Fisheries Research Institute, Chinese Academy of Fishery Sciences, Guangzhou 510300, China
4 Key Laboratory of Efficient Utilization and Processing of Marine Fishery Resources of Hainan Province, Sanya 572426, China
5 Shenzhen Base of South China Sea Fisheries Research Institute, Chinese Academy of Fishery Sciences, Shenzhen 518121, China
* Correspondence: qiugroup_bio@outlook.com (L.Q.); chpzhou@163.com (C.Z.)

**Abstract:** The present study was conducted to investigate the effects of high dietary carbohydrate levels on growth performance, enzyme activities, and gene expressions related to liver glucose metabolism and the intestinal microbiota of *Lateolabrax maculatus* juveniles. Two experimental diets with levels of carbohydrates (20% and 30%, named the NCD group and the HCD group, respectively) were designed to feed *L. maculatus* (initial weight 9.45 ± 0.03 g) for 56 days. The results showed that, compared with the NCD group, the condition factor (CF) was significantly elevated in the HCD group ($p < 0.05$). The plasma advanced glycosylation end products (AGEs), glycated serum protein (GSP), total cholesterol (TC), triglycerides (TG), high-density lipoprotein (HDL), low-density lipoprotein (LDL), and glutamate aminotransferase (AST) were significantly higher in the HCD group than those in the NCD group ($p < 0.05$). The intestinal lipase, chymotrypsin, and α-amylase in the HCD group were significantly higher than those in the NCD group ($p < 0.05$). The liver superoxide dismutase (SOD), total antioxidant capacity (T-AOC), and catalase (CAT) were significantly lower in the HCD group than in the NCD group ($p < 0.05$). The liver malondialdehyde (MDA) and hexokinase (*HK*) levels were significantly higher than those in the NCD group ($p < 0.05$). In the histopathological findings, liver cells in the HCD group appeared to have many vacuoles, and the number of lipid droplets increased. Compared with the NCD group, the relative expression of liver glucokinase (*GK*) and glycogen synthetase kinase-3 (*GSK3β*) genes in the HCD group was significantly increased ($p < 0.05$), while the relative expression of phosphoenolpyruvate carboxykinase (*PEPCK*) and glycogen phosphorylase (*GP*) genes in the HCD group was significantly reduced ($p < 0.05$). High-throughput 16S rRNA gene sequencing showed that high dietary carbohydrate intake changed the composition and structure of the intestinal microbiota. At the phylum level of the intestinal microbiota, high dietary carbohydrates decreased the relative abundance of Firmicutes and increased the relative abundance of Proteobacteria and Bacteroidetes. At the genus level of the intestinal microbiota, high carbohydrates decreased the relative abundance of Bacillus and increased the relative abundance of Photobacterium and Paraclostridium. From the results of this experiment on *L. maculatus*, high carbohydrates led to increased condition factor and liver glycogen, lipid deposition, decreased antioxidant capacity of the liver, increased relative abundance of harmful intestinal microorganisms, and disrupted glucose metabolism.

**Citation:** Zheng, L.; Wang, Z.; Zhang, B.; Yan, L.; Wang, P.; Zhao, C.; Lin, H.; Qiu, L.; Zhou, C. Effects of High Dietary Carbohydrate Levels on Growth Performance, Enzyme Activities, Expression of Genes Related to Liver Glucose Metabolism, and the Intestinal Microbiota of *Lateolabrax maculatus* Juveniles. *Fishes* **2023**, *8*, 431. https://doi.org/10.3390/fishes8090431

Academic Editor: Marina Paolucci

Received: 8 July 2023
Revised: 15 August 2023
Accepted: 16 August 2023
Published: 23 August 2023

**Keywords:** carbohydrate; intestinal digestive enzymes; liver antioxidative enzymes; intestinal microbiota; glucose metabolism; *Lateolabrax maculatus*

**Key Contribution:** This experiment investigated the effects of high dietary carbohydrate levels on growth performance, intestinal and liver enzyme activities, expression of genes related to liver glucose metabolism, and the intestinal microbiota of *Lateolabrax maculatus* juveniles.

## 1. Introduction

For living organisms, carbohydrates provide energy and organic carbon [1]. A carbohydrate source is regarded as the most economical energy source for artificial feed. In addition, including carbohydrates in the diet has a protein-sparing effect, lowers ammonia nitrogen excretion, and minimizes water pollution, all of which contribute to the feed industry's sustainable growth. Fish, however, have a restricted capacity to use dietary carbohydrates. Usually, hyperglycemia shows up after glucose loading or carbohydrate intake [2]. Carnivorous fish exhibit limited utilization of carbohydrates as a result of their shorter intestine, inadequate secretion of α-amylase, a low count of insulin receptors, and the absence of inhibition in postprandial gluconeogenesis [3]. However, the metabolic variations among fish remain inadequately characterized, thus necessitating an investigation into the dietary carbohydrates' mechanism of metabolic regulation in fish.

The liver serves as the primary site for glucose metabolism in fish, encompassing various processes such as glycolysis, gluconeogenesis, the pentose phosphate pathway, glycogen synthesis, and glycogenolysis. The regulation of glucose production and storage is influenced by hormonal and nutritional factors, which are contingent upon the expression and activity levels of crucial enzymes involved in the gluconeogenic and glycolytic pathways. Both the loss of glycemic control in coho salmon (*O. kisutch*) given a streptozotocin injection and insulin production by Atlantic hagfish (*Myxine glutinosa*) induced by glucose demonstrate the existence of a glucose homeostasis system in fish [4,5]. Glycolysis, gluconeogenesis, glycogen synthesis, and glycogenolysis are all involved in the control of homeostasis glucose [6]. As a result of increased 6-phosphofructo-1-kinase (*FBP*), pyruvate kinase (*PK*), and glucokinase (*GK*) activity, the glycolytic pathway is upregulated in the liver of many carnivorous fish species during postprandial settings. In earlier studies, carbohydrates had an impact on the gut microbiota and the transcription of genes involved in glucose metabolism. According to a previous study on *Megalobrama amblycephala*, neither the number of dietary carbohydrates nor the timing of the samples had an impact on the transcription of phosphoenolpyruvate carboxylase (*PEPCK*). However, the transcriptions of *GK*, *PK*, and glycogen synthase (*GS*) were considerably greater in the high-carbohydrates group than in the control group, whereas the transcriptions of *FBP* were the opposite [3]. High carbohydrate levels were found to enhance the prevalence of the hazardous microorganisms Vibrio, Photobacterium, and Mycoplasma in *Trachinotus ovatus* studies [7]. Glycogen synthase (*GS*) and glycogen phosphorylase are two important enzymes that control the accumulation of glycogen with regard to glycogen metabolism; increased hepatic glycogen after dietary carbohydrate consumption has been documented in certain fish. According to previous studies, after consuming a high-carbohydrate diet, the hepatic glycogen levels were increased by inducing glycogen synthesis and inhibiting glycogenolysis [8,9]. However, the exact mechanism by which a high-carbohydrate diet affects crucial glucose metabolism enzymes and thus results in abnormalities of glucose metabolism in *Lateolabrax maculatus* is still not fully understood and warrants further study.

Animals' bodies require intestinal microflora, which also plays a crucial physiological function in the host's immunological antagonism and the absorption of nutrients [8]. The consumed feed served as the major source of gut flora [10]. Numerous studies have demonstrated a connection between diseases and metabolic disorders and excessive carbohydrate intake [11]. Studies on the impact of dietary carbohydrate levels on the health of fish guts, however, are few [11].

The spotted sea bass (*L. maculatus*), a carnivorous fish that belongs to the family Moronidae (Perciformes), has distinctive conspicuous black spots on the side of its body [12]. Since the creation of the genus *Lateolabrax*, it has been regarded as a congeneric species alongside *Lateolabrax japonicus*. The *L. maculatus* has a larger geographic range than *L. japonicus*, extending from the Bohai Sea to the Indo-Chinese peninsula [13]. Due to its broad adaptability, quick growth, and high market demand, the cultivation of *L. maculatus* is becoming more widespread. According to the 2022 China Fisheries Statistical Yearbook, 199,106 tons of this fish are produced in China [14]. In light of this, the authors of this study examined the effects of increased carbohydrate intake on growth performance, plasma biochemical indices, intestinal and liver enzyme activities, liver glucose metabolism gene expression, and intestinal microflora in *L. maculatus*.

## 2. Materials and Methods

### 2.1. Experimental Design and Diet Formulation

The formulation and chemical composition of the experimental diets are presented in Table 1. The protein sources mainly include fish meals and casein. The lipid source is mainly fish oil. The carbohydrate source is mostly cornstarch. A normal carbohydrate diet and a high carbohydrate diet (NCD: 20%; HCD: 30%) were formulated as two dietary carbohydrate levels for *L. maculatus*. The ingredients were ground through a 40-mesh screen. Minerals and vitamins were mixed by the progressive enlargement method [15]. For the premixed dry ingredients, we added lipid and distilled water to a feed mixer and thoroughly mixed until homogenous [16]. The 2.5-mm-diameter pellets were wet-extruded by a pelletizer (F-26, South China University of Technology, Guangzhou, China) and air-dried. All diets in plastic bags were sealed and stored at −20 °C.

**Table 1.** Formulation and nutrient compositions of experimental diets. (%).

| Items | Diets | |
|---|---|---|
| | NCD | HCD |
| Fish meal | 40 | 40 |
| Casein | 20 | 20 |
| Corn starch | 20 | 30 |
| Dextrin | 0 | 0 |
| Glucose | 0 | 0 |
| Microcrystalline cellulose | 10.5 | 0.5 |
| Fish oil | 6.5 | 6.5 |
| Vitamin premix [a] | 0.5 | 0.5 |
| Mineral premix [b] | 0.5 | 0.5 |
| Choline chloride | 0.5 | 0.5 |
| Soy lecithin | 1 | 1 |
| Betaine | 0.2 | 0.2 |
| Antioxidant | 0.1 | 0.1 |
| Carboxymethyl-cellulose | 0.2 | 0.2 |
| Proximate analysis | | |
| Dry material | 94.3 | 94.3 |
| Crude protein | 42.9 | 43.0 |
| Crude lipid | 11.5 | 11.5 |
| Crude ash | 7.4 | 7.5 |
| Nitrogen-free extract | 20.4 | 30.2 |
| Energy | 18.2 | 19.9 |

[a] Vitamin premix (mg/kg diet): VB1 25, VB2 45, VB12 0.1, VK3 10, VC 2000, inositol 800, nicotinic acid 200, folic acid 1.2, biotin 32, VD3 5, VE 120, ethoxyquin 150, pantothenic acid 500, avicel 14.52. [b] Mineral premix (mg/kg diet): NaF 4, KI 1.6, $CoCl_2 \bullet 6H_2O$ (1%) 100, $CuSO_4 \bullet 5H_2O$ 20, $FeSO_4 \bullet H_2O$ 160, $ZnSO_4 \bullet H_2O$ 100, $MnSO_4 \bullet H_2O$ 120, $MgSO_4 \bullet 7H_2O$ 2400, $Ca(H_2PO_4)_2 \bullet H_2O$ 6000, NaCl 200.

### 2.2. Animal Rearing and Feeding Trial

*L. maculatus* was obtained from Shenzhen Long Qi Zhuang Industrial Development Co., Ltd. (Shenzhen, China). Fish were adapted for 14 days in polythene cages and fed commercial diets of *L. maculatus* (Guangdong Yuequn Marine Life Research and Development Co., Ltd., Jieyang, China). Then, similar-sized fish (average weight, 9.45 ± 0.03 g) were randomly distributed into six cages (1.0 m × 1.0 m × 1.5 m) at a rate of 25 fish per cage. One of two experimental diets was randomly assigned to fish in each cage, and each diet was tested triplicate. Fish were fed two times daily to apparent satiation (6:00 and 18:00 h) for 56 days. During experiments, the pH and the water temperature were 7.63–8.44 and 28.2–32.3 °C, respectively; the salinity and the dissolved oxygen were 25‰ and greater than 5.1 mg/L, respectively.

### 2.3. Sampling

After completing the feeding experiment, we weighed the fish after a whole day of starvation. The diluted MS-222 was used for fish that were anesthetized (Sigma, St. Louis, MO, USA). Three fish were sampled randomly from each cage for sampling. Intraperitoneal lipids and individual liver viscera were all quickly separated. Blood was collected from the caudal vein. Intestinal and liver were peeled from each *L. maculatus*, and the samples were quickly frozen in nitrogen and stored at −80 °C until analyses. Three fish were sampled randomly from each cage to obtain intestinal samples, which were immediately fixed in a 4% paraformaldehyde solution. Three fish were sampled randomly from each cage to obtain liver samples, which were immediately fixed fish liver sections in neutral-buffered formalin.

### 2.4. Measurements of Liver Glycogen

The liver glycogen content is determined by the colorimetric method using commercial assay kits (Beijing Huaying Biotechnology Research Institute, Beijing, China, Serial Number: HY-M0023).

### 2.5. Biochemical Parameters

We measured plasma glucose using the glucose oxidase method [17,18]. Insulin (INS), plasma lactate (LD), pyruvate (PA), glycated serum protein (GSP), and Advanced Glycation End Products (AGES) levels were determined by means of an enzyme-linked immunosorbent assay (ELISA) using the Huawei Delong DR-200BS Enzyme Labeling Analyzer. Determinations of total cholesterol (TC), triglycerides (TG), high-density lipoprotein (HDL), low-density lipoprotein (LDL), glutamate aminotransferase (AST), alanine aminotransferase (ALT), intestinal digestive enzymes (lipase, α-amylase, chymotrypsin), tissue protein, total antioxidant capacity (T-AOC), superoxide dismutase (SOD), catalase (CAT), and malondialdehyde (MDA) were determined using an automatic biochemical analyzer (Myriad BS-420 Automatic Biochemical Instrument). The liver lipid peroxidation product (MDA), antioxidant enzymes (T-AOD, SOD, CAT), hexokinase (HK), and phosphoenolpyruvate carboxykinase (*PEPCK*) were all determined using commercial assay kits (Beijing Huaying Biotechnology Research Institute, Beijing, China, Serial Number: HY-M0003, HY-60021, HY-M0018, HY-M0001, HY-60087, HY-NE286).

### 2.6. Intestinal and Liver Histology

The intestine samples were fixed and dehydrated in a hierarchical series of ethyl alcohol embedded in paraffin, sliced into 5-μm thick sections, which use hematoxylin and eosin to stain (H&E). The intestine villus height was measured by Case Viewer, with ten values for each group. Liver and intestinal morphology were examined using a light microscope (Olympus CKX41 microscope, Tokyo, Japan). Liver samples were fixed in 10% buffered formalin, dehydrated in a graded ethanol series, and embedded in paraffin. Sections series of 4 μm were stained with hematoxylin and eosin (H&E). Oil Red O staining was performed as described in a previous study [19,20].

### 2.7. Real-Time PCR

Bass liver RNA was extracted using the Fore gene RNA kit, RNA quality was measured with 1% agarose gel electrophoresis, RNA concentration was measured with Nanodrop 2000 (Thermo Fisher Scientific, Waltham, MA USA), and RNA was reverse transcribed into cDNA using a kit (Evo M-MLV RT Kit with g DNA Clean for qPCR II (Accurate Biotechnology)) and stored at −20 °C. The real-time PCR amplification system (10 μL) consisted of 2 × SYBR Green Pro Taq HS Premix 5 μL, cDNA template 4.5 μL, upstream primer 0.25 μL, and downstream primer 0.25 μL. In this study, *β-actin* was selected as the internal reference, and the relative expression of the target gene was calculated using the $2^{-\Delta\Delta Ct}$ method. The real-time PCR primer sequences of the gene of interest and the reference gene *β-actin* are shown in Table 2.

**Table 2.** Primers used in the present study.

| Gene | Forward Sequence (5′-3′) | Reverse Sequence (5′-3′) | Purpose |
|---|---|---|---|
| *GK* | TGACAAGGGTATCCTGCTCAACT | TTCCAACAATCATCCCAACTTCA | RT-qPCR |
| *PFK* | GGCAAACCCATCTCCTCAACT | GTCCCTGAGCGAGTGCTAACC | RT-qPCR |
| *PK* | GCTCGACTACAAGAACATTACCAAA | GTCCTGGATGTCCTTATCTGAAACA | RT-qPCR |
| *PC* | CAAACGTCCCTTTCCAGATGC | CTTAGAGCGCCGATCAGAAGC | RT-qPCR |
| *FBP* | TGACGAACCTGCTCAACTCCA | ACTGCCATACAGCGCATAACC | RT-qPCR |
| *PEPCK* | GGCAGATCGTCTCATTCGGTAG | TGCCTTGGTCGTCAAACTTCAT | RT-qPCR |
| *G6Pase* | CTCCTTCGCTGTCGGCTTCT | CTGCTGCTCTTCTTGGTCTCG | RT-qPCR |
| *GSK3β* | GAGATAAGGATGGCAGCAAGGTA | CTCTGTAGACAGTCTCGGGAACG | RT-qPCR |
| *GP* | CATTGAGAAACTCGACTGGGACA | GCAACTCCATTGACAGCGTGA | RT-qPCR |
| *G6PD* | ACGTGGTGCTGGGTCAGTATGT | TCTTGCTCATCATCTTGGCGTA | RT-qPCR |
| *β-actin* | CAACTGGGATGACATGGAGAAG | TTGGCTTTGGGGTTCAGG | RT-qPCR |

### 2.8. Intestinal Microbiota Communities

We used Hi Pure Soil DNA Kits (Magen, Guangzhou, China) to extract microbial DNA. We used PCR to amplify the 16S rDNA V4 region of the ribosomal RNA gene, using primers Arch519:CAGCMGCCGCGGTAA; Arch915R: GTGCTCCCCCGCCAATTCCT. In triplicate, we used a 50 μL mixture containing 5 μL of 10 × KOD Buffer, 5 μL of 2.5 mM dNTPs, 1.5 μL of each primer (5 μM), 1 μL of KOD Polymerase, and 100 ng of template DNA for PCR reactions. The AxyPrep DNA Gel Extraction Kit (Axygen Biosciences, Union City, CA, USA) was used for amplicons extracted from 2% agarose gels. The ABI Step One Plus Real-Time PCR System (Life Technologies, Foster City, CA, USA) was used to quantify. According to the standard protocols, purified amplicons were pooled in equimolar quantities and paired-end sequenced (2 × 250) on an Illumina platform.

### 2.9. Statistical Analysis

Statistical analysis was performed using IBM SPSS Statistics 26.0 software (IBM Corporation, Somers, NY, USA) to determine differences using an independent sample *t*-test. Normality and homoscedasticity assumptions were confirmed prior to any statistical analysis. Significant differences in values between the NCD and HCD groups ($p < 0.05$) are indicated by an asterisk above the histogram. All results are indicated as the mean ± standard error.

## 3. Results

### 3.1. Growth Performance

Compared with the NCD group, the CF and liver glycogen were significantly increased in the HCD group ($p < 0.05$) (Table 3). The survival rate (SR), weight gain rate (WGR), specific growth rate (SGR), and viscerosomatic index (VSI) in the HCD group decreased, and there was no significant difference between the two groups ($p > 0.05$). The feed coefficient (FCR), hepatosomatic index (HSI), carcass index (CI), and visceral adipose index (VAI) increased, and there was no significant difference between the two groups ($p > 0.05$).

$$\text{Weight gain rate (WGR, \%)} = (\text{final weight (g)} - \text{initial weight (g)})/\text{initial weight (g)} \times 100$$

$$\text{Specific growth rate (SGR, \%}\cdot\text{d}^{-1}) = (\ln \text{final weight} - \ln \text{initial weight})/\text{days} \times 100$$

$$\text{Feed coefficient (FC)} = \text{total feed consumption}/(\text{final gross weight} - \text{initial gross weight})$$

$$\text{Hepatosomatic index (HSI)} = \text{liver weight}/\text{body weight} \times 100$$

$$\text{Viscerosomatic index (VSI)} = \text{visceral weight}/\text{body weight} \times 100$$

$$\text{Condition factor (CF)} = \text{body weight (g)} \times 100/\text{body length (cm}^3)$$

$$\text{Survival rate (SR, \%)} = \text{terminal number}/\text{initial mantissa} \times 100$$

$$\text{Carcass index (CI, \%)} = \text{carcass weight}/\text{body weight} \times 100$$

$$\text{Visceral adipose index (VAI, \%)} = (\text{visceral adipose weight (g)}/\text{whole body weight (g)}) \times 100$$

$$\text{Feed conversion ratio (FCR)} = \text{feed intake (g)}/(\text{final weight (g)} - \text{initial weight (g)} \times 100$$

**Table 3.** Effects of different dietary carbohydrate levels on the growth performance of *L. maculatus*.

| Items | Groups | |
|---|---|---|
| | NCD | HCD |
| Initial weight (g) | 9.40 ± 0.06 | 9.50 ± 0.00 |
| Final weight (g) | 76.50 ± 2.57 | 81.33 ± 1.63 |
| Survival rate (SR, %) | 90.67 ± 3.53 | 84.00 ± 8.00 |
| weight gain rate (WGR) (%) | 641.63 ± 56.24 | 616.33 ± 51.79 |
| Specific growth rate SGR (% /d) | 3.56 ± 0.14 | 3.51 ± 0.13 |
| Feed conversion ratio (FCR) | 1.01 ± 0.04 | 1.02 ± 0.05 |
| Condition factor (CF, %) | 1.92 ± 0.07 | 2.71 ± 0.12 * |
| Viscerasomatic index (VSI, %) | 9.26 ± 0.53 | 9.03 ± 0.44 |
| Hepatosomatic index (HSI, %) | 0.98 ± 0.08 | 1.03 ± 0.10 |
| Carcass index (CI, %) | 63.07 ± 1.23 | 64.41 ± 0.60 |
| Visceral adipose index (VAI, %) | 4.46 ± 0.70 | 5.63 ± 0.86 |
| Liver glycogen(mg/g) | 9.68 ± 0.05 | 12.25 ± 0.31 * |

Data are expressed as mean ± SE. Significant differences between the values obtained in the NCD and HCD groups ($p < 0.05$) are marked with an asterisk by *t*-test.

### 3.2. Plasma Metabolites Levels

AGEs, GSP, TC, TG, HDL, LDL, and AST in the HCD group were significantly higher than those in the NCD group ($p < 0.05$) (Table 4). Compared with the NCD group, INS, LD, PA, GLU, and ALT increased in the HCD group, and there was no significant difference between the two groups ($p > 0.05$).

**Table 4.** Effects of different dietary carbohydrate levels on plasma biochemistry of *L. maculatus*.

| Items | Groups | |
|---|---|---|
| | NCD | HCD |
| AGES (mg/L) | 37.49 ± 0.98 | 40.74 ± 0.48 * |
| INS (uIU/mL) | 15.51 ± 0.46 | 16.58 ± 0.08 |
| LD (mmol/L) | 2.50 ± 0.24 | 3.23 ± 0.20 |
| PA (mmol/L) | 0.14 ± 0.00 | 0.20 ± 0.02 |
| GSP (umol/L) | 236.60 ± 7.93 | 267.56 ± 6.87 * |
| TC (mmol/L) | 4.08 ± 0.23 | 5.88 ± 0.32 * |
| TG (mmol/L) | 5.59 ± 0.22 | 6.68 ± 0.10 * |
| HDL (mmol/L) | 1.88 ± 0.02 | 2.72 ± 0.05 * |
| LDL (mmol/L) | 0.93 ± 0.01 | 1.35 ± 0.00 * |
| GLU (mmol/L) | 13.18 ± 0.14 | 13.24 ± 0.06 |
| AST(U/L) | 43.22 ± 1.23 | 56.05 ± 0.70 * |
| ALT(U/L) | 10.55 ± 0.24 | 11.25 ± 0.54 |

Data are expressed as mean ± SE. Significant differences between the values obtained in the NCD and HCD groups ($p < 0.05$) are marked with an asterisk by *t*-test.

### 3.3. Intestinal and Liver Enzyme Activities

The intestinal lipase, α-amylase, and chymotrypsin in the HCD group were significantly higher than those in the NCD group ($p < 0.05$) (Table 5). The liver SOD, T-AOC, and CAT in the HCD group were significantly lower than those in the NCD group ($p < 0.05$), and MDA was significantly higher than that in the NCD group ($p < 0.05$) (Table 6). The liver *HK* was significantly higher than that in the NCD group ($p < 0.05$). There was no significant difference in *PEPCK* between the two groups ($p > 0.05$) (Table 7).

**Table 5.** Effects of different dietary carbohydrate levels on digestive enzyme activities in the intestines of *L. maculatus*.

| Items | Groups | |
|---|---|---|
| | NCD | HCD |
| Lipase (U/mg. Protein) | 12.85 ± 0.17 | 25.46 ± 0.75 * |
| Chymotrypsin (U/mg. Protein) | 37.47 ± 0.66 | 40.72 ± 0.68 * |
| α-Amylase (U/g.protein) | 15.26 ± 0.42 | 20.87 ± 0.82 * |

Data are expressed as mean ± SE. Significant differences between the values obtained in the NCD and HCD groups ($p < 0.05$) are marked with an asterisk by *t*-test.

**Table 6.** Effects of high dietary carbohydrate level on liver antioxidant indexes of *L. maculatus*.

| Items | Groups | |
|---|---|---|
| | NCD | HCD |
| SOD (U/mg. Protein) | 4.37 ± 0.01 | 4.13 ± 0.07 * |
| T-AOC (U/mg.protein) | 0.24 ± 0.01 | 0.18 ± 0.01 * |
| CAT (U/mg. protein) | 2.56 ± 0.04 | 2.24 ± 0.03 * |
| MDA (nmol/mg. protein) | 0.27 ± 0.01 | 0.35 ± 0.00 * |

Data are expressed as mean ± SE. Significant differences between the values obtained in the NCD and HCD groups ($p < 0.05$) are marked with an asterisk by *t*-test.

**Table 7.** Effects of different dietary carbohydrate levels on glucose metabolism enzyme activities in the liver of *L. maculatus*.

| Items | Groups | |
|---|---|---|
| | NCD | HCD |
| *HK* | 0.37 ± 0.01 | 0.62 ± 0.00 * |
| *PEPCK* | 8.09 ± 0.21 | 8.35 ± 0.18 |

Data are expressed as mean ± SE. Significant differences between the values obtained in the NCD and HCD groups ($p < 0.05$) are marked with an asterisk by *t*-test.

### *3.4. Intestinal Morphology and Liver Morphology*

Compared with the NCD group, the HCD group's height uniformity and integrity of the intestinal villi were compromised, and there was a tendency for the intestinal villi to shorten and the spacing of the villi to increase (Figure 1b). The villus height in the HCD group was significantly lower than that in the NCD group ($p < 0.05$) (Table 8).

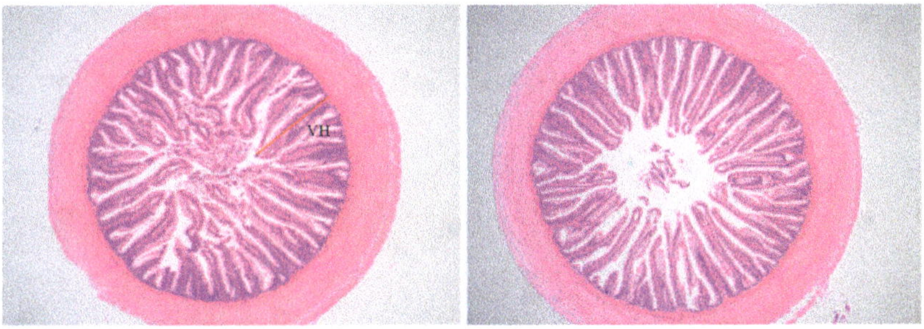


(a) (b)

**Figure 1.** Micrographs of transverse HE-stained gut sections fed diets (NCD and HCD) in *L. maculatus* for 8 weeks (×40). Normal carbohydrate diets (NCD, (**a**)), high carbohydrate diets (HCD, (**b**)). VH: Villus height.

**Table 8.** Effects of different dietary carbohydrate levels on villus height in the intestinal of *L. maculatus*.

| Items | Groups | |
|---|---|---|
| | NCD | HCD |
| Villus height | 27.85 ± 1.76 | 21.02 ± 1.68 * |

Data are expressed as mean ± SE. Significant differences between the values obtained in the NCD and HCD groups ($p < 0.05$) are marked with an asterisk by *t*-test.

The high carbohydrate effects on liver morphology and liver lipid content in *L. maculatus* are shown in Figure 2. In the NCD group, which had a complete structure and uniform shape, the density of nuclei in hepatocytes was uniform and numerous (Figure 2a). In the HCD group, the liver vacuoles increased, cell boundaries were blurred, and the number of nuclei decreased (Figure 2b). Compared with the NCD group, in the HCD group, the number of lipid droplets in the oil red O stained under high carbohydrate conditions increased, and the lipid droplets became larger (Figure 2c,d).

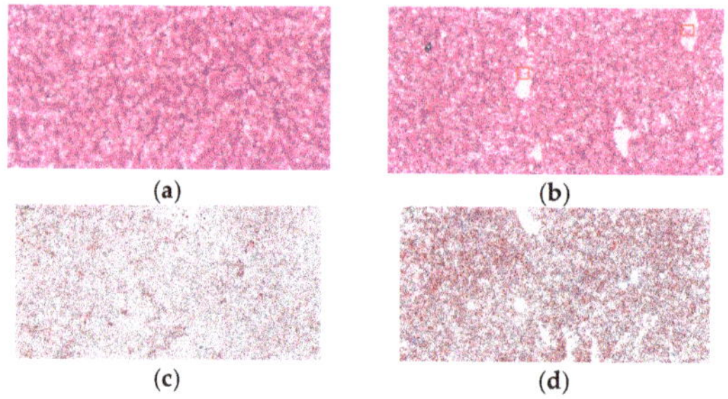


**Figure 2.** Micrographs of *L. maculatus* liver HE-stained sections and oil-red o-stained sections fed diets (NCD and HCD) for 8 weeks (×20). Red frame: The boundaries of hepatocytes are blurred. Normal carbohydrate diets (NCD, (**a**,**c**)), high carbohydrate diets (HCD, (**b**,**d**)).

### *3.5. Liver Transcription of Genes Related to Glucose Metabolism*

The mRNA levels of *GK* in the HCD group increased compared to the NCD group ($p < 0.05$) (Figure 3). The transcription of phosphofructokinase 1 (*PFK*) and pyruvate kinase (*PK*) was not affected by dietary carbohydrate levels ($p > 0.05$). The mRNA levels of *PEPCK* were significantly reduced in the HCD group compared with the NCD group ($p < 0.05$) (Figure 4). The transcription of pyruvate carboxylase (*PC*), glucose-6 phosphatase (*G6pase*), and fructose-1,6-bisphosphatase (*FBP*) was not affected by dietary carbohydrate levels ($p > 0.05$). Compared with the NCD group, the mRNA levels of glycogen synthase kinase3β (*GSK3β*) in the HCD group were significantly increased ($p < 0.05$) (Figure 5). The mRNA levels of glycogen phosphorylase (*GP*) were significantly reduced ($p < 0.05$). The transcription of glucose 6-phosphate dehydrogenase (*G6PD*) was not affected by dietary carbohydrate levels ($p > 0.05$).

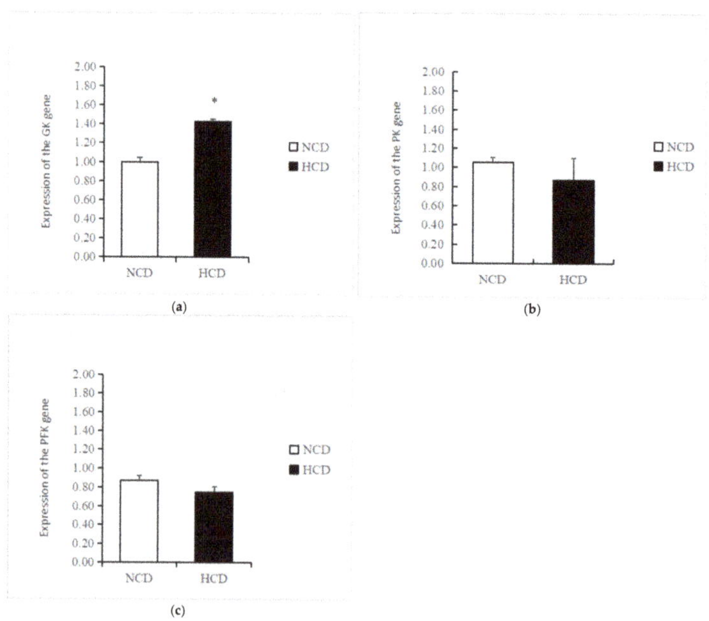


**Figure 3.** Effect of fed diets (NCD and HCD) on relative levels of liver glycolysis mRNA in *L. maculatus*. Glucokinase (*GK*, (**a**)), pyruvate kinase (*PK*, (**b**)), phosphofructokinase (*PFK*, (**c**)). Data are expressed as mean ± SE. Significant differences between the values obtained in the NCD and HCD groups ($p < 0.05$) are marked with an asterisk by *t*-test.

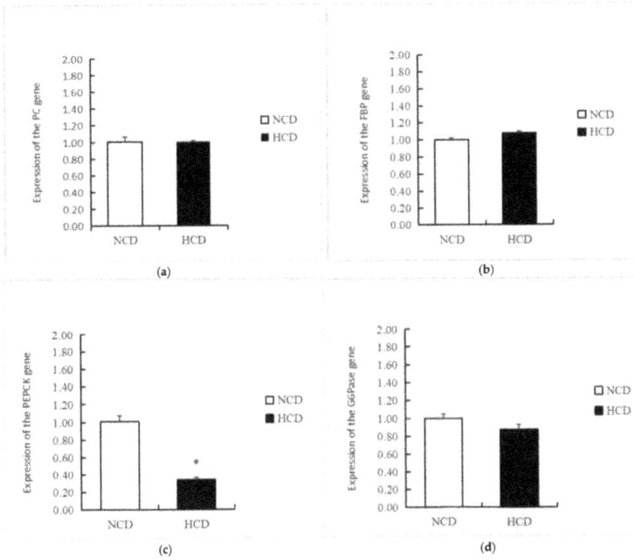


**Figure 4.** Effect of fed diets (NCD and HCD) on relative levels of liver gluconeogenesis mRNA in *L. maculatus*. Pyruvate carboxylase (*PC*, (**a**)), fructose 1, 6-bisphosptase (*FBP*, (**b**)), phosphoenolpyruvate carboxykinase (*PEPCK*, (**c**)), glucose-6-phosphatase (*G6pase*, (**d**)). Data are expressed as mean ± SE. Significant differences between the values obtained in the NCD and HCD groups ($p < 0.05$) are marked with an asterisk by *t*-test.

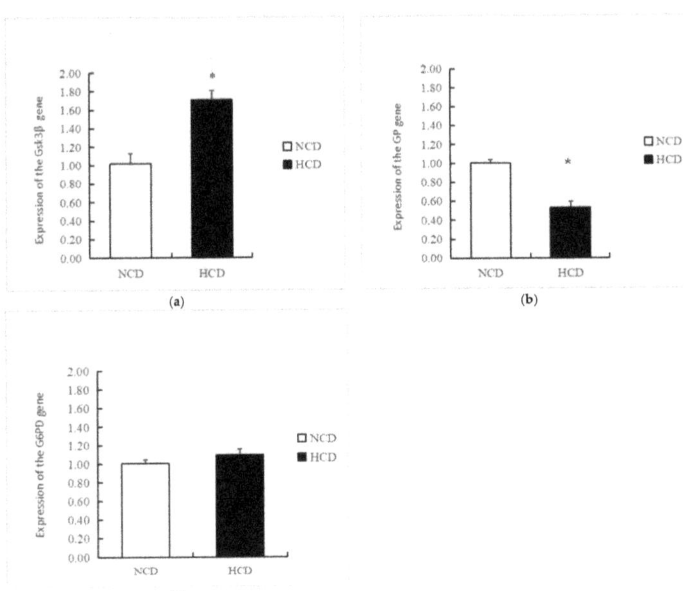


**Figure 5.** Effects of fed diets (NCD and HCD) on the relative level of liver glycogen synthesis, glycogen decomposition, and pentose phosphate pathway mRNA in *L. maculatus*. Glycogen synthase kinase-3β (*Gsk3β*, (**a**)), glycogen phosphorylase (*GP*, (**b**)), glucose-6-phosphate dehydrogenase (*G6PD*, (**c**)). Data are expressed as mean ± SE. Significant differences between the values obtained in the NCD and HCD groups ($p < 0.05$) are marked with an asterisk by *t*-test.

### 3.6. Intestinal Microbiota Community Structures

The main intestinal phylum levels of *L. maculatus* are Firmicutes, Proteobacteria, and Bacteroidetes are shown in Figure 6. The relative abundance of the intestinal microbiota varied, and the total number of OTUs increased as the level of carbohydrates increased (Table 9). Compared with the NCD group, the relative abundance of Firmicutes in the HCD group decreased. The relative abundance of Proteobacteria and Bacteroidetes increased. The genera of intestinal dominance in *L. maculatus* juveniles are *Bacillus*, *Paraclostridium*, and *Photobacterium* (Figure 7). As carbohydrate levels increased, the relative abundance of intestinal flora changed. Compared with the NCD group, the relative abundance of *Bacillus* in the HCD group decreased, whereas the relative abundance of *Paraclostridium* and *Photobacterium* increased. The principal coordinate analysis confirmed no similarity in the composition of intestinal microorganisms between the two groups, indicating that high carbohydrates affected the composition of *L. maculatus* intestinal microbiota (Figure 8).

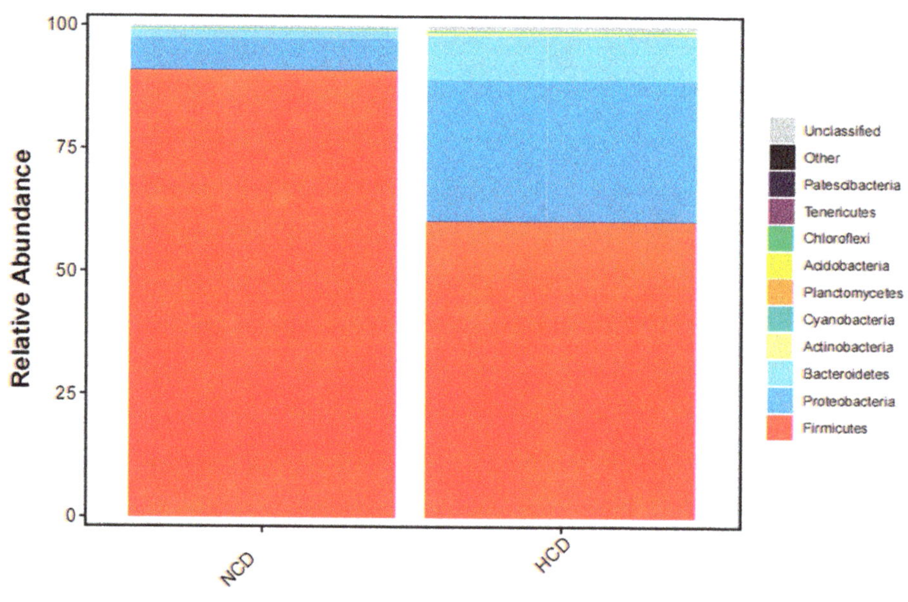


**Figure 6.** Phylum-level microbiota composition of the intestinal microbiota.

**Table 9.** Effects of fed diets (NCD and HCD) on the intestinal microbiota of *L. maculatus*.

| Items | NCD | HCD |
|---|---|---|
| OTUs | 845 | 973 |
| ACE index | 254.23 ± 45.18 | 272.29 ± 79.40 |
| Chao1 index | 240.62 ± 43.97 | 259.29 ± 77.16 |
| Simpson index | 0.42 ± 0.05 | 0.76 ± 0.03 * |
| Shannon index | 1.68 ± 0.15 | 2.76 ± 0.33 * |
| Coverage rate | 0.9995 | 0.9995 |

Data are expressed as mean ± SE. Significant differences between the values obtained in the NCD and HCD groups ($p < 0.05$) are marked with an asterisk by *t*-test.

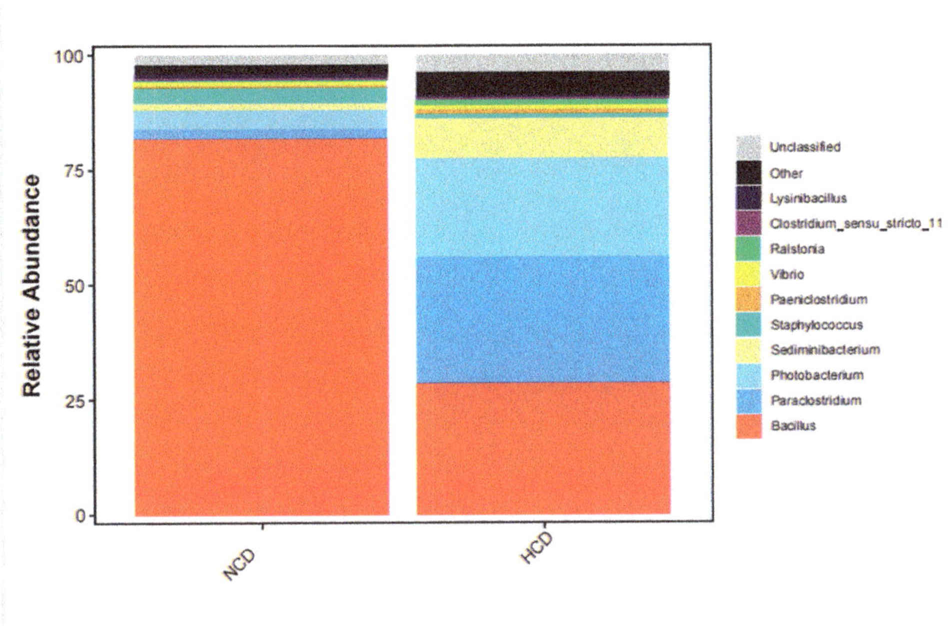


**Figure 7.** Genus-level microbiota composition of intestinal microbiota.

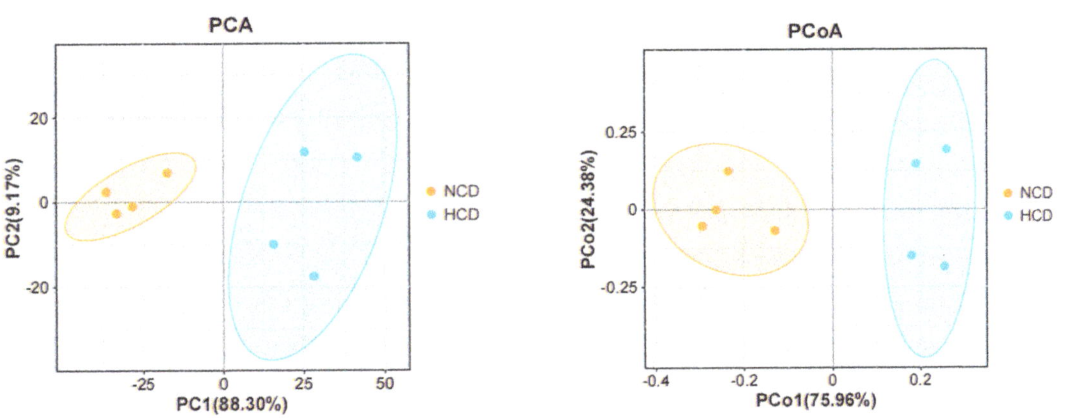


**Figure 8.** Principal component analysis and principal coordinates analysis of intestinal microbiota.

## 4. Discussion

As one of the three main sources of energy, carbohydrates are also one of the least expensive. Carbohydrates are added to the feed and utilized effectively, lowering feed costs and reducing the need for protein [21]. High-carbohydrate diets have been linked to low disease resistance, slow growth, and high mortality in fish, according to earlier research [22]. In the current study, the CF in the HCD group was noticeably higher than that in the NCD group, which was comparable to hybrid snakehead [23]. The growth balance was evaluated using the CF [24]. Therefore, it is hypothesized that the rise in CF shows that increased carbohydrate levels might promote lipid deposition. The SR, WGR, SGR, and VSI in the HCD group dropped, indicating that a 30% cornstarch level could negatively affect the growth of *L. maculatus* and cause some nutritional stress, similar to what occurs in *Micropterus salmoides* [25]. These findings suggested that the inability of fish to efficiently utilize carbohydrates may be related to their congenital diabetes [26]. Specifically in the liver or abdominal cavity, carbohydrates are transformed into glycogen through glycogen synthesis or lipid through lipid synthesis [27]. High carbohydrate intake improved liver glycogen, HSI, CI, and VAI in the current study. Similar results were seen for HSI and hepatic glycogen in other fish species, including *Nile tilapia* and *M. amblycephala* [3,28]. The steady rise in HSI, glycogen, or lipid is thought to be the outcome of either glycogen deposition or lipid deposition in *L. maculatus*.

An essential biomarker for assessing the body's health is the blood indicator [27]. Many fish species (mostly carnivorous fish) exhibit chronic postprandial hyperglycemia after being fed digestible carbohydrates, which stimulates INS secretion [27]. In the present study, high carbohydrate intake caused higher INS. Reducing sugars (like glucose, for example) and free amino groups on proteins, lipids, and nucleic acids intensify their non-enzymatic glycosylation process, which results in the formation of irreversible Maillard products AGEs [29]. In mice, AGEs disrupted the metabolism of glucolipids and enhanced the inflammatory reaction [30]. GSP is a glycated protein with ketamine bonds created by a non-enzymatic glycation process between blood glucose and the amino group at the N-terminus of albumin and other protein molecules [31]. High-carbohydrate stimulation on *M. amblycephala* results in an increase in GLU, AGEs, and GSP, which is similar to the findings of the present study [32]. The glycolysis cycle produces PA, which can be changed into lactic acid [33]. High carbohydrate intake in the current study increased plasma PA. The balance of lipid metabolism depends on the transport of TC by HDL and LDL to organs such as the liver [34]. As a result, an increase in TC levels will likewise cause an increase in HDL and LDL levels. In the present study, elevated levels of TC, TG, HDL, and LDL levels in the HCD group suggested that high carbohydrate intake increased the production of lipids. This result is in line with the previous investigation into *M. amblycephala* [3]. The liver contains the enzymes AST and ALT, which are only released into the bloodstream in the event that the cell membranes and mitochondria of the liver are damaged [35]. In this study, high carbohydrate intake led to a rise in AST and ALT values. These findings suggested that *L. maculatus's* liver health had been negatively impacted by a high-carbohydrate diet.

The organism's ability to absorb and use nutrients can be seen in the activity of intestinal digestive enzymes. Chymotrypsin is a proteolytic enzyme released by the pancreas and can reflect how an organism uses proteins [36,37]. In the present study, the chymotrypsin activity of the HCD group was significantly higher than that in the NCD group, and this finding demonstrated that raising the level of carbohydrates promoted *L. maculatus* to use protein. Similar findings were presented in a previous study on *C. carpio songpu*, which discovered that the protease activity in the low-starch group was significantly lower than that of the high-starch group [36]. The study of *Phoxinus lagowskii Dybowski* showed that both omnivorous fish and herbivorous fish have a promoting effect on protease activity after the ingestion of high-carbohydrate feed [38]. In the study of *S. meridionalis*, carbohydrates slightly changed the activity of the protease activity. According to research on *S. meridionalis*, the variations in amylase activity at different carbohydrate levels were not statistically significant [39]. In contrast, the results of the present study indicated that amylase activity was significantly higher in the HCD group than in the NCD group, which is comparable with those found in *Acanthopagrus schlegelii*, *Pseudosciaena crocea*, and carp [40–42]. According to these findings, high dietary carbohydrate promotes glucose metabolism. One of the intestinal digestive enzymes, lipase, is involved in lipid synthesis [43]. In the current study, lipase activity was significantly higher in the HCD group than in the NCD group. Given that the findings were consistent with the elevated TC and TG levels in the plasma and the lipid accumulation in the liver in our investigation, we assume that a high-carbohydrate diet may promote lipid synthesis in *L. maculatus*. A similar result was obtained on *Phoxinus lagowskii Dybowski* [38]. However, studies on *A. schlegelii* and *S. meridionalis* have demonstrated that intestinal lipase activity is not significantly affected by dietary carbohydrate levels [42,44]. There is disagreement on how dietary carbohydrates affect lipase, and there are no reliable results. This topic might be further studied in the future.

SOD and CAT are crucial enzymes in the body's first line of defense against the production of free radicals and other reactive chemicals in cells. Free radicals cause a number of diseases as they accumulate in the body [45]. MDA, the end product of polyunsaturated fatty acid (PUFA) degradation, is a marker of lipid peroxidation [46]. The body's antioxidant system can be evaluated for functionality using T-AOC [47]. In this study, SOD, CAT, and T-AOC activities were significantly lower in the HCD group compared to the NCD group, although the MDA level was the opposite. It was hypothesized that high levels of carbohydrates would limit the body's ability to produce antioxidants and result in a significant amount of the harmful chemical MDA. This result is similar to the research on *M. amblycephala* [16].

The first key enzyme in glycolytic and metabolic pathways is hexokinase (HK) [48]. In the present study, the activity of *HK* in the HCD group was significantly higher, indicating that high dietary carbohydrate intake may promote the glycolysis response. A similar result was also obtained in a previous study on *Pelteobagrus vachelli* [49]. However, previous studies on common carp and *Dicentrarchus labrax* revealed that the presence of carbohydrates has no impact on the activity of *HK* [50,51]. This could be due to the fact that hexokinase, which is virtually saturated, cannot react to slight fluctuations in blood glucose levels [52]. As a rate-limiting enzyme, *PEPCK* is a phosphoenolpyruvate carboxykinase that participates in the reaction of gluconeogenesis [53]. Similar to the findings for rainbow trout [54], the activity of *PEPCK* in this study was unaffected by dietary carbohydrate intake. However, some studies suggest that as carbohydrate levels rise, *PEPCK* activity declines [55]. When evaluating glucose metabolism, it is essential to consider not just diet composition but also feeding habits, life stage, and size, among other factors [27].

In fish species, the intestine is the primary site for digestion and absorption site [56]. The villi height and muscular layer thickness are indicators of the intestine's capacity for absorption and digestion [57]. In the present study, the intestinal villi's height was reduced, increasing the villi's spacing in the high-carbohydrate group. This indicated that excessive carbohydrate levels changed the morphology of the gut and therefore reduced its capacity for digestion and absorption. The decrease in SR, WGR, and SGR in the HCD group correlated with the shortening of the intestinal villi, which decreases the contact area of food with the intestinal villi and weakens intestinal digestion. Similar results were also discovered in previous studies on *M. amblycephala* [58] and gilthead sea bream [59]. The liver, a crucial organ for the metabolism of glucose in fish species, deposits glycogen and lipids when dietary carbohydrate intake is excessive [60]. According to previous studies, high dietary carbohydrate intake led to pathologic symptoms including lipid droplet vacuolation with displaced nuclei and cytoplasm loss [61–63]. Similar to the present study, the liver cells in the HCD group showed many vacuoles. In line with this, oil red O staining revealed a significant rise in lipid droplets in the HCD group. These findings suggested that liver injury in *L. maculatus* may occur from high carbohydrate-induced liver lipid accumulation.

In almost all types of organisms, the metabolic pathway and anaerobic energy source known as glycolysis has evolved [64]. *GK* is a special type of hexokinase that, strictly speaking, operates on liver glucose as the initial and limiting step in the storage of excess glucose. It occurs in several species [7,9]. In the current study, the expression of *GK* genes was up-regulated by high carbohydrates. Similar to this, previous studies revealed that the expression of the *GK* gene is induced by carbohydrate levels and positively linked with the rise in carbohydrate levels in *C. carpio*, *T. ovatus*, and *O. mykiss* [52,65,66]. According to the results of the current study, high carbohydrate levels had no effect on the expression of *PFK* genes, which is in line with the previous findings of *O. mykiss* [67]. However, in *Sparus aurata*, *PFK* gene expression increased as carbohydrate levels rose [68].

The liver can replenish and restore glycogen by gluconeogenesis, which can keep plasma glucose steady [69]. *G6pase* is mostly found in fish livers, where it is highly active but is inactive in the hearts and muscles of fish [70]. In previous studies, the expression of the *G6pase* gene was affected by high dietary carbohydrate intake in *Erythroculter ilishaeformis* [71]. High dietary carbohydrate intake did not, however, have an impact on the expression of the *G6pase* gene in the current study. Similarly, the earlier study found that feeding rainbow trout different starches had no impact on the expression of the *G6pase* gene in the liver [66]. The rate-limiting enzyme, *FBP*, is involved in the second step of the gluconeogenesis reaction. According to some earlier studies, dietary carbohydrates cannot regulate the activity or gene expression of *PEPCK* and *FBP* [56,72]. In the present study, high carbohydrate intake had no effect on *FBP* gene expression. Similarly, previous studies on *O. mykiss* and *M. salmoides* found that the *FBP* gene in the liver was unaffected by dietary carbohydrates [56,73]. In the current study, the expression of the *PEPCK* gene was significantly reduced in the HCD group, which is consistent with the results of *Cyprinus carpio* [67]. GSK3-β is an essential gene in glycogen synthesis, and *GP* is a key gene for glycogenolysis [74]. Glucose is stored in liver cells as glycogen, which can be then broken down and converted to glucose and released into the blood [75]. In the present study, high-carbohydrate diets promoted hepatic glycogen synthesis and inhibited hepatic glycogenolysis in *L. maculatus*. Similar results were observed in recent work in *Ctenopharyngodon idella*, where gene expression related to glycogen synthesis was significantly increased in the high-carbohydrate group [9]. In prior work on *Leiocassis longirostris*, *GP* gene expression in the high-carbohydrate group was likewise significantly reduced [9].

Animals' digestive tracts contain intestinal microbiota that are stable and can regulate the metabolisms of glycolipids [76]. Numerous factors, including fish species, physiological status, feed, and aquatic habitat, have an impact on the intestinal microbiota of fish [77]. In the present study, it was discovered that *L. maculatus*'s intestinal microbial composition was considerably changed by high carbohydrate levels. Similarly, changes in the intestinal microbiota of *T. ovatus* were caused by dietary starch levels [7]. In line with the findings of cobia [78], the number of phyla Bacteroides gradually rose as the level of carbohydrates increased in our study. Proteobacteria are common in water, soil, flora, and fauna and are members of the Gram-negative phylum. They are pathogens of the digestive system [79,80]. In the current study, Proteobacteria grew more in the HCD group. This suggests that *L. maculatus*'s susceptibility to diseases was heightened by its high-carbohydrate diet. Similarly, dietary carbohydrates dramatically increased the relative abundance of the Proteobacteria phylum in previous studies on the pearl gentian grouper and *T. ovatus* [7,81]. Gram-negative bacteria, *Photobacterium damselae*, cause septicemic diseases in aquaculture, including sea bass (*Dicentrarchus labrax*) and sea bream (*Sparus aurata*) [10]. In the current study, *Photobacterium* levels rose in the HCD group, indicating that *L. maculatus* is more prone to developing photomycosis when subjected to high dietary carbohydrates.

## 5. Conclusions

From the results of this experiment on the *L. maculatus*, high carbohydrates led to increased condition factor and liver glycogen, lipid deposition, decreased antioxidant capacity of the liver, increased relative abundance of harmful intestinal microbes, and disrupted glucose metabolism.

**Author Contributions:** Conceptualization, L.Z. and C.Z. (Chuanpeng Zhou); methodology, L.Z., C.Z. (Chuanpeng Zhou) and L.Q.; validation, C.Z. (Chuanpeng Zhou) and B.Z.; investigation, Z.W., B.Z., L.Y., P.W., C.Z. (Chao Zhao) and H.L.; formal analysis, L.Z. and C.Z. (Chuanpeng Zhou); data curation, L.Z.; writing—original draft preparation, L.Z.; writing—review and editing, L.Z. and C.Z. (Chuanpeng Zhou); supervision and project administration, C.Z. (Chuanpeng Zhou) and L.Q.; funding acquisition, C.Z. (Chuanpeng Zhou) and L.Q. All authors have read and agreed to the published version of the manuscript.

**Funding:** This work was supported by the Central Public-interest Scientific Institution Basal Research Fund, South China Sea Fisheries Research Institute, CAFS (2021XK02); the Central Public-interest Scientific Institution Basal Research Fund, CAFS (2022XT0404; 2020TD55); the Key Areas of Guangdong Province Research and Development projects (NO. 2021B0202020002); the Central Public-interest Scientific Institution Basal Research Fund of CAFS (NO. 2020TD21); and Guangzhou municipal Science and Technology (NO. 2023B03J1304).

**Institutional Review Board Statement:** The experimental design was approved by the Ethics Committee of the South China Sea Fisheries Institute (Approval Code: 20210201; Approval Date: 9 February 2021).

**Informed Consent Statement:** Not applicable.

**Data Availability Statement:** The data are available upon request.

**Conflicts of Interest:** The authors declare no conflict of interest.

## References

1. Hemre, G.I.; Mommsen, T.P.; Krogdahl, A. Carbohydrates in fish nutrition: Effects on growth, glucose metabolism and hepatic enzymes. *Aquac. Nutr.* **2015**, *8*, 175–194. [CrossRef]
2. Kamalam, B.S.; Medale, F.; Panserat, S. Utilisation of dietary carbohydrates in farmed fishes: New insights on influencing factors, biological limitations and future strategies—ScienceDirect. *Aquaculture* **2017**, *467*, 3–27. [CrossRef]
3. Adjoumani, J.J.Y.; Abasubong, K.P. A time-course study of the effects of a high-carbohydrate diet on the growth performance, glycolipid metabolism and mitochondrial biogenesis and function of blunt snout bream (*Megalobrama amblycephala*). *Aquaculture* **2022**, *552*, 738011. [CrossRef]
4. Emdin Stefan, O. Effects of hagfish insulin in the *Atlantic hagfish, Myxine glutinosa* the in vivo metabolism of [14C] glucose and [14C] leucine and studies on starvation and glucose-loading. *Gen. Comp. Endocrinol.* **1982**, *47*, 414–425. [CrossRef]
5. Plisetskaya, E.M.; Duan, C. Insulin and insulin-like growth factor I in coho salmon *Oncorhynchus kisutch* injected with streptozotocin. *Am. J. Physiol.* **1994**, *267 Pt 2*, R1408–R1412. [CrossRef]
6. Yang, L.; Qin, C.-B.; Zheng, W.J.; Lu, R.H.; Nie, G.H. Research progress on glucose perception and regulation of glucose metabolism in fish. *J. Aquac.* **2014**, *38*, 1639–1649.
7. Zhao, W.; Xie, J.J.; Fang, H.H.; Liu, Y.J.; Niu, J. Effects of corn starch level on growth performance, antioxidant capacity, gut morphology and intestinal microflota of juvenile golden pompano, *Trachinotus ovatus*. *Aquaculture* **2020**, *524*, 735197. [CrossRef]
8. Shen, J.; Liu, H.; Tan, B.; Dong, X.; Yang, Q.; Chi, S.; Zhang, S. Effects of replacement of fishmeal with cottonseed protein concentrate on the growth, intestinal microflota, haematological and antioxidant indices of juvenile golden pompano (*Trachinotus ovatus*). *Aquac. Nutr.* **2020**, *26*, 1119–1130. [CrossRef]
9. Su, J.; Mei, L.; Xi, L.; Gong, Y.; Yang, Y.; Jin, J.; Liu, H.; Zhu, X.; Xie, S.; Han, D. Responses of glycolysis, glycogen accumulation and glucose-induced lipogenesis in grass carp and Chinese long snout catfish fed high-carbohydrate diet. *Aquaculture* **2021**, *533*, 736146. [CrossRef]
10. Munro, P.O.; Barbour, A.; Blrkbeck, T.H. Comparison of the gut bacterial flora of start-feeding larval turbot reared under different conditions. *J. Appl. Microbiol.* **2010**, *77*, 560–566. [CrossRef]
11. Zhou, Y.L.; He, G.L.; Jin, T.; Chen, Y.J.; Lin, S.M. High dietary starch impairs intestinal health and microbiota of largemouth bass, *Micropterus salmoides*. *Aquaculture* **2020**, *534*, 736261. [CrossRef]
12. Liu, J. Differential population structuring and demographic history of two closely related fish species, Japanese sea bass (*Lateolabrax japonicus*) and spotted sea bass (*Lateolabrax maculatus*) in Northwestern Pacific. *Mol. Phylogenet. Evol.* **2006**, *39*, 799–811. [CrossRef] [PubMed]
13. Shao, C.; Li, C.; Wang, N.; Qin, Y.; Xu, W.; Liu, Q.; Zhou, Q.; Zhao, Y.; Li, X.; Liu, S.; et al. Chromosome-level genome assembly of the spotted sea bass, *Lateolabrax maculatus*. *GigaScience* **2018**, *7*, giy114. [CrossRef] [PubMed]
14. 2022 China Fisheries Statistical Yearbook of Mariculture Production by Region (by Species). 2022. Available online: https://www.yearbookchina.com/navipage-n3022101301000125.html (accessed on 21 June 2023).
15. Zhou, J.B.; Zhou, Q.C.; Chi, S.Y.; Yang, Q.H.; Liu, C.W. Optimal dietary protein requirement for juvenile ivory shell, *Babylonia areolata*. *Aquaculture* **2007**, *270*, 186–192. [CrossRef]
16. Zhou, C.P.; Ge, X.P.; Liu, B.; Xie, J.; Miao, L.H. Effect of high dietary carbohydrate on the growth performance and physiological responses of juvenile Wuchang bream, *Megalobrama amblycephala*. *Asian-Australas. J. Anim. Sci.* **2013**, *26*, 1598–1608. [CrossRef]
17. Asadi, F.; Hallajian, A.; Asadian, P.; Shahriari, A.; Pourkabir, M. Serum lipid, free fatty acid, and proteins in juvenile sturgeons: *Acipenser persicus* and acipenser stellatus. *Comp. Clin. Pathol.* **2009**, *18*, 287–289. [CrossRef]
18. Gutierrez, J.; Carrillo, M.; Zanuy, S.; Planas, J. Daily rhythms of insulin and glucose levels in the plasma of sea bass *Dicentrarchus labrax* after experimental feeding. *Gen. Comp. Endocrinol.* **1984**, *55*, 393–397. [CrossRef]
19. Zang, L.; Shimada, Y.; Tanaka, T.; Nishimura, N. Rhamnan sulphate from monostroma nitidum attenuates hepatic steatosis by suppressing lipogenesis in a diet-induced obesity zebrafish model. *J. Funct. Foods* **2015**, *17*, 364–370. [CrossRef]

20. Pan, Y.X.; Luo, Z.; Zhuo, M.Q.; Wei, C.C.; Chen, G.H.; Song, Y.F. Oxidative stress and mitochondrial dysfunction mediated cd-induced hepatic lipid accumulation in zebrafish danio rerio. *Aquat. Toxicol.* **2018**, *199*, 12–20. [CrossRef]
21. Fernandez, F.; Miquel, A.G. Effects of diets with distinct protein-to-carbohydrate ratios on nutrient digestibility, growth performance, body composition and liver intermediary enzyme activities in gilthead sea bream (*Sparus aurata*) fingerlings. *J. Exp. Mar. Biol. Ecol.* **2007**, *343*, 1–10. [CrossRef]
22. Dixon, D.G.; Hilton, J.W. Influence of available dietary carbohydrate content on tolerance of waterborne copper by rainbow trout, Salmo gairdneri Richardson*. *J. Fish Biol.* **1981**, *19*, 509–518. [CrossRef]
23. Xu, T.; Liu, X.; Huang, W.; Li, G.; Zhang, Y.; Xu, D.; Wang, G. Effects of dietary carbohydrate levels on growth, metabolic enzyme activities and oxidative status of hybrid snakehead (*Channa maculata* ♀× *Channa argus* ♂). *Aquaculture* **2022**, *563*, 738960. [CrossRef]
24. Du, Z.Y.; Turchini, G.M. Are we actually measuring growth? —An appeal to use a more comprehensive growth index system for advancing aquaculture research. *Rev. Aquac.* **2022**, *14*, 525–527. [CrossRef]
25. Amoah, A.; Coyle, S.D.; Webster, C.D.; Durborow, R.M.; Bright, L.A.; Tidwell, J.H. Effects of Graded Levels of Carbohydrate on Growthand Survival of Largemouth Bass, *Micropterus Salmoides*. *J. World Aquac. Soc.* **2008**, *39*, 397–405. [CrossRef]
26. Wilson, R.P.; Poe, W.E. Apparent inability of channel catfish to utilize dietary mono and disaccharides as energy sources. *J. Nutr.* **1987**, *117*, 280–285. [CrossRef]
27. Polakof, S.; Panserat, S. Glucose metabolism in fish: A review. *J. Comp. Physiol. B Biochem. Syst. Environ. Physiol.* **2012**, *182*, 1015–1045. [CrossRef]
28. Boonanuntanasarn, S.; Kumkhong, S.; Yoohat, K.; Plagnes-Juan, E.; Burel, C.; Marandel, L.; Panserat, S. Molecular responses of Nile tilapia (*Oreochromis niloticus*) to different levels of dietary carbohydrates. *Aquaculture* **2018**, *482*, 117–123. [CrossRef]
29. Rashid, G.; Benchetrit, S.; Fishman, D.; Bernheim, J. Effect of advanced glycation end-products on gene expression and synthesis of TNF-α and endothelial nitric oxide synthase by endothelial cells. *Kidney Int.* **2004**, *66*, 1099. [CrossRef]
30. Yu, W.Z. Study on Inflammation and Glycolipid Metabolism Disorders Induced by Advanced Glycation end Products (AGEs). Master's Thesis, Shanghai Ocean University, Shanghai, China, 2019.
31. Xu, X.; Chen, H.-W.; Yu, J.-P. Significance of glycated serum protein detection in diabetes surveillance. *Lab. Med.* **2006**, *21*, 136–139.
32. Shi, H.J.; Xu, C.; Liu, M.Y.; Wang, B.K.; Liu, W.B.; Chen, D.H.; Zhang, L.; Xu, C.Y.; Li, X.F. Resveratrol improves the energy sensing and glycolipid metabolism of blunt snout bream *Megalobrama amblycephala* fed high-carbohydrate diets by activating the ampk-sirt1-pgc-1αnetwork. *Front. Physiol.* **2018**, *9*, 1258. [CrossRef]
33. Zhang, X.; Jin, M.; Luo, J.; Xie, S.; Zhou, Q. Effects of dietary carbohydrate levels on the growth and glucose metabolism of juvenile swimming crab, *Portunus trituberculatus*. *Aquac. Nutr.* **2022**, *2022*, 7110052. [CrossRef]
34. Gjøen, T.; Berg, T. Hepatic uptake and intracellular processing of LDL in rainbow trout. *Biochim. Et Biophys. Acta* **1993**, *1169*, 225–230. [CrossRef] [PubMed]
35. Liu, Y. Prognostic analysis of severe hepatitis based on serum aspartate aminotransferase to alanine aminotransferase. *J. Xiang Nan Univ.* **2004**, *03*, 45–46.
36. Li, J.; Wang, C.; Wang, L.; Zhao, Z.; Luo, L.; Xu, Q. Effects of different carbohydrate types and carbohydrate levels on intestinal digestive enzyme activities, intestinal and liver histological structure of songpu mirror carp (*C. carpio specularis*). *Chin. J. Anim. Nutr.* **2016**, *28*, 3217–3224.
37. Zhao, K.; Wei, T.; Yang, F. Chymotrypsin-dependent cleavage and activation of calcineurin. *Chin. J. Biophys.* **2009**, *25*, 246–247.
38. Qu, Z.-H.; Wu, L.-F.; Zhou, K.; Yang, L.; Zu, X.-J.; Duan, J. Effects of feed carbohydrate level on digestive enzyme and glucose metabolism enzyme activity of *Phoxinus Lagowskii Dyb-Owski*. *J. Northwest AF Univ. (Nat. Sci. Ed.)* **2019**, *47*, 25–32.
39. Gao, M. Effects of dietary carbohydrates on digestive enzymes and related physiological indicators in juvenile southern catfish (*Silurus meridionalis Chen*). Ph.D. Dissertation, Southwest University, Pittsburgh, PA, USA, 2006.
40. Kawai, S.I.; Ikeda, S. Studies on digestive enzymes of fishes—II: Effect of dietary change on the activities of digestive enzymes in carp intestine. *Nippon Suisan Gakkaishi* **1972**, *38*, 265–270. [CrossRef]
41. Xing, S.; Sun, R. Effects of feed sugar level on growth and glucose metabolism of *Pseudo-sciaena crocea*. *Chin. J. Hydrobiol.* **2017**, *41*, 265–276.
42. Han, B.; Song, L.P. Effect of dietary starch level on growth and digestive enzyme activity of freshwater black seabream. *J. Shanghai Ocean Univ.* **2010**, *19*, 207–213.
43. Wang, S.Y.; Li, G.L. Advances in the study of constructive lipids and lipid enzymatic synthesis. *China Lipids* **2002**, *27*, 4.
44. Wang, M.; Zhou, P.; Huang, W.; Zhou, Q. Effects of glucose supplemental level at different protein levels on growth performance, hepatic glycolysis and gluconeogenic key enzyme activities of large yellow croaker (*Larm ichthys crocea Richardson*). *Chin. J. Anim. Nutr.* **2015**, *27*, 2431–2442.
45. Ighodaro, O.M.; Akinloye, O.A. First line defence antioxidants-superoxide dismutase (sod), catalase (cat) and glutathione peroxidase (gpx): Their fundamental role in the entire antioxidant defence grid. *Alex. J. Med.* 2017. [CrossRef]
46. Situmorang, N.; Zulham, Z. Malondialdehyde (Mda) (Zat Oksidan Yang Mempercepat Proses Penuaan. *J. Keperawatan Dan Fisioter.* **2020**, *2*, 117–123. [CrossRef]
47. Zhou, C.-P. *Effects of Different Levels of Carbohydrate Diets on Growth, Immunity and Related Sugar Metabolizing Enzymes in the Dried Bream, M. bungee*; Nanjing Agricultural University: Nanjing, China, 2012.

48. Bai, J.; Liu, L.; Han, Y.; Jia, C.; Liang, C. One-step detection of hexokinase activity using a personal glucose meter. *Anal. Methods* **2018**, *10*, 1039. [CrossRef]
49. Zhang, S.L. Effects of Sugar Structure, Sugar Level and Glycolipid Ratio in Feed on the Growth and Sugar Metabolism of Juvenile Pelteobagrus Vachelli. Master's Thesis, Ocean University of China, Qingdao, China, 2011.
50. Enes, P.; Panserat, S.; Kaushik, S.; Oliva-Teles, A. Effect of normal and waxy maize starch on growth, food utilization and hepatic glucose metabolism in European sea bass (*Dicentrarchus labrax*) juveniles. *Comp. Biochem. Physiol. A Mol. Integr. Physiol.* **2006**, *143*, 89–96. [CrossRef]
51. Panserat, S.; Medale, F.; Blin, C.; Breque, J.; Vachot, C.; Plagnesjuan, E.; Gomes, E.; Krishnamoorthy, R.; Kaushik, S. Hepatic glucokinase is induced by dietary carbohydrates in rainbow trout, gilthead seabream, and common carp. *Am. J. Physiol. Regul. Integr. Comp. Physiol.* **2000**, *278*, R1164. [CrossRef]
52. Li, J.N.; Xu, Q.Y.; Wang, C.A.; Wang, L.S.; Zhao, Z.G.; Luo, L. Effects of dietary glucose and starch levels on the growth, haematological indices and hepatic hexokinase and glucokinase mrna expression of juvenile mirror carp (*C. Carpio*). *Aquac. Nutr.* **2015**, *22*, 550–558. [CrossRef]
53. Cournarie, F.; Azzout-Marniche, D.; Foretz, M.; Guichard, C.; Ferre, P.; Foufelle, F. The inhibitory effect of glucose on phosphoenolpyruvate carboxykinase gene expression in cultured hepatocytes is transcriptional and requires glucose metabolism. *FEBS Lett.* **1999**, *460*, 527–532. [CrossRef]
54. Séverine, K.; Stéphane, P.; Lim, P.L.; Kaushik, S.; Ferraris, R.P. The role of hepatic, renal and intestinal gluconeogenic enzymes in glucose homeostasis of juvenile rainbow trout. *J. Comp. Physiol. B* **2008**, *178*, 429–438.
55. Li, J.-H. Effects of Different Sugar Levels in Feed on Growth, Hepatopancreas Tissue Structure and Metabolism of Carp. Master's Thesis, Tianjin Agricultural College, Tianjin, China, 2017.
56. Deng, J.M.; Mai, K.S.; Qing-Hui, A.I.; Zhang, W.B. Effects of soybean oligosaccharides on nutritional characters of japanese flounder (*Paralichthys olivaceus*). *Acta Hydrobiol. Sin.* **2009**, *33*, 369–375. [CrossRef]
57. Torrecillas, S.; Montero, D.; Caballero, M.J.; Robaina, L.; Zamorano, M.J.; Sweetman, J.; Izquierdo, M. Effects of dietary concentrated mannan oligosaccharides supplementation on growth, gut mucosal immune system and liver lipid metabolism of European sea bass (*Dicentrarchus labrax*) juveniles. *Fish Shellfish Immunol.* **2015**, *42*, 508–516. [CrossRef] [PubMed]
58. Yu, C.; Zhang, J.; Qin, Q.; Liu, J.; Xu, J.; Xu, W. Berberine improved intestinal barrier function by modulating the intestinal microbiota in blunt snout bream (*Megalobrama amblycephala*) under dietary high-fat and high-carbohydrate stress. *Fish Shellfish Immunol.* **2020**, *102*, 336–349. [CrossRef]
59. Castro, C.; Couto, A.; Diógenes, A.F.; Corraze, G.; Panserat, S.; Serra, C.R.; Oliva-Teles, A. Vegetable oil and carbohydrate-rich diets marginally affected intestine histomorphology, digestive enzymes activities, and gut microbiota of gilthead sea bream juveniles. *Fish Physiol. Biochem.* **2019**, *45*, 681–695. [CrossRef] [PubMed]
60. Dias, J.; Alvarez, M.J.; Diez, A.; Arzel, J.; Corraze, G.; Bautista, J.; Kaushik, S. Regulation of hepatic lipogenesis by dietary protein/energy in juvenile European seabass (*Dicentrarchus labrax*). *Aquaculture* **1998**, *161*, 169–186. [CrossRef]
61. Jiang, L.; Wu, H.; Huang, K.; Ma, Y.; Yang, Q.; Yu, D.; Zhong, L. Effects of dietary carbohydrate levels on growth performance and liver metabolism functions of juvenile tilapia (*Oreochromis niloticus*). *J. Fish. China* **2013**, *37*, 245–255. [CrossRef]
62. Miao, L. Transcriptomic Analysis of microRNAs for High Glucose Metabolism and Regulation of miR-34a on Glucolipid Metabolism in Grouper Bream. Ph.D. Thesis, Nanjing Agricultural University, Nanjing, China, 2017.
63. Essam, H.M.; Abdellrazeq, G.S. Pathogenesis of photobacterium damselae subspecies infections in sea bass and sea bream. *Microb. Pathog.* **2016**, *99*, 41–50. [CrossRef]
64. Chaudhry, R.; Varacallo, M. Biochemistry, Glycolysis. In *Stat Pearls*; Stat Pearls Publishing: St. Petersburg, FL, USA, 2022.
65. Zhou, C.; Ge, X.; Niu, J.; Lin, H.; Huang, Z.; Tan, X. Effect of dietary carbohydrate levels on growth performance, body composition, intestinal and hepatic enzyme activities, and growth hormone gene expression of juvenile golden pompano, *Trachinotus ovatus*. *Aquaculture* **2015**, *437*, 390–397. [CrossRef]
66. Panserat, S.; Médale, F.; Brèque, J.; Plagnes-Juan, E.; Kaushik, S. Lack of significant long-term effect of dietary carbohydrates on hepatic glucose-6-phosphatase expression in rainbow trout (*Oncorhynchus mykiss*). *J. Nutr. Biochem.* **2000**, *11*, 22–29. [CrossRef]
67. Panserat, S.; Plagnes-Juan, E.; Kaushik, S. Gluconeogenic enzyme gene expression is decreased by dietary carbohydrates in common carp (*C. carpio*) and gilthead seabream (*Sparus aurata*). *BBA—Gene Struct. Expr.* **2002**, *1579*, 35–42. [CrossRef]
68. Metón, I.; Mediavilla, D.; Caseras, A.; Cantó, E.; Fernández, F.; Baanante, I.V. Effect of diet composition and ration size on key enzyme activities of glycolysis-gluconeogenesis, the pentose phosphate pathway and amino acid metabolism in liver of gilthead sea bream (*Sparus aurata*). *Br. J. Nutr.* **1999**, *82*, 223–232. [CrossRef]
69. Adeva-Andany, M.M.; Pérez-Felpete, N.; Fernández-Fernández, C.; Donapetry-García, C.; Pazos-García, C. Liver glucose metabolism in humans. *Biosci. Rep.* **2016**, *36*, e00416. [CrossRef] [PubMed]
70. Yu, S.; Zhang, S. Glucose metabolism and utilization of sugar in bait in fish. *Mar. Sci.* **1986**, *4*, 58–60.
71. Wang, G. *Effects of Dietary Carbohydrate Level on Growth, Blood Indexes and Gene Expression of GK, G6Pase and HSC70 in P. upturnal*; Nanjing Agricultural University: Nanjing, China, 2009.
72. Moreira, I.S.; Peres, H.; Couto, A.; Enes, P.; Oliva-Teles, A. Temperature and dietary carbohydrate level effects on performance and metabolic utilisation of diets in European sea bass (*Dicentrarchus labrax*) juveniles. *Aquaculture* **2008**, *274*, 153–160. [CrossRef]
73. Jiang, Y. *Nutritional Physiological Response of Largemouth Bass (*Lepomis macrochirus*) to Dietary Starch Level*; Southwest University: Pittsburgh, PA, USA, 2020.

74. Fabres, A.; De Andrade, C.P.; Guizzo, M.; Sorgine, M.H.F.; Paiva-Silva, G.D.O.; Masuda, A.; Vaz, I.D.S.; Logullo, C. Effect of gsk-3 activity, enzymatic inhibition and gene silencing by RNAi on tick oviposition and egg hatching. *Parasitology* **2010**, *137*, 1537–1546. [CrossRef] [PubMed]
75. Al-Shura, A.N. Etiology of Blood Sugar Disorders. In *Metabolic Disorders and Shen in Integrative Cardiovascular Chinese Medicine*; Academic Press: Cambridge, MA, USA, 2021; pp. 11–16.
76. Yang, S.; Du, J.; Luo, J.; Zhou, Y.; Long, Y.; Xu, G.; Zhao, L.; Du, Z.; Yan, T. Effects of different diets on the intestinal microbiota and immunity of common carp (*Cyprinus Carpio*). *J. Appl. Microbiol.* **2019**, *127*, 1327–1338. [CrossRef] [PubMed]
77. Liu, C.; Zhao, L.P.; Shen, Y.Q. A systematic review of advances in intestinal microflota of fish. *Fish Physiol. Biochem.* **2021**, *47*, 2041–2053. [CrossRef]
78. Lan, K.; Wu, G.; Wang, J.; Chen, X.; Wang, Y.; Zhou, C.; Lin, H.; Ma, Z. Effects of inulin addition to feed on survival, growth and intestinal flora of *Trachinotus ovatus* juveniles. *South China Fish. Sci.* **2022**, *18*, 11.
79. Stackebrandt, E.; Murray, R.G.E.; Trüper, H.G. *Proteobacteria* classis Nov. a name for the phylogenetic taxon that includes the purple bacteria and their relatives. *Int. J. Syst. Bact.* **1988**, *38*, 321–325. [CrossRef]
80. Shin, N.R.; Whom, T.W.; Bae, J.W. *Proteobacteria*: Microbial signature of dysbiosis in gut microbiota. *Trends Biotechnol.* **2015**, *33*, 496–503. [CrossRef]
81. Textor, S.; Wendisch, V.F.; Graaf, A.A.D.; Müller, U.; Linder, M.I.; Linder, D.; Buckel, W. Propionate oxidation in *Escherichia coli*: Evidence for operation of a methyl citrate cycle in bacteria. *Arch. Microbiol.* **1997**, *168*, 428–436. [CrossRef]

MDPI
St. Alban-Anlage 66
4052 Basel
Switzerland
www.mdpi.com

*Fishes* Editorial Office
E-mail: fishes@mdpi.com
www.mdpi.com/journal/fishes

www.ingramcontent.com/pod-product-compliance
Lightning Source LLC
Chambersburg PA
CBHW061508210926
51850CB00008B/1852
* 9 7 8 3 7 2 5 8 1 2 7 3 8 *